About FIGHTING CHANCE:

2021 Next Generation Indie Book Award
Science / Nature FINALIST

2021 National Indie Excellence Award
Psychiatry / Psychology FINALIST

"A comprehensive study of the history and science of depression, FIGHTING CHANCE examines the causes, symptoms, and treatments of this disease as research and understanding have evolved over time. Providing the perspectives of clinicians, researchers, and patients, Sarah Zabel deftly illustrates the physical and emotional toll of depression on those who suffer from it. An essential addition to the library of anyone impacted by this disease."

— *INDIE READER*

"A meticulous and thoughtful scientific exploration."

— *KIRKUS REVIEWS*

"Well-researched, engagingly written and informative."

— *BLUEINK REVIEWS*

Read more at www.sarahzabel.com

FIGHTING CHANCE

For more information contact Sarah Zabel Enterprises, LLC, sarah@sarahzabel.com. First edition.

Library of Congress Control Number: 2021900941
Published in Bayview, Idaho, USA

ISBN (hardcover) 978-1-7358454-0-1
ISBN (paperback) 978-1-7358454-3-2
ISBN (eBook) 978-1-7358454-1-8

Cover design by Harry Haysom
Book design by Tom Howey
Author photo by Thomas A. Garrett

FIGHTING CHANCE

How Unexpected Observations and Unintended Outcomes Shape the Science and Treatment of Depression

SARAH ZABEL

CONTENTS

AUTHOR'S NOTE

I am very grateful to the many people suffering from depression who shared their stories with me. I changed their names to protect their privacy, but their experiences remain intact. Each named person is a unique and real individual, though the bicycling patient is a construction based on mentions in journals. A few of the patients suffer with bipolar II disorder rather than major depressive disorder. While this book was intended to be about unipolar major depression, the science applies to major depressive episodes in bipolar disorders as well.

I am not a medical professional and do not offer medical advice. I hope that readers will use the information in this book to ask meaningful questions of their medical providers and to challenge their own preconceptions. I did my best due diligence to make sure that the technical matter in this book is sound, triple-checking references and employing academic and practicing experts and graduate students in neuroscience to check as much of the material as they could. Any remaining errors are purely my own.

Introduction

AWAKENING

Carolyn's voice on the phone was shaky. "I have something to tell you...," she started, hesitantly, "I just got out of the hospital. I was admitted for suicidal ideation."

I froze; I didn't know what to say. I had never heard that term before, but it was obvious what it meant. I didn't know people were hospitalized for thinking about suicide – I wouldn't be thrown in jail for thinking about robbing a bank – but it sounded like it wasn't just stray thoughts; it was something definite and alarming. Carolyn had been grieving a loss, but I had seen her a couple months before and thought she was getting better... instead, she had become suicidal.

Carolyn and I grew up in the Air Force together, assigned to the same base as lieutenants. She was a wild child: smart, gregarious, always laughing. I was quieter, with more of an inner life than an outward-facing one, and we immediately became friends. We both lived in the base dorms, spent Friday evenings at the Officer's Club and many Saturday nights at parties. On weekends we visited nearby sights and went 4-wheeling on the beach. Military assignments separated us before long, but we stayed in contact and were reunited for a few years after she left the Air Force and moved to Salt Lake City and I was stationed in Ogden. Those Utah weekends involved skiing, nachos, and movies rather than the beach, but little else had changed.

In the summer of 2008, Carolyn was still in Utah and I had moved to a new assignment in Virginia when she called, very sad: her beloved dog, Lila, had died after a long decline. She had known Lila was approaching her end of life, but it was still devastating. Carolyn was quiet and sad on that call, and we arranged to meet in Washington D.C. for a few days to give her a break from her too-silent house. On that trip, she was as you'd expect: unhappy, but able to enjoy visits to the museums and monuments, Georgetown, and Old Town.

Over the next few months, though, her sadness didn't go away, and other symptoms of depression appeared, unrecognized as such. During a trip to Sweden in October, she began to suffer from digestive upset, not an uncommon occurrence for someone visiting a different country. When she got home, though, it didn't go away. She became increasingly affected by the dread of returning home every day to an oppressively quiet house, her constant stomach upset, and sheer exhaustion from the stress of mourning Lila. One evening in December, she picked up a bottle of Gatorade to drink, and, as she said, "It went right through me." It was too much. She didn't want to be here anymore. In a way, with the Gatorade not pausing for even a minute in her gut, it was as if she were already gone. She walked toward her garage, intending to end her life. On the way through her kitchen, she stopped, called her mother, and obeyed the sharp instructions to hang up the phone and call 911. The paramedics arrived to find her in a sobbing heap on the floor.

Up until the point she was admitted to the hospital, Carolyn's depression crept up on her, unidentified. She was able to accomplish her work on auto-pilot but had trouble eating and sleeping; problems that she assumed stemmed from stomach upset and grief. She was more affected than even she realized, though. On her first morning in the behavioral unit of the hospital, a nurse gave her a piece of leather and a much thinner leather strip and showed her how to thread the thin strip through the broad one and tie a knot. She asked Carolyn to do it. Carolyn says it was an utterly simple task she could do in her sleep, but that day she couldn't do it. The nurse showed her again; it was simple. And Carolyn still couldn't figure out how to tie that knot. She also could not read. A friend gave her a book of "vampire haiku," the story of an immortal witness to American history, told in easy-to-digest chunks of 17 syllables. Carolyn clung to that book, slowly plowing her way through the simple passages. By the end of her 7 days of hospitalization most of her cognitive skills had recovered, but it would be weeks before she could read normally again. She left the hospital with a diagnosis of major depressive disorder, antidepressant medications, and an outpatient course of psychotherapy.

When Carolyn told me about her hospitalization and diagnosis, I already had a particular view of depression based on an experience from my college days. There was a time when the world seemed entirely gray – no life in it, no hope, no enjoyment. I was sad, even though there was nothing for me to be sad about. I could remember long ago my Dad yelling "If you don't stop crying, I'll give you something to cry about!" As a wailing child I had found that threat to be outrageously illogical; years later it almost made sense. If I were sad about something I could deal with it and get over it. But there was nothing; there was just... nothing. I decided I must be depressed. After an unbearable amount of time, which was probably just a few weeks, it went away. The world came back, replete with hope and possibility.

What sticks in my memory, though, was a strong impression of differentness: a feeling that I was not just unhappy, but rather, in an abnormal state. It was centered in my mind but intimately connected with my body, and even though I knew it was abnormal I couldn't just snap out of it. From the beginning I perceived it as chemical, as opposed to situational or bacteriological. When I needed to finish a paper or prepare for a test I would load up on a wicked amount of sugar and caffeine so I could function for a few hours. It didn't make me feel normal, but it did push my "depression" to the background for a while. It seemed to me that something that could be affected by chemicals – even sugar and caffeine – must itself be chemical in nature.

Those many years later, that experience became the lens through which I viewed depression: as a physical condition, rather than a point on an emotional spectrum. Though Carolyn was sad about Lila, depression is not the same as extreme sadness. There's a lot more to it, and you can't just cheer someone up out of depression.

Carolyn felt better after her hospitalization, but not for long. Her depression and suicidality came back, again and again, worse each time. Over a period of more than a decade, she would take a variety of antidepressants, searching for one that was effective for her. She was hospitalized for suicidal ideation and attempted suicide. She had electroconvulsive therapy (ECT), which stabilized her for years, and met regularly with a psychologist and a psychiatrist. During that same time period she earned a PhD, found a new and satisfying career teaching, and moved to Georgia to help care for her aging parents. It seemed like every time we spoke there was something good ("I've defended my dissertation and will graduate this spring"), something bad ("I don't have a job and private health insurance is too expensive"), and sometimes, in the same casual tone she would use to tell me she had been out to visit her parents, "I just got out of the hospital again; this time I had closed the garage door and started the car..." To me, and to her family and other friends, on the outside with just glimpses into her world, it was as if the ground wasn't solid under our feet anymore.

Years passed, and a solution proved elusive. Carolyn practiced yoga, got another dog, and worked with a life coach. Her suicide attempts subsided following the ECT, but she was never alright. Her voice on the phone remained weak and shaky; in person she seemed timid. She was far from the girl I remembered whooping and hollering as we drove her jeep in donuts around her friend "Frank" at the beach. (We found out later from the real Frank that he wasn't at the beach that day; that's probably why "Frank" had looked so scared.) Many times I found myself stammering nonsense while speaking with Carolyn over the phone, immobilized by a fear of making her situation worse. Even a decade later, she was still sick.

The first time Carolyn attempted suicide, I was shocked and frightened. She didn't complete the act, though, and when she was hospitalized and placed in the care of professionals, I thought her crisis was over. After the second and third attempts, there could be no sense of relief. Something had her in its grip and would not let go. I kept wondering, what's doing this? How does a normal, happy person come to feel life is so bad that she must kill herself?

By the time I retired from the Air Force I had seen Carolyn's depression control so much of her life, became aware of members of my extended family who suffered from the illness, and saw too many of our Airmen and civilians take their own lives. I began a quest to understand depression: what it is, where it comes from, and how treatment makes it go away. Or, in too many cases, fails to make it go away. I had heard that no one knows what causes depression, but I wasn't satisfied with that. With so many antidepressants on the market, we must know something, right? Understanding that we don't have all the answers, I wanted to know what we do know, and think, and where we're looking. And like so many of the scientists I spoke with, I was quickly captured. The workings of the brain are fascinating; the science of depression, and the stories of the investigations and discoveries behind it, are amazing and unlikely. And it turns out that even though there is a lot that science doesn't know about depression, what it *does* know, especially the progress made in the last two decades, is actually very useful.

Perhaps because depression disguises itself so well using tools of the human condition, it took a very long time for it to be recognized as an illness. Feeling sad is a normal emotion, with an appropriate place in our lives. But depression is not just sadness; it accesses normal emotions but then reinforces them, holding them in place. It claims a physical component, so even the sufferer's body refuses to perform regular functions appropriately. This adds to the constant unhappiness, inability to take pleasure in things, feelings of guilt and worthlessness, and in many, intrusive thoughts of suicide or death, to make depression a state set well apart from normal existence. All of these symptoms, both physical and mood, are appropriate to some situation, though; they each have a legitimate seat in our bodies and brains. It's when they won't relent but instead become unshakable and create a significant disruption to functioning that we end up with a psychiatric disorder. Depression is a highly prevalent illness, striking about 15 percent of the U.S. population at some time in their lives.[1] In 2015, the World Health Organization (WHO) estimated that more than 300 million people around the world – approximately 4.4 percent of the world's population – suffered from depression. By 2017, depression had clawed its way up the WHO rankings to become the leading cause of ill health and disability worldwide.

One of the confusing things about depression is the variety of symptoms with which it manifests. Some people can't eat or sleep; others report sleeping too much and eating too much. Some feel at their worst at the beginning of the day and get better as the day progresses; for many others it is the opposite. Most have difficulty thinking or concentrating; some feel irretrievably guilty, worthless, and even become convinced they must rid the world of their burdensome presence. With none of this showing on the outside, physicians rely on the victim's self-reporting to diagnose the disorder.

Human feelings resist categorization and quantification, though, and even somatic symptoms must be interpreted through a human voice. Without standardization, a physician might know that a type of antidepressant works well on a certain type of depression, but be unable to assess whether her patient has that type. Rates and prevalence of the disorder across the world could tell us a lot about who is vulnerable and why, and help assess whether the incidence of depression is increasing, decreasing, or remaining the same over time as various environmental factors come into play: the growth of social media, air pollution, urbanization. For that information to be shared, it must be consistently measured. Such is the concept behind the two main classification systems used in psychiatric disorders. In the U.S., the Diagnostics and Statistics Manual of Mental Disorders (DSM) defines types and degrees of depression. Throughout the world, the International Statistical Classification of Diseases and Related Health Problems (ICD) contains a chapter on mental disorders, and developers work together to keep the two references consistent.

Following World War II, the American Psychiatric Association (APA) saw a need to establish a uniform nomenclature of mental disease. Multiple, conflicting standards had grown into use, making communication between agencies, schools, hospitals, and physicians difficult and error-prone. In response, the APA issued the DSM in 1952. The manual has been updated and reissued several times, with terminology and diagnostic criteria changing along the way. At the time of this writing, the DSM is in its fifth major revision.

As described in the DSM, depression ranges in severity from dysthymia, a milder but persistent form of depression, to severe major depressive disorder with psychotic or even catatonic features. It can be driven by particular challenges, as in postpartum depression or seasonal affective disorder, or recur with no apparent cause. Major depressive disorder is characterized by one or more major depressive episodes with the symptoms persisting for most of the day, nearly every day, for at least 2 consecutive weeks. The episode must be accompanied by clinically significant distress or impairment in social, occupational, or other important areas of functioning. It must include either depressed mood or anhedonia: the inability to feel pleasure. In addition, physicians look for at least four other symptoms from the following list: significant weight loss or gain, insomnia or hypersomnia, restlessness (called psychomotor agitation) or slowing of speech and movement (psychomotor retardation), fatigue or loss of energy, feelings of worthlessness or inappropriate guilt, indecisiveness or diminished ability to think or concentrate, and recurrent thoughts of death, up through suicidal ideation or suicide attempt. The major depressive episodes that occur in bipolar disorder have the same characteristics, but the mania or hypomania accompanying that disorder indicate that there are other factors at work too. Bipolar II disorder in particular leans very much towards depression, with rare hypomanic episodes to distinguish it from major depressive disorder.

The 1,500 experts collaborating on DSM-5 wanted to shift away from descriptions of symptoms toward classification schemes based on neurobiological features. By the 2013 publication date, though, science just wasn't ready for them. Despite decades of research, preclinical studies, and clinical experience, the biological processes that create these illnesses remain cloaked and debated. That's why depression, like so many mental illnesses, is classified a "disorder," not a "disease." To call it a disease, we would have to know what biological abnormality underlies its symptoms, chronicity, and recurrence. Labeling psychiatric illnesses as disorders acknowledges that we don't know what causes them.

So I came into the research for this book understanding that science doesn't know what causes depression, or even what depression really is. And it is true that very basic information about the disorder is still unknown. What I didn't expect was the value of what science *does* know. The picture is far from complete, but already it provides insight into individual risk factors, course of illness, and the process of recovery.

A little more than 60 years ago, science and medicine got their first firm footholds into depression and a slew of other psychiatric disorders, with uneven progress from there. The science of depression steers through a rear-view mirror: a substance or action unexpectedly causes the symptoms of depression to disappear, and scientists scramble to figure out why. These seemingly random leaps forward have resulted in an odd-ball collection of theories about what underlies the illness: a serotonin deficiency, stress system dysregulation, inflammatory disease, and more. Depression science took on a scatter-shot appearance, with about a dozen different theories about what causes the illness. In the last two decades, however, a new view of depression has emerged – a unifying core giving focus to all the scattered observations. That new view says that depression comes from long-term interruption of neuroplasticity: the processes through which the brain grows and adapts over the lifetime. The brain is constantly subject to abrasive forces that wear away parts of brain cells or even kill entire cells. In a healthy brain, those forces are over-matched by new growth and the development of new connections. When the forces of abrasion overwhelm the forces of growth and repair, though, the tissue damage in particularly vulnerable areas creates and maintains depression. Many of the independent observations about depression began to resolve themselves into elements of abrasion or repair. Chronic, but not acute, administration of antidepressant medications targeting the serotonin system trigger growth processes; chronic, but not acute, stress builds up particularly potent abrasive substances; ditto for chronic, but not acute, inflammation... Rather than competing explanations of depression, stress, inflammation, and other observations are revealed as facets of underlying forces that create the sense of agonized disconnect, of "otherness," that one of the people I spoke with described as living in a glass box.

With so many potential sources of abrasion and so many different factors weighing on the brain's ability to repair damage, the path to depression is an individual one. About half of the depression sufferers I talked to pointed to a precipitating event – "My depression started when *this* happened" – but for the others, it seemed to come out of nowhere, for no reason. They tended to feel it was genetic, and worried that they would never be able to shake it themselves and would pass it to their children. Family studies do indicate that there is a genetic component to depression, but other factors are at work as well. Having experienced early life adversity adds to someone's vulnerability level, and other exposures, some of them happening in the current day, have also been shown to affect vulnerability to depression.

Depression has a broad reach, but it doesn't affect all population groups evenly. There are sharp differences around the world and even within a single country on who gets depression and who doesn't. For those who do succumb, the personal impacts can be devastating. Of all potential outcomes, the most tragic and frightening – suicide – is rising in the U.S., both among young people and adults. More than half of the suicides each year are committed by people with depres-sion, either unipolar or bipolar. Even excluding suicide, people with major depression or an anxiety disorder lose an average of 8 years of life because of their disorder.[2] And with such prominent suicidality in Carolyn's experience, I had come to think of suicide as a final stage of depression, but it isn't. Suicide has its own neurobiology and risk factors; they overlap with, but don't duplicate, those of depression.

For the most part, each treatment for depression was discovered first, followed by investigations into why it worked. So it's rather ironic that scientists acknowledge that even now they're not sure *how* those treatments work. That is, they have seen in animal models that many modes of treatment cause the brain to grow (e.g. antidepressant medications, exercise, and seizure) but they don't know exactly what mechanisms are employed. There are other modes of treatment that seem to work by slowing hyperactive parts of the brain or bolstering hypoactive parts, but again, exactly how this creates lasting effects on depression remains unclear. They are still searching, though, pushing for a better understanding that will lead to quicker, more effective treatments that reach all of depression's sufferers.

And so the body of knowledge about depression continues to grow. It is *useful* knowledge, too, as it starts to reveal individual factors in resistance and recovery.

I was delighted when Carolyn agreed to be part of this project, but bad news quickly followed. Her remission was over; she had entered another major depressive episode. And like so many of her episodes before, this one was marked by pronounced suicidal behavior. Over the months that followed, she would contribute her stories and thoughts to this book as she mentally crept toward suicide. She struggled, talked with family and friends, went to her doctor, and worked her way through her anti-suicide checklist. I sent her material to try to help her find effective relief as I encountered new treatments and therapies. It felt like a race: could we find something to get her off what felt like an inevitable path to suicide before she had that last, unbearable day?

Part One

THE GLASS BOX

Chapter One

THE MONOAMINES

Theresa was floating. Tethered to her body as if she were a balloon, she looked down to see herself driving. She was navigating the road well enough this time; once before, she had watched from above as her car turned the wrong way up a one-way street.

She came back to earth with a rush. Her mother had given her an ultimatum: go to the doctor she'd found, or move out. Theresa knew she'd have to see the doctor. It was just too much: the confusion, fainting at work, feeling sick all the time... She was dying inside; her teeth turning black, falling out; the rot from within spreading. Other people couldn't see the darkness creeping out, but she did.

Theresa had started feeling different in high school, after losing a good friend to suicide. The feeling followed her to college... several colleges; she would start one then quit and head home, then off to another and another with the same outcome. She eventually graduated and moved out west with a good job, but she couldn't shake the darkness. She often felt down; didn't sleep well, if at all, and didn't eat well. She couldn't remember when the word "depression" came into the picture, but it was there too. The trip around the world was supposed to help. They say when you cross the International Date Line your depression lifts, which it did for a while. If only she could have kept on going around and around the world, she would have been fine... but she had to come home. That was when she knew she was very sick. She had the flu, and it just wouldn't go away. Doctor after doctor gave her medicine, but none of it worked.

Theresa had finally searched out a psychiatrist, one who came highly recommended. Ha! He was clueless. The medicine he gave her didn't work. The hospital he put her in was a nightmare. And when she got out, still on the medicine that didn't work, he just gave up. "You are the biggest failure in my career," he said on her last visit. She was the biggest failure in his career because the medicine didn't work. And she was just getting worse. Right now, if someone asked her name, she'd just take a guess. If someone said "What's 1 and 1?" she would say, "Uh..." And if someone told her she'd won the lottery, and she was queen of the world, and everything is great, it wouldn't matter. It wasn't about circumstances.

Moving back home after the mental hospital, she was so scared. They couldn't help her, and she couldn't function. And now her mom was insisting she see this new doctor, who she said was the best depression guy at the National Institutes of Health.

Reaching NIH in Bethesda, Theresa was at last ushered in to see the doctor, and they talked. Before long he was telling her what she was experiencing, without her telling him. "When you go to the store, and pay for something, you think you're paying with the right amount but it's wrong," he said. Theresa was amazed. *How did he know that*? And then he said, "There's a wait-list to get into the NIH depression study, but I'm bypassing it. You're going in tomorrow."

After checking herself in the next day, Theresa was escorted to some sort of auditorium where, she found, she was to be interviewed by "the guru of depression," a special honor, judging by the staff's excitement. Seated on a chair on what looked like some sort of stage, she realized the doctors and nurses were watching from behind a one-way mirror as the "guru" walked in. He looked at her; Theresa stared back, exhausted, waiting for whatever hell he was going to put her through. Instead, he asked disarmingly, "You're feeling really bad, aren't you?"

Theresa started to cry. And then they just talked; he asked questions, and she did her best to answer. Finally, he said "Listen to me. You're in the right place, and you're not leaving until you're well."

That was almost 40 years ago. Theresa had returned to her career, married, raised children, and eventually retired. Now, having heard about my project from a family support group in the National Alliance for Mental Illness (NAMI), she had volunteered to tell me her story. She is an advocate for mental health, seeking help for herself because of her son's long-standing depression, and a helper in need to others. I had come to her, and many others who suffered from depression, to try to understand the illness. What I had already learned shattered my preconceptions. There are so many different faces to the illness and so many differences in what is going on in the brain – and the body – of the sufferer, that it is difficult even to put a boundary around it. Is it a disease? Is it one point in a spectrum? Beyond that, is it inevitable? Can it be cured? To answer, science faces a key question: what *is* depression?

In the early days, Theresa herself knew nothing about depression. Her symptoms, the mixture of physical and emotional anguish, sent her repeatedly to seek medical attention but not psychiatric help. "I couldn't sleep," she said, "I couldn't eat, and I had delusions. I figured I must have a brain tumor, because what else would cause my thinking to be so distorted? I thought everything was infected, and that I was very sick, physically. I didn't have any language to understand mental health symptoms, but I had language to understand physical health symptoms, so if I felt this bad it must be a physical thing. I was just going to die. I did have some suicidal thoughts, like, well, if I'm driving my car and I just drove right into the river, everyone would be happier. But I also had a voice going, 'That's your depression talking. It's not real.' But, it was; it felt real. It was logical – your logic changes, so you don't see things like before... Depression has a loud voice.

"There was a picture on the wall in the hospital room I was in that someone, a former patient, had drawn," she continued. "It was a clear box, a three-dimensional, glass box, and there was a person in it. You could see the person with their head down, their head in their hands, looking sad. And with one foot sticking out of the box. Like you've got a foot in reality, but if you were in a room full of people, you would think all of them are on earth, and they're functioning, but I'm somewhere else. The voices are coming at me through a big tunnel, echoing; when I talk I hear my own voice far away, totally detached from the reality they are in. And how lucky they are that they get to be just alive and okay while I'm in that box. It's like a glass box; I could see through it, I had a foot in reality. I knew which day of the week it was, who was president. They ask you those questions when you first get there. I could answer those questions, but it really was a shock when they said, 'What's 5 plus 3?' and I went 'Uh...' People who don't have it, don't really have a clear idea of what it feels like, because it's not just like 'I feel crappy; I've got to go to work; what a crappy day.' Everyone has that. Say your cat gets hit by a car and you're sad. That's understandable sadness for an event; it will eventually go away. You're sad, but that is not clinical depression. I had a roommate, in the first hospital, and she was very young, 19, a newlywed, and her husband died in a freak accident on a construction job. And she went into depression and was sent to the hospital. But we both knew that they would find her medicine and that she would get out, and she'd be fine. Sad, but she would get over it and eventually not be on meds. And that's exactly what happened. But I didn't have that... for me, nothing had happened."

THE EVIDENCE

For a very long time, depression was viewed as part of normal brain function – simply an emotional extreme or a response to an event like a bereavement. It took a cataclysm the size of World War II to set the conditions for its recognition as an illness, with physiological causes and medical treatments. The immense scale of the war created an enormous demand for the products of industrial chemical companies, and the end of hostilities left those industries with extensive infrastructure but no market for their goods. Throughout the 1950s, in the scramble to repurpose plants, equipment, and workforce, those companies would turn their resources to new products and inventions, among them some new drugs that would change how the medical world perceived depression.

Geigy was one of the firms looking for new goods and markets. From its roots as a small chemist's office in Basel, Switzerland, Geigy had grown into an international enterprise with products of organic chemistry: textile dyes, industrial chemicals, and pharmaceuticals. Geigy's insecticides were in high demand through the war, supporting troops deployed around the world. As that market faded away, though, Geigy renewed its focus on pharmaceuticals.[1] Meanwhile, a short distance away in Muensterlingen, Switzerland, Dr. Roland Kuhn had grown frustrated by limited funds, shortage of medicines, and lack of effective treatments for his patients. Kuhn was on the medical staff of the psychiatric clinic in Muensterlingen, a sizable institution with an inpatient population of about 700 and growing outpatient services.[2]

They formed a relationship. Kuhn would test Geigy's new drugs on his psychiatric patients in return for access to medications. When a different pharmaceutical company's candidate antihistamine, chlorpromazine, proved to be much more effective as a sedative and antipsychotic than it was as an antihistamine, Kuhn and his Geigy contacts decided to test their presumptive antihistamine, imipramine hydrochloride, as an antipsychotic medication instead. It didn't work; in fact, some of the schizophrenia patients they tried it on actually got worse. However, the patients who also suffered from depression found their depression lifted.[3]

Kuhn then tried the new substance as an antidepressant, with great success. He reported the exciting discovery to the medical world in 1957 and within a year had treated over 500 patients with the new drug. In 1958 he was happily broadcasting his findings in multiple journals and venues:

> *The effect is striking in patients with a deep depression. We mean by this a general retardation in thinking and action, associated with fatigue, heaviness, feeling of oppression, and a melancholic or even despairing mood, all of these symptoms being aggravated in the morning and tending to improve in the afternoon and evening. From the external appearance alone it is possible to tell that the mood improves with imipramine. The patients get up in the morning of*

> *their own accord, they speak louder and more rapidly, their facial expression becomes more vivacious. They commence some activity on their own, again seeking contact with other people, they begin to entertain themselves, take part in games, become more cheerful and are once again able to laugh. The patients express themselves as feeling much better, fatigue disappears, the feeling of heaviness in the limbs vanishes, and the sense of oppression in the chest gives way to a feeling of relief. The general inhibition, which led to the retardation, subsides. They declare that they are now able to follow other persons' train of thought, and that once more new thoughts occur to them, whereas previously they were continually tortured by the same fixed idea. They again become interested in things, are able to enjoy themselves, despondency gives way to a desire to undertake something, despair gives place to renewed hope in the future.*[4]

Kuhn's results with imipramine were so impressive that scientists around the world put it through their own trials, finding that at 2 weeks it was an effective antidepressant in approximately 30 percent of their depressed patients, and by 8 weeks in approximately 60 percent.[5] Imipramine became the first tricyclic antidepressant (TCA), named for its three-ringed chemical structure. More types of tricyclic antidepressants were developed to meet a growing market demand, their method of action unknown but their effectiveness evident.

Elsewhere, several industrial chemical companies were looking for a new purpose for hydrazine, which had been used as rocket fuel during the war. Sitting on great surpluses of both hydrazine stores and production capability, they came up with imaginative applications for it. An advertisement from Mathieson Chemicals of Baltimore enthuses, "Hydrazine is the basis for many new, ultra-efficient insecticides and herbicides. One hydrazine compound promises to reduce the familiar drudgery of lawn-mowing through slowing the growth of grass. Another is used to prepare a new anti-tubercular drug. Still others are used to silver mirrors, blow foam rubber, make new plastics and disinfectants, to crease-proof textiles and stabilize fats."[6] One of those hydrazine-derived anti-tuberculosis drugs, iproniazid, would again point toward a chemical solution to depression.

In 1952, clinical testing showed iproniazid to be effective against tuberculosis, but with significant side effects that included such a dramatic improvement in mood that the test observers grumbled "So accentuated has been the feeling of the sense of well-being that disciplinary measures have been necessary."[7] Because of these mood changes, they decided to try iproniazid as an antidepressant instead. A clinical trial in 1958 tested the drug on patients with depression, anxiety, obsessional states, or anorexia nervosa. The patients with anxiety all found their anxiety worsened. Patients with obsessional states or anorexia nervosa were unaffected by the drug. However, more than half of the patients with mild depression that involved loss of weight either recovered or were much improved. A third of the weight-losing subset showed no effect, and approximately 15 percent got worse. Where there was improvement, it took up to 10 days to start to see an effect.[8] Another trial a year later reported approximately 50 percent of the patients treated with iproniazid improved, taking 2 to 3 weeks to respond.[9] Iproniazid became a market success as another type of antidepressant: a monoamine oxidase inhibitor (MAOI).

By the dawn of the 1960s, physicians had two effective types of medications to combat depression, and the pharmaceutical industry had a market demand that was unlikely to diminish over time. There were problems, though. Both TCAs and MAOIs had side effects that could prove severe, even life-threatening. TCAs like imipramine could cause blurred vision, drowsiness, a drop in blood pressure when standing up, weight loss or weight gain, constipation, sexual problems, and more.[10] MAOIs like iproniazid could cause nausea, diarrhea or constipation, headache, drowsiness, insomnia, dizziness or lightheadedness, involuntary muscle movements, low blood pressure, and sexual problems. Even more seriously, patients taking MAOIs had to avoid foods rich in tyramine, which is found in aged cheeses, sauerkraut, cured meats, draft beer, fermented soy products, and many other foods, due to a risk of dangerously high blood pressure.[11] Both medications required weeks to be effective, and were ineffective in a large swath of patients. Clinicians and pharmaceutical companies would press forward searching for other treatments, desiring something quicker to act with a more benign side effect profile.

Theresa's stay at NIH in the early 1980s was not pleasant. She had access to the most current treatments and the best psychiatric minds the country could offer, but the price was to be a test subject. "They said I could stay at NIH as a volunteer guinea pig, as long as I signed papers to agree to their testing," she said. "And they had some pretty gruesome tests. Like, one time they did a test; it was a new procedure that is pretty common now, a PET scan, which shows a picture of your brain. But it was new, so they kept me up all night with needles in my arms. And in the morning they stuck all these needles in me and put all these electrodes on my head, and they took this picture and showed me my brain and said, 'That's the depression, right there; that color. And it's big; you can see it.'"

"Did it make you feel better, or worse, to see it?" I asked.

"It made me feel better that they weren't saying, 'Oh, you're making this up,'" she said. "They were saying, 'It's real; there it is. You can see it.' Because there were other people who said, 'If you have a bad day, just look in the mirror and just say it'll get better!' As if you hadn't tried that a billion times."

Theresa continued. "They gave me an IQ test. They said 'You have a really high IQ.' But I couldn't remember my name. I was just guessing at the answers; I just figured my brain was gone. For a while I just sat in there and did jigsaw puzzles all day and didn't talk to anyone. And then they started experimenting on me; they said, 'You'll get medicine, and you'll never know which is the placebo and which is the medicine because it's an experiment.' But I took notes and I knew exactly when I was on medicine. And they took tons of blood; they took blood every day. I had freedom; I had a car, and because I wasn't suicidal I was allowed to come and go. It wasn't pretty; it wasn't a nice place to be, but there were advantages to being at NIH. They took physical symptoms very seriously. If I said, 'I think my teeth are falling out,' they would send me to the dentist and he'd say 'No, they're not.' If I said 'I think I have the flu,' they'd check and say 'No, you don't.' I had irritable bowel, and I got treated for that. But most people are on their own; they go to the doctor once a week, and they get nothing. So being a guinea pig, my price was taking all those tests, having blood drawn every day, having all those people – interns and doctors and specialists – come and talk to me all the time; that was the price."

Theresa stayed at NIH longer than anyone anticipated. "In the end, they said 'We're sorry. We thought we could help you in 3 months, but you've been here 9 months. But now we know we've got it.' And I said 'I feel exactly the same way. You're going to send me home like this, like nothing ever happened?' But they said 'Once you leave, we know that this medicine is going to work. We know it.' I had my doubts. And one day, after about 2 weeks, I woke up and I felt it lift off of me. I felt the depression just lift off of me. That doesn't mean I was free of problems. They set me up with a psychiatrist, because when the depression's gone you've got a lot of baggage left over that you need to deal with. So I was in therapy for many years." She paused to sip her drink. "The medicine they gave me was called an MAOI."

BASICS OF NEUROTRANSMISSION

Ever since a caveman first cracked open the head of a rival and peered inside, science has explored the brain and how it works. More modern methods show there are two broad classes of cells in the nervous system: neurons, which perform signaling, and glial cells or glia, which provide a variety of support functions. Though the structure of neurons in different areas of the nervous system varies greatly, a typical neuron has three regions: dendrites, which branch out and gather inputs to the cell; the cell body, which includes the nucleus; and an axon, which extends forward to the output destination where it will release a signal through presynaptic terminals.[12]

The nervous system uses both chemical and electrical signaling. Within a neuron, a signal starts in the dendrites and passes through the axon as an electric current. From one neuron to the next, though, a signal is passed by diffusing a chemical, a "neurotransmitter," across the space between the two cells, via a synapse. The neurotransmitter is released by the presynaptic neuron and binds to a receptor on the postsynaptic neuron. The principle of chemical signaling in the nervous system was suspected as early as the beginning of the 20th century, and eventually demonstrated by Otto Loewi, who famously dreamed his experiment before rising in the middle of the night to actually try it out. In the experiment, Loewi extracted beating hearts from two frogs, placing each in its own chamber. An electrical signal applied to a nerve in one of the hearts caused it to slow down, a well-known phenomenon at the time. But then Loewi took a sample of the fluid from around the now-slower heart and applied that fluid to the frog heart in the other chamber. When that frog heart slowed its beat as well, it proved that something *in the fluid* had carried a signal from one heart to the other. Loewi had proved the existence of a chemical neurotransmitter.[13]

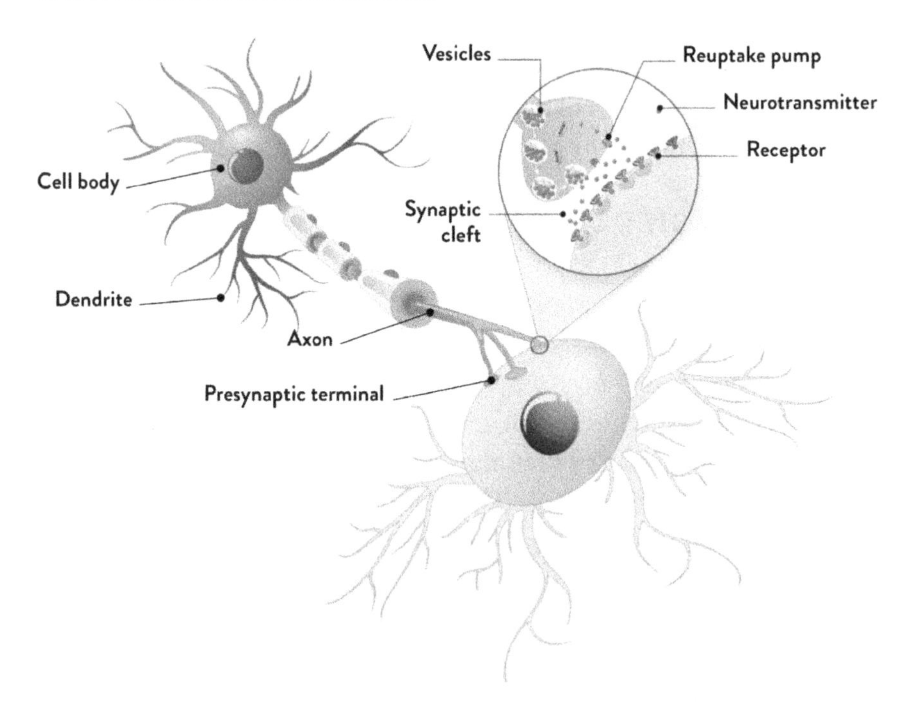

Figure 1-1: **Neuron and synapse.**

There are a large number of neurotransmitters, more than 50 known types, synthesized by cells in various regions of the body and brain. Serotonin, dopamine, and norepinephrine are all monoamine neurotransmitters (so-called because each type has one amine group in its chemical structure) and are known to be important in depression. When a neurotransmitter is released and diffuses across a synaptic cleft, it binds to its specific type of receptor; for example, serotonin binds to serotonin receptors and dopamine binds to dopamine receptors. Each neurotransmitter has a characteristic shape caused by its chemical composition and electron bonds. These molecules also have characteristic differences in polarity along their lengths, with positively- and negatively-charged portions. As an example, a water molecule (H2O) is a polarized molecule. Every water molecule has the same typical bent shape, forming an angle from one hydrogen atom to the central oxygen to the other hydrogen. The molecule's electrons are not shared evenly in a cloud around the molecule, though. The oxygen tends to pull the electrons toward itself, giving the oxygen rump at the apex a slight negative charge while the hydrogens each have a slight positive charge. The same principle holds for the larger molecules of neurotransmitters: along their lengths they fold into characteristic shapes, with some areas of positive charge and other areas of negative charge. Opposite charges attract while like charges repel, and each type of receptor has exactly the right positive/negative charge distribution and physical conformation to attract and hold its neurotransmitter.[14]

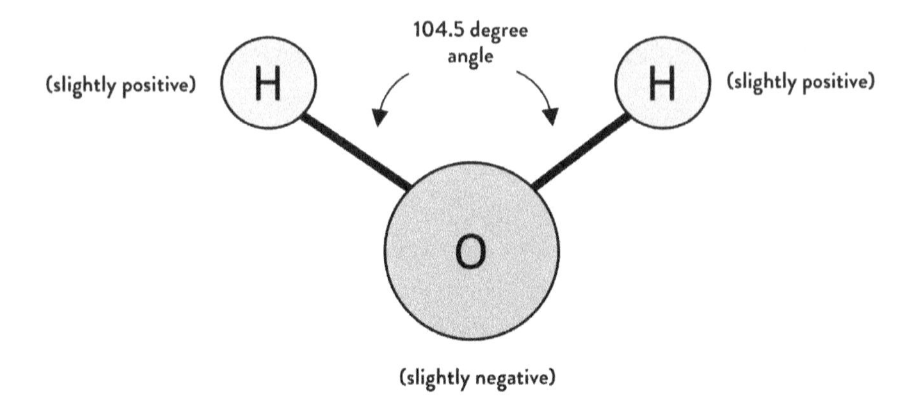

Figure 1-2: **A water molecule (H_2O), as an example of a polarized molecule.**

Signaling requires that not only do receptors bind with their neurotransmitter, but that doing so initiates action within the neuron. Neuronal signaling relies on a difference in electrical charge between the inside of a neuron and the extracellular fluid outside the neuron, created by the distribution of positively- and negatively-charged ions inside and outside the cell. That difference in electrical charge across the neuronal membrane is the neuron's resting potential. It typically ranges from -40 to -80 millivolts (mV), with the inside of the cell more negative than the outside. The neuron uses ion pumps and ion channels to maintain its polarized state. Receptors located on a neuron's cellular membrane penetrate through the membrane to offer access from outside the cell.[15]

Typically, when a neurotransmitter binds with its receptor, it changes the receptor's shape and/or electrical properties, triggering further actions. For example, picture a receptor shaped so that it provides a seat for its neurotransmitter. When that neurotransmitter molecule sits in the seat, a negatively-charged portion of the neurotransmitter comes in close proximity to a positively-charged element of the receptor. Because opposite charges attract, the positively-charged section of the receptor bends toward the negatively-charged part of the neurotransmitter, dragging open a central pore. In this example, opening that pore allows positively-charged sodium ions into the neuron, decreasing the amount of negative electrical charge inside the cell (making the interior charge less negative; i.e., closer to zero).[16] A neurotransmitter is like a key to a magnetic lock: put the right key in the right position, and the charged tumblers are pushed or pulled into the "open" configuration. The process is chemical, electrical, and mechanical, all at the same time.

As each neurotransmitter molecule binds to a receptor on that neuron, more ion channels are opened and more positively-charged ions flow into the cell. Once the charge climbs up to the threshold level (in this example, -55mV) voltage-sensitive ion channels are suddenly engaged and more positively-charged ions flow in. The positive charge increases explosively – depolarization – topping out at +40 mV. Other voltage-sensitive ion channels open, this time pushing positively-charged potassium ions *out* of the neuron – repolarization.

Just as suddenly as the charge rose to its highest value, it drops back down even below resting potential and will eventually creep back as ion concentrations are restored. The result of this sudden electrical fluctuation is an "action potential" – an electrical charge that shoots down the axon to the terminals at the next synapse. That charge causes vesicles, tiny pockets holding molecules of the neurotransmitter, to fuse to the cellular membrane and release those molecules into the synaptic cleft. Diffusing across that space, they find receptors on the postsynaptic neuron, repeating the process and moving the signal forward. Pumps are then engaged in the presynaptic neuron to restore the original balance of sodium and potassium ions – sodium ions outside, potassium ions inside – resetting for the next activation.[17] In total, the whole process takes just a few milliseconds.

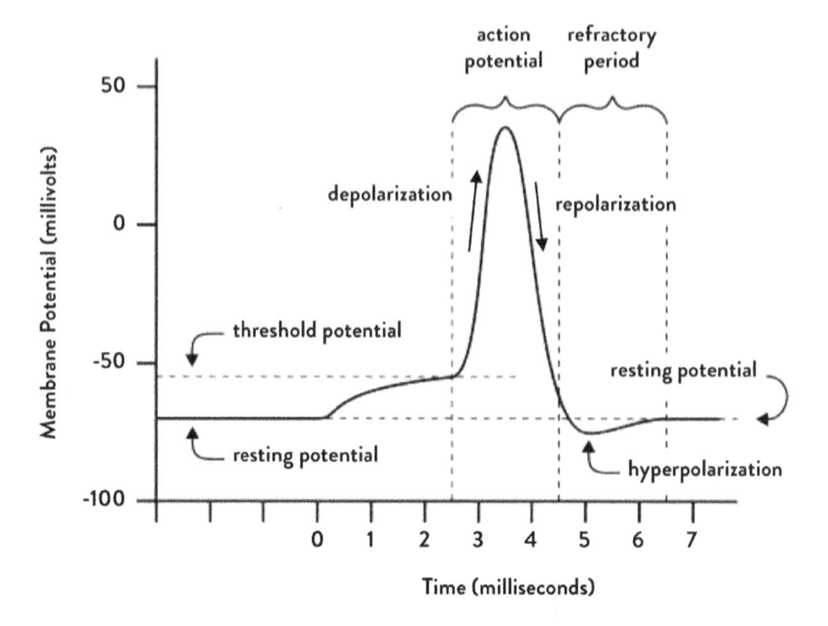

Figure 1-3: **Action potential generation.** From https://www.dummies.com/education/science/understanding-the-transmission-of-nerve-impulses/ "Transmission of a nerve impulse: Resting potential and action potential." Undated, copyright John Wiley & Sons, Inc., used with permission.

Meanwhile, the neurotransmitter whose binding started this process disengages from its receptor and floats back into the synaptic cleft, where it is swept up by a glial cell and transported back into the presynaptic neuron for reuse, a process called "reuptake." Free neurotransmitters that aren't transported back into their neuron will linger in the synapse, binding with available receptors until they are broken down into their waste products by monoamine oxidase or a similar enzyme. This is where the actions of antidepressant medications like MAOIs and TCAs come into play. By either preventing monoamine oxidase from breaking down the neurotransmitter (with a monoamine oxidase inhibitor/MAOI) or blocking the transporters used in reuptake (with a reuptake inhibitor, such as a TCA or a selective serotonin reuptake inhibitor/SSRI) these antidepressants leave monoamine neurotransmitters available in the synapse longer, increasing their signaling strength.

The type of receptor described in the example above is an "ionotropic" receptor: located directly on an ion channel, neurotransmitter binding to the receptor opens the channel. The other type of receptor is called "metabotropic"; the binding of its neurotransmitter releases a set of messengers inside the neuron that move out to take particular actions. While ionotropic receptors directly change electrical potential to cause a neuron to fire, metabotropic receptors trigger a cascade of biochemical signaling pathways inside a cell. They can open or close ion channels, activate chemical messengers for a variety of functions, cause the cell to produce various proteins, or block such protein production. With them as well, eventually the actions of the ion channels will result in signal transmission and neurotransmitter release. There are just many more activities going on in a cell as a result of a metabotropic receptor binding than the simple "fire!" of the ionotropic receptors. The proteins that are produced as a result of metabotropic receptor binding will linger in the body until they are broken down (metabolized), and so the effect of a metabotropic receptor binding can last for hours or days.[18]

All aspects of this system of neuronal signaling are important in the normal functioning of the brain. The neurotransmitters, of course, as they constitute the signaling medium. If the brain fails to synthesize enough of a particular neurotransmitter, it won't be able to keep up the necessary level of signaling. The receptors, too; if they are blocked or not present, the neurotransmitter has nothing to bind to, the neuron can't fire, and the signal can't move down the line. Glial cells are important; they sweep up used neurotransmitters to return them to the presynaptic neuron for the next firing. The transporters are important; they accept the neurotransmitter molecules and store them back in the originating neuron. The ion channels, the placement of axonal terminals of one neuron close enough to the dendrites of the next neuron that a neurotransmitter can make it across that space...[19] It's all part of a finely-tuned, complex system honed by hundreds of millions of years of evolution, but that very complexity creates opportunities for something to go wrong.

THE THEORY

By the end of the 1950s, there was considerable evidence that TCAs and MAOIs provided effective treatments for various forms of depression, but psychiatrists tended to regard each one as a specific medicine for that particular psychiatric condition, not necessarily related to one another. In the 1960s, however, American psychiatrist and researcher Joseph Schildkraut offered a bold hypothesis that would change the field forever.[20]

After graduating Harvard Medical School in 1959, Schildkraut began psychiatric training at Massachusetts Mental Health Center. Though drawn to that program by their psychoanalytic approach, once he saw the dramatic response of depressed patients to ECT, imipramine, and phenelzine (an MAOI), he became intrigued with the biological underpinnings of mental illnesses. By measuring changes in the amount of a breakdown product of norepinephrine (one of the monoamine neurotransmitters) in the urine of depressed patients before and during successful treatment with antidepressants, Schildkraut concluded that those antidepressants acted through norepinephrine. Piecing the evidence together, he proposed that depression itself could be caused by decreased neurotransmission of norepinephrine, and further, that mania could be caused by excess neurotransmission of norepinephrine and dopamine. Until this point, there had been no generally accepted scientific theory explaining how TCAs or MAOIs worked. Both types of drug were known to increase the availability of norepinephrine in the brain through different biochemical pathways, but Schildkraut was the first to propose a general theory that related the actions of those medications to the biology of depression.[21]

Published in 1965 as "The Catecholamine Hypothesis of Affective Disorders: A Review of Supporting Evidence," Schildkraut's paper launched decades of research on the relationship between neurotransmitter function and mental disorders. Even beyond the identification of the importance of those particular neurotransmitters, the implication was that depression and other affective disorders were biological diseases like diabetes, rather than emotional states – that they had identifiable causes and medical treatments. His hypothesis, and the reception it received in the scientific community, did more than suggest a mechanism for how depression worked; it placed psychiatric disorders "squarely in the realm of other medical conditions."[22]

Schildkraut's hypothesis featured the actions of norepinephrine and dopamine as the cause of affective disorders (major depressive disorder, bipolar disorder, and dysthymia). Those neurotransmitters have a direct relationship with one another. The brain synthesizes dopamine from the essential amino acid* tyrosine. Dopamine is used by some neurons, and can also be converted to norepinephrine (called noradrenaline in the U.K.). Norepinephrine can be converted to epinephrine (also called adrenaline) or broken down by monoamine oxidase. (Epinephrine is used by only a very few neurons in the brain and basically doesn't show up in discussions of brain function.)[23] Schildkraut's proposal was that depression occurs when there is not enough norepinephrine signaling to keep the noradrenergic† system working effectively. A drug like an

---

* An "essential" amino acid is one the body can't make – it has to come from the diet.

† Even though the U.S. and most of the rest of the world uses the "epinephrine" and "norepinephrine" nomenclature for the neurotransmitters themselves, the U.K. "adrenaline" and "noradrenaline" nomenclature prevails in reference to everything associated with those substances: receptors, transporters, and neurons that synthesize

MAOI would offer some relief from an insufficient norepinephrine supply by preventing any available norepinephrine from being broken down into waste products, keeping what little there is available to bind over and over again with its receptors.[24]

Two years later, a competing theory published by British researcher Alec Coppen suggested that neurotransmission of serotonin, rather than norepinephrine, was the biological basis for depression. Serotonin is also a monoamine, synthesized from the essential amino acid tryptophan. In his 1967 paper, "The Biochemistry of Affective Disorders," Coppen noted that when a patient was given tryptophan along with an MAOI, the antidepressant effects of the MAOI were heightened. The additional availability of the tryptophan translated to additional availability of serotonin, preserved in the extracellular spaces because the MAOI prevented it from being broken down. When norepinephrine's precursor was administered along with an MAOI, though, there was no impact on the antidepressant effect of the drug, implying that increased availability of norepinephrine made no real difference. Thus, Coppen theorized that antidepressant action of MAOIs stemmed from elevating extracellular levels of serotonin rather than norepinephrine.[25] Further evidence implicating serotonin over norepinephrine came when two Soviet researchers published a review that pointed out that all recognized antidepressants, plus ECT, increase brain serotonin levels, but do not necessarily increase norepinephrine levels.[26]

Whether one fell into the serotonin camp or the norepinephrine camp or thought both substances equally culpable, the implication was that a deficiency of one or more monoamines in the brain causes depression. This concept became known as the monoamine deficiency hypothesis: that the underlying biological basis for depression is a lack of neurotransmission of serotonin and/or norepinephrine, and that antidepressant treatments restore normal function by targeting this deficiency.[27]

---

and process it. So, receptors for norepinephrine and epinephrine are called "adrenergic receptors," and the system of neurons that release or respond to norepinephrine is called the "noradrenergic system."

THE SUSPECTS

While there are many neurotransmitters at work in the brain, dopamine, norepinephrine, and serotonin earn special attention because of their involvement with the wide array of symptoms of depression, from mood to cognition to physical functions like sleep and appetite. Though they are all monoamines, each of these neurotransmitters has multiple receptor types, and their effects depend on the type of receptor activated and its location in the brain.

Dopamine

Oddly, dopamine doesn't get as much attention in the depression literature as do norepinephrine and serotonin. Dopamine is a precursor to norepinephrine, of course, but it is important in its own right in the pleasure and reward systems of the brain. Dopamine neurotransmission in a small structure of the brain – the nucleus accumbens – appears to be a primary mechanism through which we experience pleasure. Anhedonia (the inability to derive pleasure from things) would seem to point directly to a dopamine dysfunction in depression.[28]

Humans have two families of dopamine receptors: D1 and D2; antipsychotic medications block D2 receptors. A dysregulation of dopamine is considered to be the cause of Parkinson's disease, the second most common neurodegenerative disorder (after Alzheimer's disease). This association with Parkinson's disease has revealed that dopamine affects growth and survival of neurons as well as motivation, reward, cognition, and emotion.[29]

Norepinephrine

Norepinephrine used in the body is primarily synthesized in adrenal glands on top of the kidneys, while norepinephrine used in the brain is synthesized in a brain structure called the locus coeruleus. Norepinephrine controls the sympathetic nervous system, mobilizing the brain and body for action through the "fight-or-flight" response. Release of norepinephrine serves as a general alarm signal for the whole body. Norepinephrine signaling reduces the urge to eat or sleep;

it focuses the attention by enhancing long-term storage and retrieval of negatively-toned emotional memories; it makes the pupils dilate and heart rate increase; it helps direct blood flow to the skeletal muscles at the expense of the gastrointestinal tract. It is a tool for survival.[30]

Norepinephrine binds with two basic classes of adrenergic receptors: alpha-adrenergic and beta-adrenergic, each with many subtypes. The actions resulting from norepinephrine binding depend on the receptor type and location, and also on the levels of norepinephrine in the brain region. The parts of the brain where these receptors are found indicate that they play a role in arousal, the stress response, memory consolidation, immune response, endocrine function, sleep and wakefulness, and pain-threshold regulation. The direction that they affect them – positively or negatively – is more complicated. The alpha-2 adrenergic receptors have the highest affinity for norepinephrine (i.e., they can bind with it even at the lowest concentrations), and tend to be inhibitory: they slow down signaling in the noradrenergic system. When the level of norepinephrine rises, however, the alpha-1-adrenergic and beta-adrenergic receptors are drawn in, and these receptors tend to increase norepinephrine signaling. So, under normal circumstances it is as if we have a foot gently on the brakes of the sympathetic nervous system, and then, when something happens, we step down on the accelerator. As researchers have experimented with various receptors to try to achieve certain effects – improving memory, for example – they have found it to be a delicate balance between blocking one type of receptor and activating other types in the same region.[31]

Serotonin

Serotonin is an ancient signaling molecule; there is evidence that the first serotonin receptors may have developed more than 700 million years ago. It is probably because of its long evolutionary history that serotonin is involved with so many functions, from gastrointestinal and cardiovascular physiology, to sensory perception, mood, behavior, and memory. In 1943, Dr. Albert Hoffman of Sandoz Laboratories discovered – and experimented with – lysergic acid diethylamide (LSD). Experimentation showed LSD's mind-altering effects, and further research found that the LSD molecule is built on a scaffold of tryptamine. Serotonin, also built on tryptamine, was discovered just a few years later, and soon there were reports that both serotonin and LSD could alter the state of mind.[32]

Most of the body's serotonin, 95 percent of it, is synthesized by cells in the gut. The last 5 percent, the brain's serotonin, is primarily synthesized in cells of the raphe nuclei from the essential amino acid tryptophan. Serotonin (officially named 5-hydroxytryptamine, abbreviated 5-HT) can be converted to a variety of products depending on which enzymes it encounters. Some serotonin is converted to melatonin (used to control the sleep-wake cycle), but most of it is broken down by monoamine oxidase to become a waste product excreted through the kidneys. There are 14 types of serotonin receptors; they are named 5-HT1 through 5-HT7, many with subtype designations (e.g. 5-HT1A, 5-HT1B, and 5-HT2A). Serotonin in the brain is broadly understood to affect mood, cognition, sleep, anxiety, appetite, body temperature, sexual behavior, movements, and gut motility. But even with so much study directed its way, there is a lot still unknown about the serotonergic* system.[33]

5-HT1A receptors were the first to be characterized, and their function is inhibitory – they slow down the firing rate of the serotonergic system. Called "autoreceptors," they sit on serotonin-releasing neurons where they catch some of the serotonin that is released. When that serotonin binds with a 5-HT1A autoreceptor, it opens a channel to allow some positively-charged ions out of the neuron, so the inside of the cell is made more negative than before. That means the neuron must bring in even more positively-charged ions to crawl up to its activation threshold – it is inhibited from firing again for a while. Thus, activation of 5-HT1A autoreceptors results in *reduced* serotonin release.[34]

While the 5-HT1A autoreceptors affect how much serotonin is available throughout the brain, the other types of serotonin receptors tend to have more behaviorally-oriented effects. The 5-HT1B receptor

---

* The "serotonergic system" is the collection of neurons that release or respond to serotonin.

regulates aggression and impulsivity. Activating this receptor decreases aggression, but also decreases learning and memory capabilities. The 5-HT2A receptor is the primary target of LSD and other psychedelics; its activation creates hallucinogenic effects. 5-HT2C receptors inhibit dopamine neurotransmission, so they could be involved with the loss of pleasure so common in depression. They are also found in the amygdala, where their activation can produce feelings of anxiety. Additionally, activating these receptors reduces food intake and blocking them produces weight gain, even obesity.[35]

The higher-numbered serotonin receptors are generally newer discoveries and in many cases their full functions are not known. Activating the 5-HT4 receptor enhances learning and memory. Little is known about the 5-HT5 and 5-HT7 families of receptors, but from their locations in the brain, they are likely involved with circadian rhythms, mood, and cognitive function. Scientists have not been able to pin down the functions of the 5-HT6 receptor, though it seems to be involved with dopamine neurotransmission, cognitive abilities, and obesity.[36]

In all, dopamine, norepinephrine, and serotonin are deeply involved in all of the functions that are affected in depression. However, the way in which those monoamines affect such functions varies widely depending on the type of receptor activated, its location in the brain, and, in some cases, the ambient level of that neurotransmitter. It's not a simple "too much" or "too little."

Both Schildkraut and Coppen admitted upfront in their very influential papers that they, and scientists and physicians in general, did not have anything approaching a full understanding of the mechanisms that caused or remitted depression. They did have enough evidence to challenge the prevailing opinion that depression was an emotional state rather than a physiological one, though. In his 1967 paper, Coppen acknowledges that he is promoting a provocative idea: "It must be admitted that it is paradoxical that a psychiatric syndrome such as a depressive illness could be considered biochemical in origin. After all, depressive reactions are a part of everyday life, and it has been argued that a depressive illness differs only by its severity from these common experiences." He saw the actions of antidepressants and ECT as proof that the biochemical changes that accompany depression and its treatment are actually responsible for the development of the disorder, not secondary to it. "In my view," he continued, "the weight of evidence, although it is by no means conclusive, suggests that biochemical changes are most important in the aetiology of affective disorders. A biochemical aetiology implies that there are certain biochemical changes in the brain which need to be restored to normal before the patient's clinical condition will improve."[37]

RECONSIDERING THE HYPOTHESIS

There was a problem with the monoamine deficiency hypothesis, though. Even in its early days there was evidence that while monoamines play an important role in the actions of antidepressants, they are not the *cause* of depression. One of the first observed issues was the placebo effect. While some drug trials reported 60 to 70 percent of test subjects recovering under antidepressant treatment, approximately 30 percent of the subjects unknowingly taking a placebo instead of the active drug also recovered.[38] If recovery depended on the actions of antidepressants on monoamine concentrations, then what could prompt a patient to recover on her own? The miraculous 60 to 70 percent effectiveness of antidepressant medications faded away to 30 to 40 percent efficacy over placebo. Still high, and the medications were still very useful, but less compelling in support of the theory. Another contradictory observation was the lag time between start of antidepressant treatment and its effect in patients. Though animal studies indicated that the drugs were active in the brain within hours of treatment, it usually took 2 to 3 weeks to see an effect in patients. What was going on between the time the drugs reached the brain and effects were seen? Lastly, by the 1980s several selective serotonin reuptake inhibitors (SSRIs) were in common therapeutic use and shown to be effective. SSRIs block the transport of serotonin back into a neuron, with the result that serotonin lingers in the synapse longer to compensate for an overall deficiency in that neurotransmitter. However, other substances also known to block serotonin reuptake, such as cocaine and amphetamine, were ineffective in treating depression.[39]

Other studies examined the effect of artificially lowering tryptophan as a route to reducing brain serotonin levels in order to induce a depressive state experimentally. When test subjects are given a diet that includes all other essential amino acids but not tryptophan, the liver rapidly produces a load of enzymes to break down those products, including the ones to break down tryptophan. Those enzymes go to work on any existing free tryptophan in the body, reducing its levels drastically. Tryptophan can't get to the brain to be converted to serotonin, and as a result, brain levels of serotonin drop quickly. This effect was seen in tryptophan-depletion studies attempting to determine if that sudden unavailability of serotonin in the brain would cause depression in healthy volunteers. Hundreds of subjects were put through this test, with the results showing that tryptophan (and by extension, serotonin) depletion did not induce depression in healthy people. It didn't even reduce their mood. In people who were previously depressed and in remission, tryptophan depletion could prompt a relapse. Reducing norepinephrine and dopamine levels in the brain through similar studies also failed to produce depression in depression-naïve subjects.[40]

Another problem with the monoamine deficiency hypothesis is that it assumes that what gets someone out of depression is necessarily the reverse of what got them in that state, but that's a bad assumption. Life is full of one-way functions, transformations that are simple and straightforward in one direction but intractably difficult in the reverse direction. Putting a spoonful of sugar in a cup of tea is simple, but getting that sugar back out is practically impossible. The actions of TCAs and MAOIs may have been the equivalent of diluting the tea: they don't undo the sugaring, but they help the patient get by in spite of it.

THE MONOAMINE HYPOTHESIS TODAY

For a disproved theory, the monoamine deficiency hypothesis has been remarkably durable. With very few exceptions, all of the antidepressant medications available today are marketed for their effects on monoamine systems. In addition to TCAs, MAOIs, and SSRIs, there are serotonin-and-norepinephrine reuptake inhibitors (SNRIs), norepinephrine reuptake inhibitors (NARIs), and many more. And what's great for depression patients but frustrating for someone trying to figure out what causes the disorder is that once they find the right combination, about 50 percent of depression patients do achieve remission with them.[41]

I spoke with Dr. John Mann of Columbia University, a leading expert in the relationship of serotonin with depression, to see where the monoamine theory stands today. Dr. Mann is currently the Director of Research and Director of Molecular Imaging and the Neuropathology Division at the New York State Psychiatric Institute, and for decades he has worked to elucidate the pathways of depression. Though the monoamine deficiency hypothesis of depression has gone by the wayside, monoamines are clearly related to depression, even if not the direct cause. Both serotonin and norepinephrine are actively involved in mood and somatic symptoms of major depression, and monoamine-based antidepressants, particularly SSRIs, are still its first line treatment.

I asked Dr. Mann what he tells his students about the monoamine deficiency hypothesis. "We used to tell our depressed patients that they probably had a deficiency of serotonin in their brains," he replied, "and we were going to correct that, indirectly, by giving them SSRIs, which worked by slowing the removal of serotonin from the synaptic clefts. If you have less serotonin but you leave it around for longer, signaling in the brain, then it amplifies the signal. So you can either have a lot of serotonin in there for a short time, and whisk it out quickly back into the serotonin terminals, or you can have less serotonin and leave it there for a longer time and thereby get an equivalent signal, by giving an SSRI.

"So that's what we told our patients. And then our group started counting the number of serotonin neurons in the brain in people who were depressed. And instead of finding fewer neurons, which was one possible explanation for a serotonin deficiency hypothesis, we found that there were normal or more neurons. So then we said, maybe the neurons aren't functioning properly; maybe they aren't making enough serotonin. And so we looked at the enzyme that makes serotonin, tryptophan hydroxylase, and we looked at the expression of the gene for that enzyme, and we found that, overall, across all the serotonin neurons, there was more tryptophan hydroxylase gene expression and protein. So there's no evidence at all that these neurons are damaged or under-functioning; they seem to have actually increased capacity. So, that was a bit paradoxical, and we weren't sure what to make of that."

Advancements in structural and functional neuroimaging technologies provide opportunities to investigate what is going on, in a living depressed brain, in much greater detail than ever before. The results of many of these studies drew attention to the 5-HT1A autoreceptors on serotonergic neurons. "A long time later, when we started doing PET scans, we discovered that there were more of these autoreceptors in the brain of depressed patients, not only while they were depressed, but after they've gotten better, between episodes of depression, and even in their offspring," Dr. Mann continued. "All of that indicates that this autoreceptor may be very important.

"And then a colleague of mine here at Columbia made a mouse with more autoreceptors and another with fewer autoreceptors – two flavors of this mouse, one with the gene over-expressing and one with the gene under-expressing. The mouse with the over-expression of autoreceptors had lower firing rates of serotonin neurons and released less serotonin. In mouse models of depression, this mouse behaved like a depressed mouse. So we came up with this model, and it's the old story: 'Water, water everywhere, but not a drop to drink.' There's plenty of serotonin in the brainstem neurons of depressed patients. But the firing rate is set artificially low because of too many autoreceptors, and they release less serotonin, so they effectively have a serotonin deficit in signaling, instead of not generally having enough serotonin overall."

Dr. Mann's team has poured out dozens of studies of the behavior of serotonin and its receptors in depression, helping to clarify the time-based actions of antidepressants, and why they take so long to show effect. With SSRIs, for example, they act by inhibiting serotonin transporters on the serotonin-releasing neurons right away. Normally, that serotonin would be recycled back into the presynaptic neuron, but the SSRIs plug that route. As the serotonin present in the synapse gradually builds up, initially, it binds to 5-HT1A autoreceptors and inhibits firing (and thus, serotonin release) from these neurons. During these early stages, the brain releases less serotonin: it is strongly inhibited because these autoreceptors are so highly active. Eventually, those autoreceptors become desensitized and even go down in number. Without the inhibition from too many autoreceptors, the brain goes back to releasing the full amount of serotonin, which remains in the synaptic cleft longer to trigger its other receptors more often.[42]

But SSRI function shows a way *out* of depression, not how someone got *in*. As Dr. Mann noted, we hadn't even discussed the mechanisms that disturb the equilibrium of which autoreceptors are an important part. People have the same genes throughout their lives, but most develop depression later in life. Under what circumstances would those genes start to over-express 5-HT1A autoreceptors? "And," he added, "We don't know at the moment just what proportion of depressive illness out there is related to this specific [autoreceptor over-expression] problem. We believe it is clearly not all depressions because SSRIs only work for about half of the patients. This suggests that serotonin-related abnormalities don't involve all depressed patients. There are other things wrong with depressed patients that we are studying, energetically."

So, the monoamine systems control key functions affected in depression – sleep, appetite, arousal, memory, cognition, and mood – and putting pressure on a monoaminergic system over a significant period of time is enough to promote remission in about half of depression's sufferers. But, those monoamine systems seem to be a lagging part of both the development and recovery from depression. Something affects them, before they affect an individual.

Theresa was born about the same time that TCAs and MAOIs were undergoing testing; the same time that science and medicine started making inroads into depression. She is a direct beneficiary of those early advances, and has been taking MAOIs since her release from NIH. "An MAOI is not the drug of choice that they give just anyone, because it comes with a diet," she said. "They sent me out of the hospital on the highest dose you can get. I'm still on it, and it's been, maybe 40 years?

I'm still on it, but I'm on the lowest dose. They have other medicines now, and doctors have said 'You should get off what you're on, with that stupid diet, and get on this. It's new, and it's better, and it will work.' But my original doctor says, 'It's not broken; don't fix it. It took you years to find it; just stick with it.'" She shrugged and smiled. "So I wrote a cookbook." And the MAOIs continue to work. "There's been a couple of times when I thought 'Uh oh, it's starting again…' But I knew what it was, and I really believed it would stop, and it did. I've been under huge amounts of stress, and I never went back into it. In the end, my doctor out in Bethesda, instead of 'You're the biggest failure of my career' says 'You're the biggest success of my career.' I had a life after that. I have three children who are grown now."

But that was how Theresa and I had met. One of her children had suffered from depression for years before being successfully treated. "If anything good came out of it, maybe I had to go through this to help my son," she said. "I can tell him, no matter how bad it gets, we are going to find you help, and you're not always going to feel like this." It is advice she gives others, too. "You have to be really strong, and you have to fight off the voice that's telling you you're better off dead. You have to not give up when you think you should. Just push and push until somebody helps you. My husband, all these years, he didn't understand. He thought my son was just slacking off. But in the mental hospital, he was in really bad shape. And when they found his medicine, you could see the difference. My husband went, 'Oh my God!' If you don't know, you just think 'Get over it,' but you try and try and can't. If someone has chicken pox, you see it on them, and they've got it. Depression is a horrible disease that people can't see."

The legacy of Kuhn, Schildkraut, and many other scientists and physicians goes well beyond the modern availability of antidepressant medications. It starts to answer the question, what is depression? Theresa experienced these developments intimately, and has a clear view of the meaning of this progress. "I was lucky to be born at a time when there were things that could help," she reflected. "I would say nobody understood it except the doctors who were studying it. The first one who told me what it was, he didn't really know much; it was a new thing. Nowadays, any doctor could diagnose it, and people need to listen to that. One time I had a guy come to my house to fix a broken heater and he was talking to me about his life and telling me that he was very unhappy, he had a new baby but didn't know why he couldn't have any joy in it, and he didn't know what to do. I said, 'I know what's wrong with you.' It turned out it was: it was depression. He didn't know; people, a lot of times they think they're going insane and they don't know. I was lucky there was a name for what it was."

And she is willing to call it out by name. "My mother always told me never to tell anyone I have depression; they'll think I'm crazy." She laughed. "I tell everyone."

Chapter Two

STRESS

The car careened toward the edge of the road. In the back seat, Julia gripped the armrests, stunned into silence. Her husband was still yelling, trying to wrestle control of the vehicle from Michael – their son, Michael, who appeared intent on killing them all. Dragging on the steering wheel, shouting at Michael to pull over, he finally brought them to a shuddering halt at the side of the road.

It had all started when Michael went to Europe for the summer... something had happened there. After he returned and went to work on a marijuana farm in Mexico, he started having psychotic episodes. They had finally gone to pick him up, to convince him to come home. They were happy when he agreed... and now, at the side of the road only half-way home, they were witnessing for themselves a psychotic break.

That was more than 10 years ago. For Julia, the time since had seen repeated hospitalizations for Michael, crises as he disappeared from time to time, dealing with the financial impact of mental illness, and coping with her husband's growing depression as well.

It is very common for caregivers to develop depression. Under chronic stress and without adequate time for self-care, they are among the most vulnerable populations. "I didn't know what depression really was until my son got it," Julia said. "I was home with my husband who had recently retired, and I couldn't handle it. He went through a period of major depression. It was a serious break, not psychotic, like my son's, but he was totally depressed." Her husband had retired from the Florida Air National Guard a year before their son's crisis. "He had reached the mandatory retirement age. He couldn't fly jets anymore, and he also got disgusted with his regular job, in government. It was very difficult living with him. He was depressed; he was majorly depressed. And it wasn't until my son's diagnosis that I could get him to do something about it. That's when I started reading a lot, and learning a lot. It's an awareness we have now... I know now that there were people I worked with who were depressed. They had the symptoms and they also had the behavior, but they never said it."

With husband and son both ill, extra responsibilities fall on Julia. "I'm the person who does most of the stuff, makes most of the decisions in the family," she said. "I get grouchy. When I'm grouchy, the whole house is grouchy, even the dog cringes. It was after all this came to a head that I said, 'I just can't deal with it all.' I started to feel depressed."

She went to her nurse practitioner, who prescribed antidepressants and the anti-anxiety medication Xanax. "I take the Xanax very seldom, but the anxiety was definitely there as well as the feeling, 'What am I going to do?' I love both those members of my family; it's just that it was so bad. I'd always been a high-energy person, pretty positive about things, and versatile and flexible, because we've lived in lots of places. But to be at this point in my life, my retirement, and thinking, 'This is what I have to look forward to.' My husband, looking back, he was depressed during most of our marriage, but I was able to nudge him to go to the gym, and I do think that being a member of the military helps a lot because he had to get his hair cut, he had to maintain weight, he had to exercise, all those things. Leaving the military was a big point in his life."

"What helped you cope with your depression?" I asked.

"My nurse practitioner, she has me on Cymbalta [an SNRI]. I've been on some other ones, and sometimes they wear off, lose effectiveness after a while. I started with anxiety and depression, but now it's more depression. I also exercise quite a bit and go to therapy, pastoral counseling. I knew we needed something, and I took my husband in, but he wouldn't cooperate, so the counselor said he'd work with me. That has been the biggest help in my life in coping with the members of my family. He was a wonderful counselor, but he retired. I never thought I'd go to someone like him, but I found him, and for 11 years he was my good friend. He helped me through a lot of things. He helped me to get rid of some of the guilt. When dealing with people who are mentally ill, it's the guilt. Get rid of the guilt, and cope."

She smiled. "Today, I feel fine. I make sure I go out with my friends. If my husband doesn't want to go out, and my son doesn't want to go out, I go out. That's one of things I learned in therapy, that I do have control over my life."

STRESS AND STRESSORS

Stress is the next piece of the puzzle of depression. Rather than stress at the macro level – the load of financial, professional, personal, and social problems that anyone with or without depression can point to – the processes in the body that make up the stress response, from the micro level on up, are deeply implicated. This came to light in the 1970s with such strength that depression is now commonly referred to as a stress-related disorder, and the mechanisms underlying the body's stress system show up repeatedly in different pathways to depression.

In 1936, Hans Selye, a Hungarian physician who had immigrated to Canada, wrote a letter to the editors of the journal *Nature* describing a "general adaptation syndrome," he had observed in test animals. The animals had been put through different sorts of shocks to their system: exposure to extreme cold, injury, excessive muscular exercise, or treatment with sub-lethal doses of a variety of drugs. In these animals Selye saw consistent changes in internal organs that had no apparent relationship with the site of injury, as well as a general loss of fat tissue and other organism-wide alterations not immediately explainable by the injury itself, no matter what the source of harm.[1] By the 1950s he had renamed the phenomenon to "stress," which in the physical sciences referred to the interaction between a force and resistance to that force. Selye applied the concept to physiology as the body's resistance to the application of a force, whether that force came from physical or psychosocial factors. That latter concept was key in the continuing contribution that Selye's work would make to psychiatry: the fact that the body could react to psychosocial events in the same way it did to physical injuries opened the door to understanding much of the two-way communication between the brain and the body.[2]

The full stress response goes beyond the "fight or flight" reaction originally described by physiologist Walter Cannon in 1915. "Fight or flight" is the body's initial reaction to an acute stressor, in which the sympathetic nervous system interacts with the adrenal glands to release loads of epinephrine and norepinephrine, preparing an animal for violent exertion on its own behalf.[3] Cannon's concept of "fight or flight" is a short-term reaction to an acute threat, involving only the release of those substances. In contrast, Selye's "stress syndrome" describes longer-term changes in the physical state, in which the body first goes through an initial alarm reaction, then a resistance phase, and then on to exhaustion and death. Selye's description of the stress response involved a characteristic pattern of involvement by specific glands, organs, and chemical signals that created predictable physiological changes and clearly went beyond epinephrine and norepinephrine. He called it "the syndrome of just being sick," and over three decades he would trace the specific actions and effects of stressors on the body.[4]

Selye's letter to *Nature* described a three-phase syndrome. In the first phase, from 6 to 48 hours following the injury, certain organs had shrunk in size, the body temperature fell, small erosions were seen in the stomach and digestive tract, fat tissue disappeared, and other effects were observed. After 48 hours, a second phase began with enlargement of the adrenal glands, changes in the pituitary and thyroid glands, cessation of general body growth, atrophy of the gonads, and cessation of milk secretion in lactating animals. If the insult or injury continued, the animal built up enough resistance that the appearance and function of their internal organs returned to normal. After 1 to 3 months of continuing slight injuries, however the animals lost their resistance and succumbed to the symptoms of the first phase.[5]

Selye recognized it wasn't only laboratory animals in which this pattern of responses presented themselves, though. As a medical student, he had seen similar reactions to different illnesses and injuries in patients, a standardized response to a variety of harmful experiences. During rounds, Selye noticed that people suffering from distinctly different ailments – cancer, injury, infections – often had numerous complaints in common: loss of appetite, weight loss, looking tired, and not being in a mood to go to work. Years later, new experiences would bring these observations back to mind.[6]

By then at a fellowship at McGill University in Montreal, Selye was tasked to isolate and identify various female sex hormones. For this project, he injected extracts of cow ovaries into female rats and measured their responses. Necropsies of the rats yielded surprising findings: enlargement of the adrenal glands, atrophy of the lymphatic system, and peptic ulcers in the stomach and small intestine. Notably, every injected noxious agent produced the same results; it was not a reaction to one particular hormone. He continued his experiments by placing the rats in various stressful situations: on the cold roof of the medical building, or a mechanized treadmill that required they run continuously. Each experiment produced the same findings: hyperactivity of the adrenal glands, lymphatic atrophy, and peptic ulcers. Selye recognized it as an expression of the body trying to return to its original condition under challenge; his insight was to link the hypothalamic-pituitary-adrenal axis to the process of a body coping with stress.[7]

In his letter to *Nature*, Selye recounted the reactions he had observed in the rats, calling it the general adaptation syndrome because it seemed to represent "a generalized effort of the organism to adapt itself to new conditions." He also commented that "the symptoms of the alarm reaction are very similar to those of histamine toxicosis or of surgical or anaphylactic shock; it is therefore not unlikely that an essential part in the initiation of the syndrome is the liberation of large quantities of histamine or some similar substance, which may be released from the tissues either mechanically in surgical injury, or by other means in other cases."[8] Years later, he would go on to name and identify the effects of the key substances responsible for the body's stress response: glucocorticoids and mineralocorticoids.[9]

THE HYPOTHALAMIC-PITUITARY-ADRENAL AXIS

Today, researchers continue discoveries detailing the stress response and the hormones involved. (Hormones are chemical messengers like neurotransmitters, but created in the body's endocrine glands and secreted into the bloodstream from which they can act on distant organs.) The stress system is an exquisitely balanced system-of-systems, with both positive and negative feedback loops. It is carefully orchestrated to stir up a massive response to a threat, mobilizing the body and the brain for immediate action and rapid repair... but it also contains the mechanisms to stop those actions and return to a normal state once the danger is past.[10]

The core of the stress system is the hypothalamic-pituitary-adrenal (HPA) axis. A part of the brain, the hypothalamus is considered the master integrator of many vital functions of the body: blood pressure, hunger and feeding, the sleep-wake cycle, sexual behavior, body temperature, and the "fight or flight" response. When activated by a stressor, the hypothalamus releases corticotrophin releasing hormone (CRH). CRH binds to receptors in the adjacent pituitary gland, causing that organ to release adrenocorticotropic hormone (ACTH) into the bloodstream. When ACTH reaches the body's adrenal glands, they release glucocorticoids: cortisol in humans and other primates, corticosterone in rodents. Some of those glucocorticoids are transported back across the blood-brain barrier to the brain where the cycle originated. Binding to receptors in several structures, they inhibit further CRH release and stop the cycle. So, the HPA axis is controlled by a negative feedback loop: products of the stress response itself swing back around to turn that response off.[11]

Hormones in the bloodstream don't reach just one target, though; they reach *all* targets, even those outside the HPA axis. Wherever a hormone finds its compatible receptor and binds to it, just as with neurotransmitter binding, something happens as a result. In addition to activating the pituitary gland, CRH released in the brain near the amygdala directly promotes anxiety and fear-related behaviors. It activates the sympathetic nervous system, causing the secretion of norepinephrine and epinephrine. It inhibits the thyroid, sex and growth hormone production, appetite, and sleep. Finally, it activates the immune system, preparing it to deal with an attack. Circulating norepinephrine, meanwhile, causes the release of inflammatory agents, further activates the amygdala, and also suppresses growth hormones, sex hormones, appetite, and sleep. It is a self-reinforcing, accelerating cycle of hormone production and physiological effects that focuses the attention on the threat, disables normal functioning that could be a distraction, and gets the body in a state to deal with the likely outcomes: wounding, infection, and running away.[12]

Once the danger is past, though, the acute stress response must stop and let the body return to normal. This is the function served by the original, negative feedback loop of the HPA axis, where the glucocorticoids go back to the brain to inhibit CRH release, and everything consequent to it winds down. When the threat can't be defeated or avoided and the stress response is not allowed to subside, though, it becomes "chronic stress." Chronic stress creates the long term changes that Selye observed in rodents that were kept under a repetitive stressor for months. Where a healthy, short-term stress response increases chances of survival, its long-term sustainment creates problems in multiple systems.[13]

Like Julia, Henry has experienced the detrimental effects of chronic stress. He is also a caregiver; his wife suffers from schizophrenia. "It started with extreme sadness, and almost panic, back in '98," he said. "I'd say, starting in '95 my wife was starting to act a little strange, but she was still active, still getting out. So it all started with my wife coming down with what turned out to be paranoid schizophrenia, or schizo-affective disorder. Even before, right after we were married, she was having trouble getting along with the people at work. She was dealing with some prior issues from before I met her that I had known about but thought were resolved. When you don't know anything about mental illness, you think things can get resolved pretty easily, but not so. They stay under the surface, and bubble and regurgitate up, and that's kind of what happened."

Henry first experienced one of his wife's psychotic episodes about 20 years ago. "It was mostly an outburst that I had ruined her life," he said. "Then she became reclusive; she wouldn't go out at all. Eventually, after an arduous struggle I came to the conclusion that she just wasn't herself; she wasn't the person I had known for 10 or 15 years. There are all these decisions you have to make at that point. It's a real struggle, with mental illness. Do you know how difficult it is to get someone detained against their will? Get them into a hospital, but against their will?"

I admitted that I didn't. It made sense to me, though, and I commented, "You can't just lock someone up because her husband says she's crazy." That would probably happen a lot.

He nodded. "It's probably the most difficult thing for someone who is on the outside, trying to help someone who is ill. It's been set up because of the patients' rights philosophy; their ability to say 'no.' But what they don't realize is that these people aren't thinking straight. Of course they think they're well; we all think we're well, even when we're doing crazy things, standing on our head naked in public, or something like that."

Henry began to develop symptoms of an illness himself. "About the time of her first hospitalization, I was getting sick," he said. "I got sick at work with symptoms involving weakness, nausea, dizziness; then I started having trouble with my muscles, twitching, involuntary muscle contractions, that sort of thing. I couldn't even sit in a chair because I was just antsy, extremely antsy. My doctor said it might be Parkinson's disease, so he referred me to a neurologist who ran some tests. He said, 'There's no positive test for Parkinson's disease. It's a brain disorder, a movement disorder.' So I had an MRI, and there wasn't a stroke or a brain tumor. They were ruling out things like multiple sclerosis and ALS. They gave me some medication. But what a lot of people don't realize is that it is normal for the medication to immediately wipe out the symptoms, and when the symptoms go away they don't say 'You're cured,' because there's no cure for Parkinson's disease. They say 'Keep taking the medication, because your symptoms will return and they'll get worse. But you can live a pretty good life for a long time if you follow the regimen.' Which is medication and exercise."

Henry was almost ready to give up. "I was sort of blaming myself for my wife's illness because she put me on this guilt trip with the outbursts," he said. "And her concerns and claims and accusations were accurate. I couldn't deny it. Every relationship has arguments; you make some mistakes and you apologize for it and think everything's going to be OK. But she still continued to cloister herself away, living as a recluse. Once I got her treatment, though, she stabilized, and I could get her out. It doesn't cure things; she didn't return to the same old personality. But because I was on this guilt trip and I started calling around to get some help for myself, I ended up with some pastoral counseling. He kind of restored my faith in myself, that I wasn't the cause of her illness.

"She had this one hospitalization; she was stable for a while, and I made the determination that I was going to try to help her as best I could. I loved her, and this wasn't her fault. None of this was her fault; she didn't bring it on herself. Just the struggle of care-giving, and the fact that I had to work too – though work was kind of a vacation at that point, getting out of the house and the social type of atmosphere at work. But on weekends, when there was no work, the sadness would set in. Sometime before that, when I was first diagnosed with Parkinson's, I was thinking about suicide. What did I have ahead of me? But any thoughts of suicide always came back to what's going to happen to her? I'd be fine; I'd be gone, but she'd be left with no connection to anybody else."

Henry's stress continued, taking a toll. "At some point I realized I just can't get my old zest for life back. It was quite a ways down the road. I had committed to helping her, and was having some success getting her out for different activities, but it was still a struggle for me. I couldn't travel, because I was afraid she couldn't survive if I were gone even 2 or 3 days. I was going to sessions, and I went to see her psychiatrist maybe every 3 months at that point. The psychiatrist would call me in for the last 5 minutes. So one day I said 'I'm not feeling so good myself; I think I might have depression. Is there some sort of medicine you could suggest for me?" I wasn't asking him to prescribe it because he wasn't treating me. So he suggested I ask my doctor about Zoloft [an SSRI]. At my next visit with my primary care doctor

I asked about it. She didn't hesitate; she'd been treating me for many years. She gave me a low dose of Zoloft, and I took that for a number of years. It did help; it blunted the sadness, and basically the suicidal thoughts pretty much went away.

"Then I felt like the Zoloft wasn't working. I was having trouble getting out of bed. And when I got out of bed and had two cups of coffee, it was still like, I want to go back to bed. I wasn't getting things done; I wasn't living other than keeping my wife going, what I had to do for her. I wasn't able to do basic things like housekeeping, running errands, the general things we all have to do. I didn't have too much interest in entertainment, or anything like that. By then I had retired, in 2008, so that allowed me more time to concentrate on her and myself. And when my wife first got sick, she was saying, 'No phone calls. I don't want you answering the phone.' I didn't have to buy into that, but I remembered how much my family likes to ask questions, and they couldn't help, so I cut everybody off. They live out of state, and I cut myself off from them. Maybe I was mistakenly buying into what my wife was saying. 'Don't have any contact with them; they haven't done much good for you.'"

It was difficult to resist her mindset. When you are living with someone, they can have an overwhelming influence on you, much stronger than others far away. "They didn't do anything bad, but in my mind they weren't going to be able to help in my crisis: the Parkinson's and my wife's illness. So I cut them all off. I didn't answer the phone; I didn't call them, didn't visit with them. And this was all that was left of my family. That went on for 12 years. So when I retired and I had that extra time, I decided to reconnect with everyone. Of course, there was a lot of explaining to do, a lot of apologizing. I was pretty sure they weren't going to get it, as to why I did it. So I was trying to repair those connections."

After the Zoloft seemed to lose effectiveness, Henry's physician switched his medication to Pristiq, an SNRI, and it has helped. Supportive family and social connections are a stress reliever, and Henry is working hard at repairing them. He also got another bit of good news. "Parkinson's turned out to be not the correct diagnosis. The reason I got un-diagnosed is that I was going to support groups for Parkinson's, thinking I had Parkinson's, and someone came up to me and said, 'You're not like us.'" He laughed. "After about 3 years they had taken me off medications, as an experiment, and 6 months, a year later, the symptoms never came back. So someone said 'You're not like us, and you're not on medication. Did you ever think about getting a second opinion?' I did, and the neurologist, a different neurologist, said 'You know, Henry, you ran me through all the tests the first neurologist did, and I don't really think you have Parkinson's. Again, there are no guarantees with this stuff, but knowing your history and that you're off medication with no symptoms, just go live your life.'"

STRESS AND DEPRESSION

Dr. Philip Gold, a longtime associate of the National Institute of Mental Health (NIMH), has spent most of his career searching out and exposing how stress turns into depression for some. With an undergraduate degree in English, Gold was always drawn to understanding meanings in words and ideas and decided to leverage that interest by becoming a psychiatrist. Though his initial focus was psychoanalysis, Vietnam wartime service brought him to the NIH where he was tasked with depression research. It was immediately evident to him that there were biological factors behind the disorder. "One of the premises of my work over the years has been that depression represents a stress response that has gone awry," he told me. "We need the stress response to survive. When we're stressed we become anxious; we focus on the danger at hand, so we turn down our capacity to be distracted by pleasurable phenomena because that promotes survival. We lose the propensity to eat, sleep, and participate in sexual activity because that would also diminish the likelihood of survival."

Chronic stress, however, leaves a permanent mark, and when crossed with the vulnerabilities imposed by genetics and early life experience, can set up the conditions for depression. It shows up physically in damaged or destroyed brain tissue, a neuropathic condition. "We're beginning to localize anatomically in the brain specific areas that confer specific symptoms of depressive disorder," Dr. Gold commented. "There is a loss of tissue in the prefrontal cortex. The subgenual prefrontal cortex does many things with relevance to depression. It is smaller and hypoactive in some types of depression, which makes it unable to effectively regulate cortisol and the norepinephrine system. It restrains the amygdala, so when it is smaller anxiety is exaggerated. It promotes the activity of the reward center, so when it's smaller, it's difficult to anticipate or experience pleasure."

Just before Gold joined NIH in the mid-1970s, a suspected link between chronic stress and depression was confirmed by Edward J. Sachar when he examined cortisol production in depressed patients. Several previous studies of the subject had found inconclusive results, a fact that Sachar attributed to issues in the control of experiments. In experiments to measure levels of stress hormones, methods that increase the immediate stress of the experimental subjects (like watching a researcher poke a needle in their arm) tended to interfere with a clear result.[14]

Sachar tried something different: inserting a catheter in the vein of a forearm, he took samples every 20 minutes for 24 hours with the blood collecting in another room and the patients unaware of when that collection occurred. All the patients were initially unmedicated, though after the initial 24-hour baseline test the depressed subjects received medication in order to measure its effect. Eight healthy controls and six psychotically depressed patients were tested in this way. The results were striking: in their unmedicated state, the depressed patients secreted almost twice as much cortisol (the primary glucocorticoid in humans) per day as did the controls, including through the night as they slept. After antidepressant treatment, cortisol secretion diminished in all six of the depressed subjects; in four of them cortisol secretion approximated normal and they recovered completely.[15]

Sachar noted that the depressed patients who were over-secreting cortisol did not appear to be doing so because they were over-stressed or over-anxious; after all, the excess cortisol production continued while they were asleep. Instead, he observed, it could be a "central limbic system dysfunction," a disturbance in the way the depressed subjects produce cortisol or a loss of the feedback inhibition that was conjectured to take place.[16]

Under normal conditions, cortisol is secreted along a daily pattern, with the highest amounts in the morning around awakening, then falling during the day and through the night to reach its lowest point about 2:00 or 3:00 am, with minor rises at the usual mealtimes of noon and early evening interrupting its decline. That morning awakening surge is dramatic, with concentrations reaching nine to ten times the level of the nighttime low in just an hour or so.[17]

Further research of cortisol secretion often failed to find a consistent pattern across a broader population base of people with depression, but did find clusters of commonality. There were groups that over-produced cortisol (hypercortisolemia), groups that under-produced it (hypocortisolemia), and groups in whom the daily cycle was flattened: not as high upon waking as in healthy controls, but tapering off less over the day and evening, so that through the night they maintain levels significantly higher than do healthy controls. In some depressed subjects, other HPA abnormalities have been detected, with the usual cycle advanced in time by several hours from normal, or small bursts of CRH or ACTH during the day unmatched by an echoing cortisol release. Though scientists initially referred to depression as involving "HPA axis hyperactivity," over time that designation changed to a more general "HPA axis dysregulation."[18]

And so scientists conjectured that perhaps the root cause of depression is a faulty stress response, leading to disturbances in monoamine systems. It is easy to see how chronic or simply ill-timed activation of the HPA axis would cause many of the problems seen in depression, starting with activities like sleep and appetite that are under hormonal control by the hypothalamus. A constantly activated amygdala could be responsible for the pervasive anxiety and negativity that characterizes depression, and there are clear paths to other symptoms of depression as well. Animal studies show that excess glucocorticoids inhibit the growth of new neurons in the brain, decrease formation and turnover of synapses, and can also damage neurons, possibly leading to cognitive and memory deficits or the damaged brain tissue Dr. Gold mentioned. Other animal research has shown that glucocorticoids regulate tryptophan hydroxylase availability, and thus serotonin synthesis, in the brain. Additionally, reducing glucocorticoid levels in rats resulted in an increase in some serotonin receptors, and restoring those glucocorticoid levels decreased the number of serotonin receptors, tying glucocorticoid levels to serotonergic activity and thus to mood, sleep, appetite, and more.[19]

As Dr. Gold remarked, HPA axis dysregulation resulting in either too much or too little cortisol compared to what the body needs and expects at that time of day would lead to the contrasting symptoms sets seen in depression, particularly in melancholic versus atypical depression. In melancholic depression, the sufferer loses appetite and sleep, and experiences a loss of pleasure in all, or almost all, activities or a lack of reaction to usually pleasurable stimuli, and even something good happening cannot improve his mood. Melancholic depression is at its worst in the morning and can improve as the day goes on. In contrast, a person suffering from atypical depression has increased appetite or weight gain, sleeps too much, experiences a feeling of leaden paralysis, and has a long-standing pattern of extreme sensitivity to perceived interpersonal rejection. Atypical depression involves mood reactivity – the capacity to be cheered up when something good happens.[20] Confusingly, many people experience episodes of depression that involve features of both subtypes. Statistics show that 25 to 30 percent of people with depression experience purely melancholic features, 15 to 30 percent experience purely atypical features, and the others experience a mix. Additionally, the characteristics of a person's depression can change as their depressive episode continues. A person in a major depressive episode may sleep too much one week, then be unable to sleep the next week, and so on to a confusing mixture of symptoms that emphasizes the complexity of the disorder.[21]

Those collections of symptoms could reflect the results of either too much or too little cortisol secretion. "When you look at the phenomenology of the stress response and you look at the phenomenology of melancholic depression," Dr. Gold noted, "they're virtually identical except that the changes are more pronounced and prolonged in melancholic depression. Melancholics are anxious, don't sleep as much, don't eat as much, sexual activities diminish, and so forth. They have activation of cortisol and norepinephrine secretion, increased inflammation and clotting and so forth. And so, melancholic depression looks like a stress response that's stuck in the 'on' position.

"Atypical depression is the antithesis: it looks like it's stuck in the 'off' position. Anxiety is not generally a prominent part of atypical depression, but patients with atypical depression often say they feel out of touch with themselves, with their pasts, and with others. They feel fatigued, they sleep more, they eat more, and so forth. In the 1920s, when children were put in orphanages early and neglected, after about 3 to 4 months they stopped protesting, or crying, or trying to get anybody's attention, even when they were hungry or thirsty; they shut down. That looks like it may have been the onset of atypical depression."

A link between the contrasting symptoms and the stress system was confirmed in 2013, when over a hundred each of patients with chronic, severe, melancholic depression and chronic, severe, atypical depression were compared to hundreds of control subjects. The patients with melancholic depression had significantly higher cortisol levels than the patients with atypical depression or controls, especially upon waking – they were hypercortisolemic. The patients with atypical depression had significantly lower awakening cortisol than patients with melancholic depression and controls, showing hypocortisolemia, but their cortisol levels did not drop as much during the day.[22]

Though a strict under- or over-engagement of the stress system wouldn't explain the many people whose depression shows mixed features, I would think that these differences would provide important clues as to what lies beneath this disorder. It is startling, though, how little discussion there is of the differences behind the subtypes. (For a while I thought that "atypical antidepressants" were designed to treat atypical depression, but not so. In fact, since their side effects include weight gain and drowsiness, they seem more likely to give someone symptoms of atypical depression than relieve them.) The difference does matter, as Dr. Gold pointed out. "The tricyclic antidepressants, which were the mainstays before the SSRIs, work with melancholic depression but not with atypical. TCAs reduce the activity of the CRH system and HPA axis; a drug that does that is not suited to treat atypical depression."

GLUCOCORTICOIDS AND THEIR RECEPTORS

Doing the research for this project, I was surprised by how actively we use our genetic information. As the body develops and grows, our genes determine if we have blue eyes or brown, and a whole lot more. After a cell is built, though, its genetic information is in constant use to operate the cell. Genes encode proteins, and proteins basically carry out all the body's functions. They form the building material for cellular structures; they are messengers; they transport substances into and out of cells – they are the "apps" of the cell. Because proteins get broken down by metabolic processes, they have to be replenished frequently. Liver cells are in constant demand to produce enzymes – which are also proteins – to break down food; membranes and other structures need replacement in all cell types, and so on. Depending on the type of cell and the signals it gets, it is in a perpetual state of expressing genes; i.e., using the information encoded in the gene to produce the corresponding protein.

Receptors are proteins, too. There is a gene that encodes for each type of receptor – a serotonin 5-HT1A receptor gene, a dopamine D2 receptor gene, a glucocorticoid receptor gene, etc. When that gene is expressed, one of those receptors is produced. It takes its position either inside or spanning the outer membrane of the cell, waiting for its ligand (like serotonin, dopamine, or cortisol) to come along. If an insufficient number of receptors are produced or insufficient ligand molecules synthesized, or something interferes in the binding process, then the functional effect won't take place.

Selye had identified two types of substances that are involved in the stress response: glucocorticoids, which, among other functions, travel back to the hypothalamus to turn off the stress response when it is no longer of use, and mineralocorticoids, which were named for their association with controlling absorption of minerals like sodium and potassium. Both mineralocorticoid receptors (MRs) and glucocorticoid receptors (GRs) play roles in the stress system. Though cortisol is a glucocorticoid, it has a higher affinity for MRs than for GRs. As a result, in places where both types of receptor exist, MRs are constantly occupied with normal levels of cortisol and GRs really only get busy when the MRs are saturated. This happens at certain times of the day starting with the early morning peak of cortisol production, or when a heightened stress response floods the body with cortisol.[23]

Once bound, either type of receptor initiates a chain of activity. GRs have been closely studied, so a lot is known about their structure and function. They position themselves in the cytoplasm of a cell, surrounded and held in place by chaperone proteins. When a glucocorticoid such as cortisol diffuses from the bloodstream across the cell's membrane and binds to the GR, the physical conformation of the receptor complex changes, causing the GR to break away from its chaperones and migrate into the nucleus of the cell. Once inside the nucleus, the GR attaches to a glucocorticoid response element on a strand of DNA, and the target gene is transcribed. Alternatively, a GR can also attach to a part of the DNA where it represses transcription of a gene that would otherwise take place. A common target of this repressive action are genes that produce inflammatory agents; that's one way cortisol gets its anti-inflammatory quality. When gene transcription is finished, the GR is recycled out of the nucleus back to the cytoplasm, to rejoin some chaperones and wait for the next suitable hormone to arrive.[24]

Of course, all of this is what happens in a normal, healthy stress response. The point of HPA axis dysregulation is that something is *not* happening as it should. It could be a problem with the H, the P, or the A; it could be a problem involving CRH, ACTH, cortisol, or any of their receptors. By devising tests to probe along the path of the HPA axis, scientists were able to pin down where the breakdown occurs: in feedback inhibition, or the property of cortisol to turn off CRH release. They eliminated MRs as the culprits and turned to the GRs, finding that GR response to glucocorticoids is delayed in atypical depression and impaired in melancholic depression. The faulty response of GRs in the presence of high concentrations of cortisol is called glucocorticoid resistance, and is now considered one of the most consistent characteristics of depression.[25]

So, chronic stress can impact the normal functioning of the stress response by interfering with its feedback loop, while also affecting levels of norepinephrine, expression of serotonin receptors, and more. It is no wonder that chronic stress has a strong role in depression.

Both Julia and Henry became depressed later in life, subjected to a slow, grinding, chronic stress. For both of them, the way through it has been individual counseling to help them deal with their still-ongoing stressors, and a monoamine-based antidepressant medication to treat its neurophysiological effects. It is a mix of therapeutic modes that many physicians, including Dr. Gold, endorse. Though he has dedicated his career to teasing out the biochemical details of those systems and targets, he never lost his focus on the human element of this disorder. "We still need psychotherapy in the optimal treatment of depression," he emphasizes. "When we're stressed, the emotional memories of that stress change the structure of the stress system, and are consolidated as emotional memories that might emerge at any time we're under great stress or depressed again. The burden of stress and the exposure to stressful stimuli, based on how we're wired to respond, have a great deal to do with precipitating depression and influencing its natural history. A person with a lot of stressors imprinted on the brain is not likely to respond to antidepressants as well unless those psychological and existential elements are addressed at the same time."

Chapter Three

INFLAMMATION

Dr. Charles Raison was on his way to a meeting addressing psychedelics – LSD, psilocybin, and such – in the treatment of depression when we spoke. A University of Wisconsin-Madison professor and medical doctor, he also directs research in an organization working for permission to study psilocybin in the disorder. And for many years, Raison has studied the role of inflammation and the immune system in depression.

Immunotherapy is an exciting and promising field, turning the body's own defenses against diseases like cancer and hepatitis. Early studies showed an unexpected drawback, though, when many of the patients treated with this new therapy developed classic symptoms of depression, including "anhedonia, helplessness, and dysphoria, in addition to fatigue, apathy, and mental slowing."[1] Certainly, someone who needs experimental therapy to treat a life-threatening disease could be assumed to develop depression from long-term stress, but the number of cases and the timing – soon after initiating these therapies – raised flags. Up to 50 percent of the patients receiving interferon-alpha (IFN-α) for the treatment of cancer or infectious diseases developed behavioral symptoms strikingly similar to depression. Plus, their depressive symptoms responded to treatment by traditional antidepressants. By all appearances, the immune factors used in their therapy could cause depression.[2]

And these were not just mild cases of depression. Raison and co-author Dr. Andrew Miller commented in a 2011 paper,

> *The power of IFN-α to induce psychic misery in patients taking it for hepatitis C virus infection or cancer is not subtle, as any clinician who has treated these patients can attest. At high doses, fully 50% of patients without pretreatment depression will meet criteria for major depressive disorder within 3 months of commencing therapy, and a far higher percentage–perhaps 90% or more–will endorse at least one or two significant depressive symptoms (with fatigue being the most common)... Except for the rare induction of mania, no one feels better during IFN-α treatment. Some people slog through with only malaise and fatigue to report, whereas others develop suicidal major depressive episodes that require IFN-α discontinuation and psychiatric hospitalization. Between these two extremes, one observes every possible level of depressive outcome.*[3]

Immunotherapy uses cytokines – small proteins that act as chemical messengers, like hormones; the only difference seems to be that cytokines signal to elements of the immune system. The cytokines relevant to depression fall into two basic groups: "pro-inflammatory," telling the immune system to initiate a state of inflammation, and "anti-inflammatory," telling it to turn off an inflammatory response. The cytokines used in immunotherapy tend to be pro-inflammatory as they are intended to bolster the body's defenses. Though there are many pro-inflammatory cytokines active in the body and brain (including IFN-α), the ones that have earned the most discussion in studies of depression are interleukin-6 (IL-6) and tumor necrosis factor (TNF).

After clinicians reported that people treated with immunotherapy were getting depressed, several studies confirmed an association between depression and the pro-inflammatory cytokines used in that sort of treatment: TNF and IL-6 were each found to be significantly higher in people with depression than in control populations.[4] Then, putting immunotherapy aside and looking at patients with major depression from any source, researchers found that depression is associated with higher levels of C-reactive protein (CRP) and IL-6. (CRP is produced by the liver in response to inflammation. Though not itself a cytokine, its presence indicates a recent or ongoing inflammatory response.) And the depressed patients used in the study had been carefully selected so that they didn't have any sort of medical illness that could itself explain the inflammation.[5] Somehow, inflammation and depression were linked.

Angela is simultaneously one of the luckiest and unluckiest people I've ever come across.

We met near Baltimore on a scorching day in July. Driven from the cafe by the noise and crowd, we found a shady spot and settled to talk. "I've probably been suffering from depression since the age of 13," she began. "I've been suicidal since that age also. Nobody picked up on it. I was extremely high-functioning: I was a star athlete, a star student, went to Cornell, ran Division 1 track; nobody ever picked up on it. In 2001, my father passed away very suddenly of a heart attack. He was only 53. Two weeks later, I had just gone back to work, and 9-11 happened and I lost some friends from the towers. So there was a very serious period right there where I could easily say I was depressed, and people just felt it was situational. Two years later I ended up breaking up with my boyfriend. After that, I was feeling really bad, and I went to Faculty and Health Services. I was working at Johns Hopkins at the time. I said 'I'm not feeling great about things, and I think maybe I need some help.' So they sent me to a psychiatrist who put me on Celexa [an SSRI], a pretty low dose, and some trazodone [an atypical antidepressant], and sort of followed me for a couple years. She wasn't too terribly concerned, though, and I sort of fell off the radar. I think I was probably very good at concealing it."

As the years passed, Angela's mental state didn't improve, and she started experiencing other, strange symptoms. "I wasn't on the Celexa anymore, and I wasn't doing any worse, but I wasn't doing any better. I had managed to cope and get through that period of the breakup. I met my husband in 2005. Then shortly after my 30th birthday, all of a sudden I developed this extreme pain in my abdomen. So I was in the hospital for that. They took out my appendix, they did two exploratory surgeries. I had a colonoscopy; I had an endoscopy; I had all these things, and no one could quite figure out what it was. I had some of the symptoms of Crohn's disease, but not all of them. No one could help, but at least no one denied that it was there."

For 2 years, Angela went to various specialists, trying to get a handle on her physical symptoms. She was still depressed, and also had abdominal pains, swelling joints, migraines, and night sweats. As an employee of Johns Hopkins, a premiere medical institution, she had access to the top doctors in the world, but they couldn't figure out what was wrong. "So I saw the top rheumatologist; I saw the top GI people; nobody could figure it out. And none of my markers were positive for anything. You start to feel like you're crazy. They put me on steroids and then I was being seen by the pain clinic at Hopkins. So I was also on opioids, to manage the pain. And then finally, in 2009, I walked into my doctor's office and said 'I can't do this anymore. I'm done. I just don't want to be here anymore.'"

There was more to that last visit to the doctor, as she would tell me as we continued. Angela's suicidality had increased along with her depression, and she had made plans to kill herself during an upcoming business trip. "It was always there," she said about her suicidality, "though there were some ebbs and flows. It was always this sense that somehow there was something inherently flawed about me. That I was not a lovable person; that I somehow just didn't fit. I needed to escape. It was always there, and I always had plans. When I was younger they were more vague plans. I would occasionally try to hurt myself, to see how it would go. What landed me in the hospital was that I had a completely well-thought-out plan. I had a trip planned, because I traveled a lot for work. I had some preconditions, like I didn't want my husband or my mom to find me. I didn't want them to have to deal with that. I was traveling to Cincinnati, and at that time I was taking so many medications, it was going to be very easy."

The doctor's visit derailed those plans. "When I walked into my doctor's office, it was a regular appointment. I don't know that I was preparing to say anything," she said. "I don't know what it was; it might have just been her questions, her pointed questions at the time. She, being the adept doctor that she was, called up the Hopkins mood disorders clinic and they got me in there to be seen. I stayed in the mood disorders floor for 3 months straight, and then another 3 months plus at the day hospital. And during that time I had over 50 bilateral ECTs."

"Fifty?" I asked. "Isn't that a lot? I thought the usual number was in the teens."

"Fifty," she replied. "Bilateral. I lost my memory. I don't remember the period when I was depressed at all. When I was in the hospital, I have vague flashes of that; but the day hospital, I don't really remember that. Or certain things that happened right after, while I was still getting the ECTs. But once I stopped ECTs, it's been fine." The ECT sessions took place over a year, starting with three sessions a week, then once a week, then every 2 weeks, and so on, pacing down to a maintenance course by the end.

"Did it work?" I asked.

"It did," she said. "I've been, ever since..." she paused. "It's been a gradual process. But in my mind it's the best thing that could ever have happened to me. That hospitalization, all that treatment, everything, because I wouldn't have the life I have today. Not even close. And I wouldn't have the understanding and the ability to cope. At this point I feel like I'm healthier than the majority of the mass public, just because I have that ability to look inside myself, analyze my emotions, and deal with things. And I've been extremely fortunate to have been followed by really, really great doctors."

Her pain diminished after the ECT, but even as her mental state improved, many physical symptoms remained. "I was suffering from chronic migraines, and we now think that the original stomach pain was an abdominal migraine. And a lot of the things they were doing just exacerbated it. Of course I saw a neurologist, and I saw some doctors about migraines, and a lot of it is diet and this and that. I worked with my psychiatrist ultimately to help find a course to manage the migraines. So it got better, and then things started to get worse. That's when we found the MAP."

MAP is *Mycobacterium avium paratuberculosis*, an infectious organism usually found in farm animals like cows and sheep. In people infected with the organism, it is called 'human paratuberculosis.' "We think I've had it – MAP – since I lived in Poland as a kid. We moved there when I was 8 years old. They think I might have got it from unpasteurized milk."

Angela had apparently been carrying the infection since she was a child, though usually, her immune system kept it in check. That implies long-term immune system activation, though, and with her growing depression and the particular stresses she had endured, the MAP finally broke out. It was a chance meeting in 2016 that would eventually provide relief. "We discovered my condition when my husband was in Russia for a conference; he's an MD-PhD. I'd been sick for years and years. He heard about MAP from a colleague giving a talk, talked to him, and said 'I think my wife might have something like this.' So we sent my blood to New Zealand, and it came up positive. At the time there were only two doctors in the U.S. studying it; one happened to be in Maryland."

Seeing the similarity of his wife's symptoms to the MAP infections he heard about at the conference, Angela's husband searched among his colleagues and made a connection to that physician, Dr. Harry Oken. For most of the last decade, Dr. Oken has spent a lot of time and effort working with patients suffering from Crohn's disease or other similar conditions who have not been successfully treated. He has become an expert in MAP infections and has great success in treating these patients. It involves a load of powerful antibiotics, over a long period of time.

I went to see Dr. Oken to try to understand how a MAP infection could interact with the immune system and the brain. "Outside the body, MAP is a thick-walled microorganism that lives in soil, the udders of cows, and milk," he explained. "But in the body, it sheds its cell wall and it becomes what we call a sphereoplast, or cell wall-deficient bacteria. In that form, it can live many places, even inside our cells, and be very difficult to detect."

They were able to find MAP in Angela's blood by isolating white blood cells and culturing the MAP microorganism residing inside them. "Angela was likely exposed to MAP during her childhood in Poland, where she recalls having suffered from almost constant diarrhea," Oken said. After weighing the options, they decided to treat her with a course of three antibiotics. Within 2 months, Angela noticed an improvement in her headaches and energy level, and progress slowly continued. Less than a year after starting her antibiotics, Angela stated, "My life has dramatically and astonishingly improved."

THE IMMUNE SYSTEM

The human immune system evolved over millions of years, driven by continually changing threats from biological pathogens. It has multiple layers of protection and defense, starting with physical protections such as the skin and the acidic environment of the stomach. Invading organisms that survive the physical layer meet with a variety of innate immune defenses to slow them down while adaptive defenses are prepared and deployed. Physical defenses exist all the time, under attack or not; innate immune defenses are called into place when a threat is realized. They include inflammation, phagocytic cells (which consume invaders), natural killer cells, and more.[6]

Detecting damage or evidence of an invader, a cell sends out cytokines as a "help" signal. The first set of cytokines are interferons, a type of pro-inflammatory cytokine that calls cells of the immune system to the site of damage. This initiates inflammation, an environment simultaneously inhospitable to the invader and supportive to the body's immune defenses. Blood vessels in the area start to leak, allowing cells of the immune system to exit the bloodstream and go to the site of damage; this leakiness causes inflammation's redness, swelling, heat, and pain. More cytokines are released to mediate the rest of the immune response. Natural killer cells arrive and put out another cytokine signal, this time interferon-gamma, to orchestrate more defenses. Phagocytic cells are drawn in to ingest and destroy the pathogens.[7]

When one of the varieties of phagocytes detects material indicating a bacteria or virus, it binds to it, destroys it, and carries pieces of the pathogen (antigens) back to the adaptive immune system to craft a targeted response to the new attacker (antibodies). After the attack or damage has been dealt with, anti-inflammatory cytokines are released to marshal the forces necessary to restore order and terminate the inflammatory response.[8] Chronic inflammation, in which the inflammatory response keeps going well past its usefulness, causes tissue damage and can result in conditions like heart disease. It can be detected in higher levels of circulating pro-inflammatory cytokines as well as other markers like CRP. And evidence shows that chronic inflammation in the body can lead to inflammation in the brain, causing changes in the brain's biochemistry associated with depression.

THE BRAIN'S IMMUNE RESPONSE

The brain benefits from the body's immune system and adds layers of its own, starting with the blood-brain barrier. Endothelial cells lining the brain's capillaries form a tight shield, helped by astrocytes (a type of glial cell) that wrap their end feet appendages around those blood vessels. Together, they regulate what comes in and out of the brain from the bloodstream: specialized transporters carry energy substrates and essential amino acids into the brain and metabolic (waste) products out. Toxic substances and circulating neurotransmitters like serotonin or norepinephrine from the body are kept out.[9]

The brain's next line of defense comes from microglia, another class of glial cells, specialized to orchestrate and execute inflammatory, protective, recuperative, and toxic processes for the brain. Microglia account for about 10 percent of all brain cells. They are very mobile, typically resting quietly in place but quickly activated to meet a variety of threats: infection, injury, or neurodegeneration. In their activated state, microglia direct their extensions toward the site of infection or injury, proliferate, and change their shape and size. A fully activated microglial cell has retracted its extensions and assumed the shape of an amoeba. As a phagocyte, it envelops and consumes invading cells or cell parts, and also secretes either pro- or anti-inflammatory cytokines and other mediators, depending on how the microglia was activated.

Microglia can remain in an activated state for several months, releasing more cytokines all the while. Microglia, and the neuroinflammatory processes they trigger, are essential parts of the immune response of the brain to pathogenic challenges, but their immune response can also go too far and damage neurons and glia.[10] The other two classes of glial cells, astrocytes and oligodendrocytes, also participate in the brain's immune system; the astrocytes support the blood-brain barrier, and both they and oligodendrocytes can release pro- and anti-inflammatory cytokines.[11]

It was long thought that the blood-brain barrier ensures a complete separation of the brain's immune system from that of the body, but now scientists know that inflammation from the body can cross into the brain through several mechanisms. First, there are "leaky" regions in the blood-brain barrier that cytokines can cross, despite being relatively large molecules, and some inflammatory agents are actively transported across that barrier as well. The vagus nerve, extending between the gut and the brain, has been shown to relay information to cause release of cytokines in the brain. Recently, an extension of the lymphatic system was discovered in the tissue immediately surrounding the brain, providing another potential avenue for immune system cells to cross in. Other mechanisms are suspected; the blood-brain barrier is not impenetrable after all. Influence goes the other way as well: the brain affects the body's immune system through hormonal control of the autonomic nervous system, regulated by the hypothalamus. Norepinephrine released from nerves of the autonomic nervous system diffuses to distant sites in the body to interact with immune cells.[12]

Now, 2 years after initiating her treatment for MAP, Angela is still improving but has not been able to get off the antibiotics. "It feels like I am, probably, 75 percent? Every time we've tried to reduce the antibiotics, it starts to come back," she said. And, after 3 years of learning through active practice, Dr. Oken is still confident that Angela had been exposed to MAP and that the long course of antibiotics is effectively treating her, but cannot say with confidence that her MAP exposure – rather than other infectious agents – specifically caused her symptoms. Or whether changing her microbiome with the antibiotics may have made her more immunocompetent; nevertheless, Angela has improved with the intervention.

"When I first started working with her, I thought, 'Ah-ha! She's got MAP," he said. "I do think that she had MAP, and it could be a manifestation of MAP, but you can grow MAP from normal people." He compared MAP to *Mycobacterium tuberculosis*, the microorganism that causes tuberculosis. "There are about 7 billion people in the world. Two billion have been exposed to tuberculosis; we would call them PPD positive," he said. "But 2 billion people don't have tuberculosis; about 10 million a year do. We could actually demonstrate immune response to tuberculosis in those 2 billion people, but they're not sick. What's happened is the body has learned how to quarantine these tuberculosis bacteria so they don't cause a problem, and only a small percentage per year come down with active tuberculosis, typically when there is a weakening of their immune system." MAP is also thought to have widespread exposure throughout the world's population, but only in some people, likely those who are immune-compromised in some way, does it become active as Crohn's disease spectrum. "My conclusion is lots of people are exposed to MAP," Dr. Oken continued, "but in general they have the immune system to make it quarantined and a passenger." In Angela, the early symptoms, then the surgical investigations, steroids, and other medications used in an attempt to treat it would have taxed her immune system. Possibly very early on, the MAP broke loose and intensified, causing a cascade of harm.

Or was it another infectious organism? "I think, as time goes on," Dr. Oken said, "we're going to find out that lots of illnesses, depending on our microbiome and depending on our immune state, that we are going to see lots of things that we will decide have an infectious origin." Including conditions like schizophrenia and Alzheimer's disease. The former has shown an association with *Toxoplasma gondii*, another microorganism that the body "safely" quarantines, and in the latter, the amyloid plaques characteristic of advanced Alzheimer's disease could be a product of microglial anti-bacterial activity. "We think amyloid is a byproduct of the microglia trying to defend themselves," he said. He spoke of a colleague, who repeatedly found another cell wall-deficient bacteria, *Chlamydia pneumoniae*, in brain tissue of Alzheimer's disease patients. "Most people are aware that syphilis can cause 'madness,'" he commented. "So why is it so hard to figure out that maybe there's an organism responsible for dementia too?"

It was a long and painful journey for Angela and her family, but in the end it brought healing. When I asked her how she felt, she responded confidently, "I feel great. This sounds crazy, but it is the best thing that ever happened to me, because it enabled me to live my life with confidence, and awareness, and a greater sensitivity." She is very positive about the psychotherapy she received, cognitive behavioral therapy. "It was amazing. All of my psychiatrists are Hopkins-trained therapists and part of their practice involves CBT as well. It's been fantastic in terms of being able to stop and think, 'Why am I being upset about this?' I think it's a good life habit for everybody."

Angela talks with a lot of people about depression. It hasn't recurred since her hospitalization in 2009, but it is still a vivid memory. I asked what she would advise people reading this book. "When you're in the middle of it, it just feels like you are unique, that you are the only person in the world that ever felt this way, and everything feels completely insurmountable. There's no way out; there's not even a pin-prick of light. It alters your mindset so much that you can't figure it out until you're out... that your thinking is not clear. Being treated for it, what do you have to lose? You have nothing to lose, and everything to gain. You're already at the point of no return in your mind, so why not?"

INFLAMMATION AND DEPRESSION

It shouldn't be surprising that inflammation and depression are related. After all, iproniazid – the first MAOI – had been developed as an anti-tuberculosis drug, intended to fight a particular type of infection. Iproniazid does fight tuberculosis, but not as well as its sister drug, isoniazid. Many inflammatory conditions are known to be co-morbid with depression, such as cardiovascular disease and irritable bowel syndrome. So, while the experience with immunotherapy and depression really caught the attention of scientists and clinicians, it wasn't the first evidence that depression and inflammation might be closely linked.

Research has uncovered mechanisms that link inflammation and depression together, starting with the stress response itself. One of the functions of CRH in an acute stress response is to cause the release of pro-inflammatory cytokines, in preparation for possible wounding and fighting an infection. It is a mild inflammatory state, though, just preparation. Farther along the process, glucocorticoids swing around to turn off the stress response, in part by inhibiting production of pro-inflammatory cytokines while initiating the production and release of anti-inflammatory cytokines. But this latter action happens when GRs are activated, and depression is often marked by glucocorticoid resistance. So in a way, we had already seen that depression and inflammation are closely related: the stress response initiates a mild inflammatory state and terminating it also terminates that inflammation. That means chronic stress with glucocorticoid resistance implies chronic inflammation too.

Microglia present another mechanism relating inflammation and depression. A state of inflammation in the body allows pro-inflammatory cytokines to be carried across the blood-brain barrier into the brain to activate microglia, which then produce their own pro-inflammatory cytokines to initiate neuroinflammation. When they do that, they also produce an enzyme called indoleamine 2,3-dioxygenase (IDO), which initiates the breakdown of the brain's tryptophan supply along a non-serotonin pathway. Specifically, IDO converts tryptophan to kynurenine instead of serotonin. Kynurenine undergoes further chemical reactions into either kynurenic acid or quinolinic acid. Both substances impact the survival of neurons: kynurenic acid has an overall protective effect on neurons, while quinolinic acid is toxic to them. Either way, pushing tryptophan down the IDO pathway deprives the brain of serotonin. This means pro-inflammatory cytokines can activate microglia to hijack the brain's serotonin production and produce a toxic substance in its place.[13]

IS DEPRESSION AN INFLAMMATORY DISEASE?

That particular question has been asked often since the immunotherapy experience caught the attention of scientists and clinicians, and human and animal studies provide both confirmatory and contradictory evidence of such a relationship. In depressed patients without an illness or injury that would cause inflammation, some show elevated inflammatory markers, but many do not.[14] The Whitehall Study, in which over 3,000 British civil servants were monitored over a period of 12 years, found that inflammation predicted the onset of depressive symptoms.[15] Scientists have also tested the theory that neither depression nor inflammation directly causes the other, but that increases in CRP and IL-6 concentrations could be explained by behavioral changes associated with depression, instead. The 5-year Heart and Soul Study, involving 667 people with coronary artery disease, seems to support this. In this study, having inflammation at baseline did not predict the development of depression, and though the people who started with depression at baseline typically showed higher levels of inflammation 5 years later, that rise in inflammatory markers was no longer statistically significant after adjustment for poor health behaviors associated with depression: higher body mass index, smoking, and physical inactivity.[16]

If inflammation causes or supports depression, then anti-inflammatory drugs should have antidepressant qualities, and overall, they do reduce symptoms of depression. However, studies that looked at the anti-depressive qualities of non-steroid anti-inflammatory drugs (NSAIDs), like aspirin, produced mixed results: some supported an anti-depressive effect and some showed no difference. Additionally, studies of people being treated with NSAIDs for pain actually found cases where those drugs appear to have induced symptoms of depression – symptoms that went away when the patients stopped taking the NSAIDs. NSAIDs used in conjunction with antidepressant treatment have sometimes been shown to worsen depressive symptoms.[17]

Researchers also looked at the situation the other way around: do antidepressants have anti-inflammatory effects? Yes and no. Some

SSRIs have been shown to be anti-inflammatory, while others are potently pro-inflammatory. ECT, though a very effective treatment for depression, has been shown to prompt a brief rise in pro-inflammatory cytokines.[18]

Evidence supporting a role of microglia in the development of depression also goes both ways. Some PET imagery studies show strong indications of microglial activation in patients with depression, while others do not. One postmortem analysis found microglia activation in only one of six patients with affective disorders, while three other postmortem studies found no difference between people with depression and controls. A small number of studies using drugs to either promote or inhibit microglia in humans and mice also found conflicting results: in one case, microglia activation in humans relieved their depression for 24 hours. However, in animal studies, different circumstances under which microglia were activated could either promote or relieve depression-like symptoms.[19]

Speaking with Dr. Raison as he drove to his meeting, I asked for his assessment of the relationship of inflammation and depression. In 2011, Raison and Miller published a paper titled "Is Depression an Inflammatory Disorder?" and they had answered their question with "a resounding 'no.'"[20] Time had not altered Dr. Raison's opinion. "Depression is not an inflammatory condition. In fact, many people with depression have low levels of inflammation," he stated firmly. "We know that inflammation is not necessary to cause depression; we know that not everyone who is inflamed gets depressed."

Though not a direct cause, there is a relationship, he clarified. "It's a probabilistic thing. What Miller has shown over the last several years is that if you take people who are otherwise healthy but depressed, and measure their level of inflammation and then you go looking in their brains, as the inflammation increases you begin to see different patterns of brain functioning. Basically, people who are depressed and have increased inflammation have this sort of disconnection of an area called the ventral striatum with a core other area of the brain, the ventromedial prefrontal cortex. They're less connected: they have decreased activity in the ventral striatum, which is sort of a pleasure center of the brain." The ventral striatum holds the nucleus accumbens, considered the home of reward and motivation. "Inflammation seems to induce this anhedonia, the depleted-dopamine, flat, tasteless feeling about life."

In their 2011 paper, Raison and Miller had suggested that inflammation contributes to depression in a subset of patients – it's a risk factor to some. "So, it's pretty clear that the way to think about it is not that depression is an inflammatory condition. What is true is that in general, inflammation, especially when it is chronic, predisposes towards depression, and some people are more vulnerable than others. Where we've been, with IFN-α as a model system, when people started taking this drug it just jacks up the inflammation and some of them get depressed. But the people who get depressed are the ones with a lot of added risk factors for depression. They have abnormalities in their HPA axis; they're already a little bit depressed. Inflammation is one of the ways people get depressed, and a subgroup of the people who are depressed have usually mildly elevated inflammation markers. That subgroup of people probably have different brain functioning than people who are just as depressed but don't have inflammatory conditions."

Dr. Raison also commented that the higher levels of inflammation in a person with depression are not even close to the degree of inflammation seen in an acute response to infection. The levels of inflammatory markers in someone with depression are often two to three times that of normal levels,[21] which, though it sounds high to me, is not nearly up to infection levels. "Take CRP, which is one of the most standardized of the inflammatory markers. If you are a healthy person you want your CRP to be below 1 milligram per liter. The Heart Association says that above 3, you enter a risk zone for increased risk of long term issues like a heart attack or stroke or dementia. When you get the flu, your CRP goes up to 100. There's no comparison."

It's not an inflammatory response *per se* that is necessarily on the pathway to depression, he explained. "The problem seems to be more the chronic wear and tear. It's not that depressed people are not average – what we call 'inflamed' is often at 'high normal'; it's just that they are there all the time. Thom McDade – the anthropologist at Northwestern – measured CRP levels of foragers-horticulturist tribes down in the Amazonian rain forest, who lived in a more traditional kind of environment full of pathogens, so infectious disease was the primary cause of death. And what he found there was that, unlike us, their CRP was almost zero, until the people get sick, then it shoots up. But if they live, then once they've fought off the infection it drops back down to about zero. For us, it tends to be fixed so if you run your inflammation high, it will tend to always run high. It's sort of the chronic grinding of that inflammation over time that really seems to get to people."

IS DEPRESSION AN EVOLUTIONARY ADAPTATION?

The association of depression with immune response has some scientists wondering if depression itself has been preserved in the human race through natural selection as part of an elaborate disease-fighting strategy.

We all know what it feels like to be sick from a cold, flu, or anything else: fatigue, achiness in muscles and joints, sleepiness, reduced appetite, as well as whatever discomforts the particular disease adds to our state of being. We look beyond the common symptoms to figure out precisely what we have. What if those common aspects of illness play a bigger role in survival, though? Along with fever, the changes seen in behavior and psychological state when the body is fighting off a pathogen may be part of a well-organized strategy for fighting infection. This strategy is called "sickness behavior," and is initiated by pro-inflammatory cytokines. While unpleasant, sickness behavior may have evolutionary value through encouraging sick people to conserve energy and avoid unnecessary social contact. In a deeper sense, our perceptions and motivations change when we are sick, making us want to take different actions than when we are feeling good, which might help our bodies fight off infection or illness.[22]

One of the theories of an evolutionary purpose for depression is the pathogen-host defense theory of depression, published by Raison and Miller in 2012, proposing that genes that promote an inflammatory bias have been positively selected for over evolutionary time. This is in part because fever creates an inhospitable environment for the infectious agent, but also because it promotes the development of depressive symptoms as part of an immune response. Depressive symptoms aid survival by conserving energy and encouraging social avoidance and hypervigilance in the host.[23] Dr. Raison explained. "The idea here is that, at the end of the day, depression evolved; it evolved because it is a way of managing relationships. It's not always the best way, but it's sufficient enough that the risk factors it makes us avoid maintains the population. When you think about relationships that are most relevant to the survival and reproduction of humans, they really fall into two camps: other humans, and bugs. Those are the things that were most likely to kill you or that were most necessary for reproduction."

Disease, rather than being trampled by a buffalo or eaten by a lion, is what has historically claimed life before it can procreate and raise its young. The evolutionary history of mammals involves bacteria mutating to become potent disease agents, then the mammalian immune system adapting to defeat it, then it mutating again, and so on; a race to the death. When the enemy tries something truly new and spectacular, somewhere in our genome are forgotten chunks of DNA holding an answer. And sometimes the cure is just a little less burdensome than the disease, so it survives in the gene pool.

Raison and Miller's proposal goes beyond tolerating an unhappy effect in order to survive the evolutionary arms race; they propose depression as having been positively selected for. As Dr. Raison put it, "One of our large relationship domains is our interaction with the microbial world. In that context, depression evolved as sort of an elaboration of sickness behavior. It is very clear now that sickness symptoms are an evolved defense strategy. We get sick because it helps us survive infection. There's a lot of overlap between the symptoms of sickness and depression, which really can't be accidental. And that's another part of the argument that depression has something to do with inflammation, because inflammation is what induces sickness. It's not the bugs themselves; it's the inflammatory response to infection that makes us feel sick. And so, the idea here is that one of the pathways whereby what we call depression evolved is that it evolved out of sickness. Depression just extends the basic purpose of sickness."

The extension, of course, is necessitated by human nature. "Depression has cognitive elaborations to it," he continued. "If you think of depression evolving as a way of navigating relationships with other humans, well, relationships with other humans was a risk factor for infection in many ways in ancestral environments. Humans are a risk factor for infection, from physical proximity, sexual activity, but also from violence. Until very recently, violent, threatening interactions were often associated with wounding, and wounding was the fastest way to die of infection. One of the reasons stress may cause depression over time is because it activates inflammation, and the reason stress activates inflammation is because across evolutionary time it was a reliable enough warning signal that you were about to be injured. Stress then signals the immune system to activate preemptively, to be ready for the increased infection risk."

"That doesn't explain why depression would involve feelings of guilt, or suicidal ideation," I commented.

"This the one of the places in the theory where there's something we don't understand," Raison agreed. "Years ago, Miller and a colleague of ours did a study where they looked at symptom patterns of people who were getting IFN-α, and were inflamed and depressed, versus symptom patterns of people who were just depressed. A lot of the symptoms were very similar, but where they diverged was you were more likely to see guilt, like, 'Do you feel crappy about yourself?' you were more likely to see that in regular depression. But I just don't know. There does seem to be sort of a specific link between suicide and inflammation. It seems that one thing inflammation does is drive people to kill themselves, for reasons unknown. There are postmortem brains studies of normal people who died of something natural, depressed people who did not commit suicide, and depressed people who did commit suicide, and it's the people who committed suicide that had increased inflammatory markers in their brains. And so whether you are in so much stress that you decide to kill yourself, that activates inflammation, or whether inflammation drives suicidality, nobody knows."

And it is also not the case that something that may have aided survival in ancient times still serves a beneficial purpose. When I asked whether depression might instead be a modern *mal*adaptation rather than an evolutionary inheritance, he gave me a firm "no." "Depression must be ancient," he said. "If depression wasn't ancient, how could we use rodents in antidepressant studies? But that doesn't mean the modern world is not making it worse. We all have an inflammatory bias that comes from the modern world. There's a sort of sad irony here, because we need inflammation less than ever. But we've done a number of things in the modern world that exacerbate it."

It seems that chronic inflammation and chronic stress reverberate to destructive effect, with impacts linking to the monoamine systems, depression, and suicide... but not in all cases. It helps build a picture of a changing physical environment in the brain, though, with long-lasting – and sometimes deadly – consequences.

Chapter Four

PANDORA'S BOX

Please, please, please don't fire me, Dinah thought. *Please just bear with me through this episode and then I'll be an excellent employee again.* But she was pretty sure it was too late. It wasn't the accommodations she'd had to request, or the occasional missed day. She could make up for the days that she just couldn't get out of bed. No, it wasn't that so much as the unexplainable and unmerited bouts of anger. It was a combination of issues that had lost her jobs in the past, and would likely lose her this one as well. She had read that 80 percent of people with mental illness were unemployed, the largest demographic segment of unemployment in the country.

It was incredibly stressful, for herself and her family. Much of the time she was fine, but about every 12 to 18 months she would descend into a major depressive episode, and it was hell. There were minor episodes, too, in the intervening time, where she'd have to pull out of work a couple times a month. But it was the major episodes that were so toxic. And getting accommodations at work was never easy; they didn't understand why she couldn't just snap out of it. Like, "Do you really need accommodations? So what: you're feeling sad. Just get up!"

It was at home that she felt the most worrisome impact, though. Whenever she was depressed, her husband would have to take care of her and their son, and run everything around the house; tasks made even more difficult by her loss of income. It was like having a partner with a chronic physical ailment. It had played a hard role in their marriage, particularly with the agitation, irritability, and anger. And when she was non-functional she was also ignoring their 3-year-old son; not the best parent she would like to be. *People don't seem to talk about that a lot,* she reflected. *They talk about how depression affects the individual who is experiencing it, but it affects the family life. Which in turn can re-affect the individual, because you feel like you're hurting everyone; you're letting everyone down. It's really hard when family gets angry at you too. They're frustrated because they're overwhelmed; sometimes they don't understand. They grow numb and start lacking empathy because they have been dealing with it every month or year or whatever.*

It had all started about 10 years earlier with her first major depressive episode. She doesn't know why it happens to her. Like they tell their son, Mommy has a broken brain.

Several of the people I spoke with said they felt relieved and validated when their doctor first told them they had a depressive disorder, characterizing it as a chemical imbalance in the brain. After loving family members and fond friends had told them to cheer up! and think happy thoughts! it was good to have acknowledgment that they were suffering from illness, not attitude. Relief wore thin, however, when no one could tell them what chemicals were out of balance, or precisely how to get them back in balance.

Imaging studies of depression are shifting the focus from chemicals to brain tissue, revealing the neuropathic condition that Dr. Gold commented on. Depression affects both the structure and function of the brain, in region-specific ways. Someone with depression has lost brain mass, including brain cells – a rather upsetting prospect. And because these changes primarily occur in areas of the brain responsible for control and regulation, their impacts ripple through multiple functions. Depression stirs up anger as well as sadness, and irritation as well as guilt. It has a broad reach, affecting not only sleep and appetite, but even how someone processes perceptions about their world.

THIS IS YOUR BRAIN

When scientists first started exploring, identifying, and naming different parts of the brain, they did so based on its visible features. Regions of the brain are named simultaneously under a system of coordinates and one based on clumps of gray matter containing distinguishable cell types. Gray matter forms itself into common types of structures, including a cortex (a thin outer layer, like a sheet) and nuclei (clusters of neuronal cell bodies).[1]

The coordinate-based system orients us along three major axes. The first goes from the direction of the nose, "rostral," toward the tail, "caudal," and is sometimes referred to as "anterior" and "posterior." The next axis goes from the top, "dorsal," toward the belly, "ventral." Because humans walk upright, "dorsal" refers to the top of the head and "ventral" is towards the jawbone rather than the belly. Dorsal and ventral are sometimes referred to as "superior" and "inferior," respectively. The last axis goes from the center, "medial," outward to either side, "lateral."[2] Together, these naming systems identify areas such as the ventromedial prefrontal cortex or the dorsal raphe nucleus in a reproducible way.

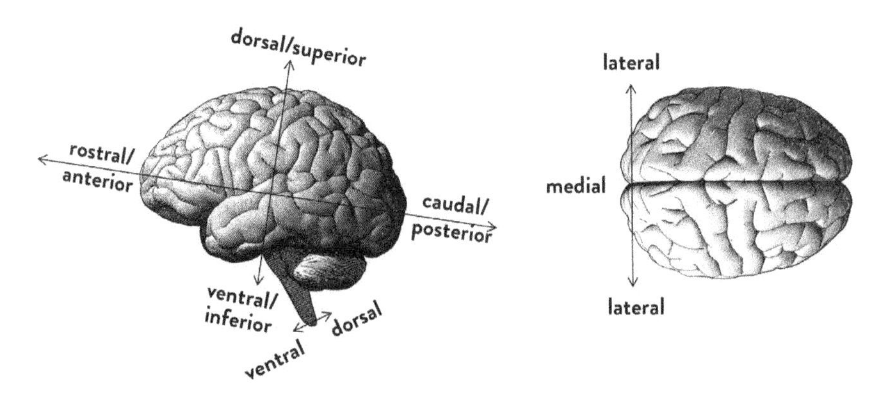

Figure 4-1: **Coordinate system of the brain.**

The main parts of the brain are the brainstem, diencephalon, cerebellum, and cerebrum. Each of those parts is associated with an overall type of function and several structures. Sitting at the end of the spinal cord, the brainstem holds structures that bring sensory information in and send information relating to arousal and awareness out. One of those structures, the pons, is of known relevance to depression because it holds the locus coeruleus, where the brain's norepinephrine is synthesized. Another structure in the brainstem, the midbrain, holds the raphe nuclei, where the brain's serotonin is synthesized. The next part of the brain, the diencephalon, is dorsal (or superior) to the brainstem; it holds the thalamus and hypothalamus. The thalamus processes most of the sensory information from the rest of the central nervous system, directing it to areas in the cerebral cortex, and the hypothalamus regulates autonomic, visceral (gut), and endocrine (hormonal) functions – including the stress response. The next part, the cerebellum, lies posterior to the brainstem, modulating the force and range of movements and the learning of motor skills. The largest part of the brain, the cerebrum, is divided into frontal, parietal, occipital, and temporal lobes. It holds the cerebral cortex and three deeper structures: the hippocampus, amygdala, and basal ganglia.[3]

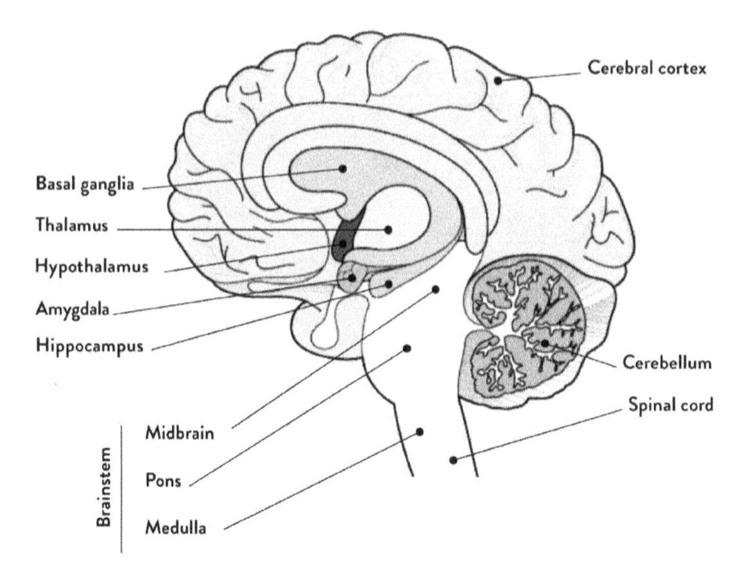

Figure 4-2: **Parts of the brain.**

The frontal lobe of the cerebrum is largely concerned with executive control, short-term memory, speech, the planning of future actions, and control of movement. The parietal lobe is involved with somatic or body sensation, forming a body image and relating it to extra-personal space. The occipital lobe is concerned with vision, and the temporal lobes with hearing and, through the hippocampus and amygdala, with learning, memory, and emotion. The cerebral cortex, a thin, wrinkly sheet comprising the outermost part of the brain, is the primary location for the operations responsible for our cognitive abilities. The portions of the cerebral cortex are named for their lobe: frontal cortex, parietal cortex, occipital cortex, and temporal cortices. The cerebral cortex also includes two deeply buried structures: the cingulate cortex in the medial region, and an insular cortex on each side. The neocortex – the structure closest to the skull bones – is the last part of the brain to develop, both evolutionarily and in a maturing human being. It won't reach full maturity until an individual reaches 23 to 25 years of age.[4]

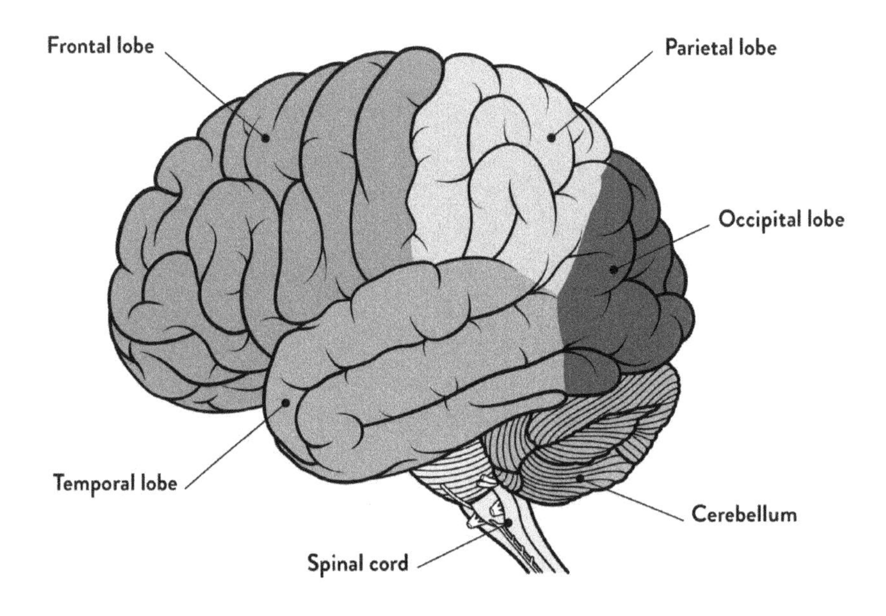

Figure 4-3: **Lobes of the brain.**

FROM LEFT TO RIGHT

We say that there's a difference between the left and right sides of the brain: that left-handed people use the right side of their brain and are more creative, and right-handed people use the left side and are more logical. While I don't know if that particular bit of urban wisdom is true, it is true that the left and right hemispheres of the brain are not the same. Though structurally very similar, there are often different functions between a structure in the left hemisphere and its equivalent in the right hemisphere.

Eighty years ago, scientists proposed that the left and right sides of the prefrontal cortex (the anterior portion of the frontal cortex) provide different types of inputs with regard to motivation. Observing the behavior of patients with damage to the left versus the right side of their brain, scientists created a model in which the left side of the prefrontal cortex is more involved in positive approach motivation and the right side more involved in negative withdrawal motivation. To prove it, they used a test in which they injected a barbiturate derivative into one of the internal carotid arteries, left or right, to suppress the activity of that hemisphere only. With the left side of the brain suppressed, the subjects showed depression. With the right side suppressed, the subjects were euphoric. These effects were interpreted as releasing one hemisphere from inhibition by the other.[5] So, the left side of the brain normally keeps the "unhappy" right side in check, and when that control fails, the unhappiness comes out. And vice versa: the right side of the brain normally keeps the "happy" left side in check, and when that control fails, euphoria comes out. This left-side versus right-side effect on mood and emotion was also seen in studies of more than 100 patients who had suffered lesions to one side of the brain from surgery or trauma. The patients with damage to the left side sometimes cried spontaneously, while the ones with damage to the right side tended to show spontaneous laughter.[6] And it is not just happiness or unhappiness, as the author of the study noted. "Damage to the left side of the brain is associated with a dysphoric reaction. Patients display feelings of despair, hopelessness, and anger. They show heightened tendencies toward self-blame, self-deprecation, or fits of crying."[7]

So maybe that is how major depression gets its character: an impaired left side of the brain can't keep the Pandora's box of the right side from flying open and letting out anger, guilt, hopelessness, worthlessness, pestilence, disease, and death. Maybe this is why Dinah's depression holds so much anger and irritation as well as sadness.

IMAGING DEPRESSION

For quite a while, "broken brains" were a mainstay of scientific research into the structure and function of the brain. Lesion studies used surviving victims of stroke, epilepsy, or head trauma to see what functions changed when an injury happened in a particular place, and postmortem studies showed physical features after death. The living brain, however, remains a black box. We can apply certain inputs and measure outputs, but we can't see exactly what goes on between. Starting in the late 20th century, however, *in vivo* imaging technology has opened a window into much more detailed structural and functional characteristics of the brains of living subjects. Magnetic Resonance Imaging (MRI) is the primary technology used to assess differences in the physical structure of the brain because it can differentiate between types of tissue and detect even very small structures. Using this technology, scientists can look at the brains of people with depression and compare them to the brains of people without depression to assess structural differences.[8]

In the early 1990s, researchers noticed that MRI sensors could also measure changes in energy usage resulting from brain activity, and used that principle to invent functional Magnetic Resonance Imaging (fMRI). Active neurons consume more oxygen than neurons at rest, and the change in oxygen level in the blood affects MRI's magnetic field. Another imaging technology, Positron Emission Tomography (PET), uses a radioactive tracer labeled with an isotope of carbon, oxygen, fluorine, or nitrogen that is quickly absorbed into molecules containing that element. Depending on the type of element marked with the tracer, PET imaging measures consumption of oxygen or glucose, or blood flow, throughout the brain. PET and fMRI are functional imaging technologies. Scientists use them to look at activity levels in different areas of the brain in people with depression while they are either doing an assigned task or simply resting, and compare them to the activity levels in those areas of the brain in people without depression under the same conditions.[9]

As the pace of structural and functional imaging studies in depression and other disorders picked up in the last two decades, those studies repeatedly show that certain structures and pathways are involved in depression. In every case there are some studies with conflicting results, often attributed to differences in study populations, actions of antidepressant medications, and other variables. Overall, though, structural studies are showing that depression involves loss of volume in certain areas, implying loss of brain tissue. Functional imaging studies are showing areas of increased activity (hyperactivity) and areas of decreased activity (hypoactivity) in the depressed brain compared to the brains of healthy subjects, implying that the depressed brain functions differently too. And the particular areas that are showing these structural and functional differences trace depression's course and symptoms.[10]

The prefrontal cortex is responsible for aspects of executive control like remembering and acting on our intentions. Two main subregions of the prefrontal cortex commonly show differences in people with depression compared to those without: the dorsolateral prefrontal cortex and the ventromedial prefrontal cortex.

The dorsolateral prefrontal cortex is associated with executive control, working memory, and inhibitory control over behavior. It selects the appropriate response to a situation, inhibits things we know we shouldn't do, and monitors actions to ensure they match intentions. Structural imaging shows that the volume of the dorsolateral prefrontal cortex is often reduced in people with depression compared to healthy controls, and some studies have shown that volume reduction to be corrected with successful antidepressant treatment. Functional imaging studies show reduced activity in dorsolateral prefrontal cortex overall in people with depression. Further, the studies looking for differences in the left side compared to the right side in depression tend to show hypoactivity in the left dorsolateral prefrontal cortex, and/or hyperactivity in the right dorsolateral prefrontal cortex. The ventromedial prefrontal cortex, in contrast, is part of the "emotion circuit": it modulates activities in several areas of the brain involved in generating emotions. Functional imaging studies show evidence of increased activity in the ventromedial prefrontal cortex in people with depression compared to controls, even while they are in a resting state.[11]

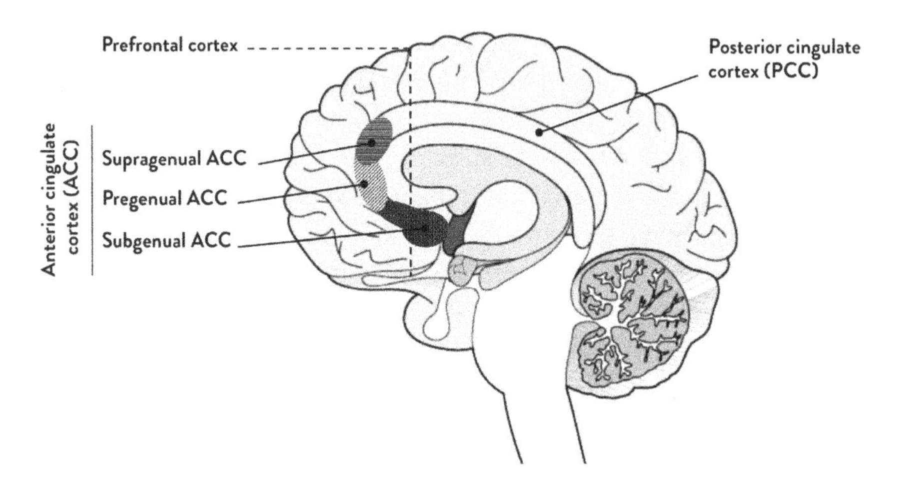

Figure 4-4: **Subregions of the cingulate cortex.**

Moving on to the next area with significant differences in depression, the cingulate cortex is part of the limbic system, an ancient part of the brain with functions tied to emotion and instinct. As part of the semi-modern cortical structure, though, the cingulate cortex has regulatory and cognitive roles; rather than *generating* emotions or instincts, it *regulates* them. Within the anterior cingulate cortex (ACC), the subgenual ACC has become an area of particular interest in depression research as more and more studies implicate it as a center of dysfunction. The subgenual ACC is activated in the memory of negative events and sadness tasks. Numerous structural imaging studies of depression show gray matter volume loss – up to a whopping 48 percent – in the left side of the subgenual ACC, and some, but not as much, volume reduction in the right side. After correcting for that reduced volume, however, functional imaging studies indicate that activity levels in the subgenual ACC are actually *increased* in unmedicated depression.[12] The pregenual ACC, on the other hand, is activated during happy events and self-relevant tasks. Not surprisingly, it also shows abnormal metabolic activity in major depression, with hypoactivity observed in the "happy" pregenual ACC, in contrast to hyperactivity in the "sad" subgenual ACC.[13]

Continuing in the emotion and instinct-driving limbic system, the amygdala is commonly called the "fear center" of the brain. It is a complex structure that is involved in evaluating the emotional significance of our perceptions. Though it can assess an input to be on the "good" or "happy" side, it tends to work mostly with tones of fear, sadness, and anger; perhaps indicating that, over eons, our instincts were shaped more by bad experiences than by good ones. There is inconsistent evidence of volume change in the amygdala in depression, but consistent evidence of its increased reactivity to negative stimuli in depression. For example, amygdala responses to negative words could be seen in depressed patients for at least 25 seconds on average, but were no longer visible in healthy controls after 10 seconds. Patients with depression remember negative words better than positive words, correlating with increased activity of the right amygdala.[14]

The hippocampus – also part of the limbic system – is deeply implicated in depression. It plays a critical role in the consolidation of new memories, and damage to the hippocampus results in anterograde amnesia: the inability to form new memories, though old ones remain intact. The hippocampus interacts with the amygdala to provide input regarding the context in which stimuli occur, and encodes emotionally-relevant information into memory. It also modulates both the amygdala and the HPA axis. In the stress response, for example, a stimulus is perceived from any of many modes; the hippocampus interacts with the amygdala to assess "stressor!" and then signals the hypothalamus to produce CRH while initiating norepinephrine production.[15] Imaging and postmortem studies show a loss of volume in both sides of the hippocampus in depression, with the left hemisphere of the hippocampus reduced by an average of 8 percent in patients with depression, and the right hemisphere reduced by an average of 10 percent. Further, the right-side reduction was more severe in patients who had experienced more episodes of depression.[16]

So there are differences in both structure and function between a depressed and a healthy brain that show up in multiple areas. It is also interesting to consider what these studies *don't* show. First, they don't show reduced volume – with its implication of damage – in the brain structures that produce some of the chemicals strongly associated with depression. These areas include the hypothalamus (the "H" of the HPA axis), where the stress response begins and ends, and the brainstem, where the brain's serotonin and norepinephrine are synthesized. The studies don't show if the lower volumes were pre-existing risk factors that made someone more likely to develop depression, or if they are a result of having depression. The studies also don't show increases in volume of any areas of the brain. They show increases in activity levels, but not in volume. Depression is characterized by loss, not growth.

Quantifying the differences between the depressed brain and the healthy brain remains a challenging task for researchers. The subjects in these studies tend to be few in number; there are confounding variables like whether depressed subjects have been medicated or not; image resolution in the earlier studies is not as good as in the more recent ones; and there are a lot of individual differences in brain size. Even with those considerations, though, the more-consistent differences between depressed and healthy brains stand out in a meaningful way. In terms of structural differences, depression is associated with volume loss in the dorsolateral prefrontal cortex, the subgenual ACC, and the hippocampus. Often, that volume loss differs between the left and right sides of those structures. In terms of activity levels, reduced activity (hypoactivity) is seen in the dorsolateral prefrontal cortex (particularly on the left side) and the pregenual ACC. Increased activity (hyperactivity) is seen in the ventromedial prefrontal cortex, the subgenual ACC, the amygdala, and in some studies, the right dorsolateral prefrontal cortex.

What do these differences mean? Bringing the functions of the affected structures into account, reduced tissue and/or activity levels are seen in the areas of the brain responsible for executive control, memory, and the experience of happier emotions over darker ones.

Increased activity levels are seen in the areas of the brain responsible for the experience of the darker emotions. The data begins to draw a picture of a brain in which various control structures are not working well, and possibly as a result, some of the structures they are supposed to control are running wild.

BUILDING A NEW MODEL

With these findings, a new model of depression is coming into prominence. This new model emphasizes what is happening to brain tissues and structures rather than the state of the chemicals (like serotonin, norepinephrine, glucocorticoids, and cytokines) in the brain. This doesn't mean that the chemicals are unimportant. Excess glucocorticoid signaling and chronic inflammation are known to damage brain tissue; when traditional antidepressants promote more serotonin and/or norepinephrine signaling, some of the damage is remediated. Each of these chemicals exist for a reason, though; none of them are good or bad in themselves. The inflammatory response, managed with pro- and anti-inflammatory cytokines, is an important part of the immune system; the stress response, within which glucocorticoids and norepinephrine both have key roles, is important for survival. Even serotonin can produce good mood or bad, depending on what type of receptor it hits and where. These chemicals are tools, and there are many others as well. So, this new model of depression places the focus on structures in the brain and what is happening to them, layering a tissue-level view over the top of the chemical view. In doing so, it opens the door for a richer understanding of how depression comes about, and how people recover from it.

Chapter Five

NEUROPLASTICITY

I had met Dinah on a trip to Georgia, luckily on one of the cooler days as we had to wander quite a way to find a quiet place to talk, and she told me about her 10-year struggle with depression. "I kind of distinguish between little 'd' depression and big 'D' depression, and I've always struggled with little 'd' depression, from as early as elementary school. I can remember being worn down more than other kids."

Something traumatic happened to her in 2009, something that she didn't want to tell me about, but it kicked her from little 'd' to big 'D.' "It started with me losing motivation and energy," she said. "I was in grad school at the time, and suddenly I was having trouble working. That was highly unusual for me because I was a very motivated person. I was a triathlete; I was just a very motivated person. Then I started having trouble with sleeping. I was over-sleeping, then I was under-sleeping, and my sleeping schedule was just all over the place. That was unusual for me as well. The next thing that happened to me was that I started dropping weight. I dropped like 10 or 15 pounds over the course of a month. I just wasn't hungry. I didn't connect all the dots until the depressed mood hit, because the mood thing came after the other symptoms. When everybody thinks of depression the first thing they think of is depressed mood. So when that hit, I connected the dots and I thought 'I must be depressed.'"

In her early 20s, doing well in school and athletics, Dinah felt she had command of the situation. "I went to a therapist, who recommended that I get on some meds right away, asked me about some family history and stuff like that. I said 'I want to do this on my own; I think I can do this.' The reality was that I spiraled further and further down. Long story short, I ended up in the hospital, and I think it wasn't until that point that I realized I really need some medicine; I can't do this on my own." Over the years, she went through many different combinations of medicines and a few more hospitalizations. "It's up and down," she said. "Things happen. I think I'm with the waves of what happens in my environment, like I feel like I don't have as much control as I would like. I'm at the whim of what happens."

Dinah does currently have a job that she enjoys, in an up-and-coming area of technology, and she and her husband work hard to keep a steady home life for their family. "We weren't even going to have kids because of my illness, and my son was actually a happy surprise. Best thing that's ever happened to me, there's no question there. But there was a conscious choice that this thing should not be passed on. So now we're in a situation where he knows Mommy has a broken brain, and that it makes Mommy really sad sometimes, and that it's OK for Mommy to be sad; it's not his fault." They look to the future for their son as well. "One of the things we want to do is have him in therapy as he is growing up so that he can learn techniques that can prevent that sort of thing happening to him. I just cross my fingers and hope more of my husband's genes got in there. And even if he doesn't get it, it's going to cause him issues that Mommy has it. We want him to be able to grow up knowing it wasn't his fault; there's nothing he could do; stuff like that."

Dinah and her family have learned the hard way just how complex such a condition, and its treatments, can be. In addition to depression, Dinah suffers from anxiety, to the point of occasional panic attacks. Early attempts to treat the anxiety had a poor result. "I'm not allowed to take benzodiazepines, the most common treatment for anxiety, because it causes me to go crazy, for lack of a better word," she said. "I have attacked people. I've been just out of my mind on benzos, so I'm not allowed." None of her medications have proved very effective for her depression, either. "I feel like none of them get rid of the symptoms, but there are some that make me functional, able to be a contributing member of society, whereas I wouldn't be able to get out of bed if I didn't have them. I still struggle with days I have to call into work, for example, on my medicines. It's relative. The current cocktail is...," she paused. "I have lots of weird side effects."

Psychotherapy has also helped, but with limits. "There have been times when I've given up on psychotherapy," she said. "For probably 6 to 8 months over the last year I gave up; I just wasn't seeing anybody. Sometimes I just feel like I'm treading water and it doesn't change anything. But sometimes it's helpful. I think that my depression has a pretty strong biological component, so sometimes I feel like psychotherapy only helps if I'm doing OK. The meds have to be working, and I have to be at a certain level or threshold before the therapy will start working. If I'm in the depths of it, therapy isn't going to help. It's a medicine thing, it feels like, that needs to happen."

It does have a place in her coping mechanisms, though. "I think the psychotherapy has been helpful for me in a preventative way, more than it has been in the ability to pull me out of things. Because I think once I hit there it doesn't help. But it helps when you are starting to go down, and sometimes you can't prevent going down but you can make it as graceful as possible and try to prevent it from disrupting your life. Usually for me it's just a matter of getting through it without making any decisions while I'm in it. And that's usually my best goal, right there."

With all the challenges Dinah faces, among the most difficult is the lack of popular understanding of major depression. "One of the hard things is when people think they understand what depression is. So many people have told me, 'Oh yeah, I've been depressed' or 'I've had depression.' And they may have even been to a therapist. But they don't really understand it; they think that your way of thinking just needs to be fixed. That it's all about learning how to be positive and seeing the positive in things and you can pull yourself out of it if you really tried. I think that's the most frustrating thing that I face, that lack of understanding. People offer advice all the time. They say, 'Have you tried meditation; have you tried exercise; have you tried this or that,' and I'm thinking, 'Don't you think I've tried everything?' At this point I don't even say anything anymore; I just nod along and say 'thank you' because it gets to be pointless to keep trying to get it out there. It's not like they don't want to understand or don't believe. If someone's open and they ask questions then I will go into it, but too often they already have their preconceptions, so I just let it go."

The extent of tissue loss in the brain of someone with major depression is alarming, but one thing science has learned over the years is that the brain changes. A lot. There's a term for it – neuroplasticity – referring to the whole set of adaptive changes seen in the brain, including the birth of new brain cells, changes in cell function, reduction or creation of synaptic connections, and changes in structure of cells.[1] There is actually a lot that is changing all the time.

Dr. Ronald Duman of Yale University has been in the forefront of neuroplasticity research for decades, and he talked with me about how stress can affect the actual structure of the brain. His work, and the idea that neuroplasticity is closely related to depression, ascended in the 1990s as researchers observed the physical changes that stress causes in the brain. "There was a lot of work going on looking at how stress affects the processes of neurons in the brain, and studies had been published showing that repeated stress exposure could cause retraction or atrophy of the dendritic processes of the brain, both in non-human primates and also in rodent models," he said. Evidence grew on various fronts that stress, and the synaptic changes that occur with stress, were involved in precipitating those changes. And with areas like the prefrontal cortex and hippocampus – both involved in the control of mood and emotion – physically changing in depression, it is hard not to see a relationship between those changes and the illness.

The neuroplasticity hypothesis proposes that impaired mechanisms of neuroplasticity are a core feature in the development of depression, that chronic stress is an important factor causing this impairment, and that antidepressant treatments act, at least in part, by mitigating damage or repairing the mechanisms of plasticity. The mechanisms involved include the creation of new neurons (neurogenesis), reduction, loss, or death of cells (neurodegeneration), and changes in the shape, conformation, or function of existing neurons (synaptic plasticity). Together, these elements describe an adaptive brain and factors that work it into or out of depression.[2]

ADULT NEUROGENESIS

In the 1960s, Dr. Joseph Altman published a series of papers in respected journals showing that new neurons are created in a few, restricted parts of the brains of adult mammals (rats, for his research). Altman was a self-taught postdoctoral fellow working on his own in the Massachusetts Institute of Technology's psychology department. Despite the quality of the school and the journals that published his findings (including *Science* and *Nature*), his discoveries were simply ignored. It was the immovable opinion of the scientific community that when we are born, our brains have our entire lifetime total of neurons already. Other types of cells can divide and proliferate, but not neurons. His work became an occasional footnote in textbooks, and Altman moved on to other topics of research.[3]

It wasn't until 30 years later that his discovery was revived and extended. Studies using new ways of labeling cells in a living brain showed adult neurogenesis in various rodents, then primates, and finally, humans. In particular, the olfactory bulb and the hippocampus continue to generate new cells throughout the lifetime. There is some evidence from animal models that new cells can be born in the cerebral cortex too, but that picture has been slower to develop.[4] Many of the new cells die before they reach maturity and about 20 percent become glia rather than neurons, but it is now estimated that the adult human hippocampus grows about 500,000 new neurons each year.[5] Of course, since our brains have somewhere around 100 billion neurons in total, that's a small fraction, even over a lifetime.[6]

But these new neurons are thought to be especially important because they specifically help the brain integrate new information into pre-existing contexts. It's a very advanced sort of learning, supporting reassessment, reconsideration, and even forgetting – capabilities that seem to be lacking in depression. And since the hippocampus is part of the limbic system, this sort of learning involves affective behaviors[7] – regulating our emotional state. Dinah had noted that, for herself at least, psychotherapy is effective only after she starts seeing the effects of antidepressant medication. Maybe that is what is happening to her when neurogenesis is interrupted: her brain literally cannot change its emotional state because it takes new neurons to do that.

In 2000, Dr. B. L. Jacobs proposed a neurogenesis hypothesis of depression, suggesting that the off-again, on-again progress of neurogenesis in the hippocampus may be the deciding factor in the development and remission, respectively, of major depressive episodes. This hypothesis was prompted by observations that stress reduces adult neurogenesis, and antidepressant medications help restore it.[8] Mice and rats show higher than normal adult neurogenesis when they experience physical exercise, learning, environmental enrichment, and even acute, controllable stress. On the other hand, exposure to excessive glucocorticoids, as seen in periods of chronic stress, reduces neurogenesis. When rats are given high levels of corticosterone (their equivalent of cortisol), neurogenesis decreases in the hippocampus. Conversely, when the adrenal glands of rats are removed to prevent corticosterone production, neurogenesis increases over baseline levels.[9]

Other animal studies showed that common antidepressant treatments, including antidepressant medications and electroconvulsive shock, increase adult hippocampal neurogenesis. Some studies indicate that that increased neurogenesis is necessary for antidepressants to work. In rodent models of depression, when the hippocampus is irradiated to reduce neurogenesis, antidepressants become ineffective in reducing some anxiety-related and depressive behaviors, and the rodents are unable to return a hyperactive HPA axis to normal after a stressor. The time factor also favors the neurogenesis hypothesis: the time required for a progenitor cell to migrate, differentiate, grow an axon and dendrites and form connections to other neurons matches the time lag seen in the start of antidepressant administration to when they have noticeable effect: 2 to 3 weeks.[10]

But neurogenesis alone is insufficient as an explanation of depression. The number of new neurons produced in a lifetime is minuscule when compared to the total number of neurons in the brain, or even just in the hippocampus; halting that new growth can't account for the volume loss measured in multiple areas of the brains in victims of depression. Structural and functional imaging, as well as postmortem tissue analysis, show the effects of depression to extend well beyond the hippocampus, to include the prefrontal cortex in particular; how the growth or lack of growth of new cells in the hippocampus would cause these effects is unclear. So, while adult neurogenesis has an important place in the development and remission from depression, it is not seen as the only factor.[11]

NEURODEGENERATION

Going from a stem or progenitor cell to a neuron is only part of the story. Surprisingly, about half of the neurons our brains generate throughout our lifetime die a programmed cell death, making survival of new neurons an important part of neurogenesis. Every cell in our body comes with a self-destruct program called "apoptosis." When apoptosis is triggered, the cell rounds up its genetic material, breaks apart its nucleus, and releases materials that can be recycled for use in other cells. This program can be initiated under various circumstances, including when the cell's defenses detect that it, or its DNA, has suffered significant damage and should be taken out of action. The brain also has other tools to reduce the number of cells or remove connections by trimming off dendritic branches or terminals. In fact, the developing brain first over-grows, then trims away a lot of unnecessary connections as it matures. There are planned and appropriate uses of these tools, but they can also be activated where such losses are not beneficial. Necrosis – when a cell is killed by a pathogen or some means other than apoptosis – leaves behind debris that can be mistaken for an invading pathogen, prompting an autoimmune response. Another common way for neurons or glia to suffer damage is through excitotoxicity, in which calcium ions accumulating inside the cell activate mechanisms that can rip apart cellular structures. Excitotoxicity can shave off dendrites or even kill entire cells.[12]

Structural imaging and postmortem studies of depression emphasize losses – not increases – in various areas of the brain, so a neurodegeneration hypothesis of depression has also been proposed, suggesting that prolonged glucocorticoid exposure increases susceptibility of neurons to damage from multiple sources, in turn increasing the rate of cell loss from toxic challenges or even ordinary cell attrition.[13] It piles on: damage from excess glucocorticoids reduces the brain's ability to stop the HPA axis, resulting in more glucocorticoids and more damage.[14] And in addition to glucocorticoids, pro-inflammatory cytokines and oxidative stress can damage cellular DNA to the point of triggering apoptosis. Pro-inflammatory cytokines can also activate microglia in the brain to divert tryptophan to produce a neurotoxic substance instead of serotonin.[15] With increasing losses in key control areas of the brain like the hippocampus and parts of the prefrontal cortex, it is left with fewer resources to exert control over negative emotions and the stress system... The sufferer spirals down into a major depressive episode.

Newer discoveries have shifted the view of neurodegeneration, though, showing that the volume reductions seen in depression come from loss of glia and reduction in the size of neurons, not from the loss of whole neurons. Neurons lose dendrites and synaptic connections and the generation of new neurons through neurogenesis is interrupted. Glia can be damaged or killed, but depression does not mean the death of neurons.[16]

NEUROTROPHINS AND SYNAPTIC PLASTICITY

Once a neuron is born, it must migrate to its position, grow an axon and dendrites, and form synaptic connections with other cells in order to become functionally effective. As the new neuron matures, the direction and degree it grows its axon and dendrites will determine what signals will activate it and what targets it can signal. Mature neurons undergo changes too: they manage synaptic connections and prune and/or grow new dendritic spines. Those factors are determined both by internal (to the neuron) and external factors. Once dendrites are extended, they grow spines: small protrusions that will receive the majority of excitatory inputs. On those spines, receptors for various neurotransmitters will grow and can change throughout the lifetime of the neuron, also based on both internal and external factors. Dendritic spines can be pruned away, removing the capability to signal through associated synapses. Each of these processes can be halted and reversed. Altogether it shows a high degree of potential for changes in the basic communications paths of the brain.[17]

Brain-derived neurotrophic factor (BDNF) was first isolated in pig brains in the 1980s, joining a family of nerve growth factors that guide the growth path of new neurons while preventing their self-destruction. Researchers found that BDNF is produced in response to seizure activity and by sensory stimulation, and that it turn it provides support and protection for growth, differentiation, survival, and functional programming of neurons. BDNF itself is a protein, produced in neurons from the *BDNF* gene, then released outside the cell to go about its work. BDNF secreted in the area of the dendrites as they form seems to attract neighboring neurons to start growing dendrites its direction, with growth effects far from the source. Also, and very importantly, BDNF has been shown to play a major role in the process the brain uses to strengthen synaptic connections through repetition, called long term potentiation (LTP), which is key to learning and forming memories.[18]

Whether they developed before we were born or well into our adult years, neurons form and modify connections with other neurons throughout their existence. Synaptic plasticity refers to changes that the brain creates at the level of individual synapses, enhancing actions across some and inhibiting actions across others in response to repeated stimuli. LTP ensures that recent or frequent stimuli create a faster synaptic activation, while the counter process of long-term depression tends to turn down the firing capacity across a synapse. While adult neurogenesis seems to be restricted to a few parts of the brain, synaptic plasticity occurs throughout. Not every neuron or synaptic connection has an equal chance of firing. Instead, as we experience the same stimulus over and over again, the synaptic connections involved are strengthened, so that they are more likely to fire again on the next such stimulus.[19]

Another type of plasticity is gaining support: neurotransmitter switching. It has long been thought that once a neuron develops – for example as a serotonergic neuron – it keeps that identity throughout its existence. Recent animal studies have shown that electrical activity can cause a neuron to add, lose, or change neurotransmitters, including its receptors and transporters, even after the brain is fully mature. Adult rats exposed to a longer-than-normal day swapped some of their dopamine neurons to use a different neurotransmitter, and switched back to dopamine when they were exposed to shorter days. These swaps had behavioral impacts as well: the long-day rats with reduced dopamine neurotransmission showed increased anxiety and depression; this was reversed when they went back to a shorter day and gained back dopamine neurons. While a lot remains to be discovered about mechanisms and consequences of neurotransmitter switching, at the least it shows that the brain can exercise levels of plasticity going much further than once believed.[20]

In the early 2000s, Dr. Duman proposed a neurotrophic hypothesis for depression. "The neurotrophic hypothesis was based on data from a number of labs, going back to the early 90s, when it was discovered that these very important neurotrophic factors in the brain that were known to be important for development of the brain and survival of neurons were also very highly regulated by activity of neurons and by other conditions. It was also discovered at that point that stress exposure caused a down-regulation of one of these key neurotrophic factors, the brain-derived neurotrophic factor, or BDNF. That was pretty interesting because around that time also there was a lot of work going on looking at how stress affects the [dendritic] processes of neurons in the brain, and studies had been published showing that repeated stress exposure could cause retraction or atrophy of the dendrite processes of the brain, both in non-human primates and also in rodent models. So there was sort of a potential link between the reduction or loss of the growth factor and this retraction of processes of neurons."

Duman's neurotrophic hypothesis of depression incorporated the observations that the volume losses observed in depression correspond to a loss of BDNF production, and also that the reverse is true: restoring BDNF production restores lost connections and helps new cells to survive and grow. "In postmortem tissue from depressed patients there was a decrease in levels of BDNF, but also some evidence that people who were on an antidepressant at the time of death had reversal or did not show this decrease in BDNF," he said. Then, in looking at mechanisms of action of chronic treatment with antidepressants, his team found that different classes of antidepressants, including SSRIs and MAOIs, could increase levels of BDNF.[21]

The neurotrophic hypothesis proposes that depression results from decreased neurotrophic support; i.e., from decreased BDNF and other nerve growth factors. That loss of neurotrophic support leads to the atrophy of existing neurons, decreased hippocampal neurogenesis, and loss of glia. Further, antidepressant treatment gains its effectiveness from blocking or reversing such atrophy and cell loss.[22] BDNF is critical to the growth, survival, and protection of neurons. Beyond that, research has shown that BDNF is an important part of neuronal response, even after those neurons have fully matured. When BDNF fuses across neural tissue and lands on its receptor, it activates a number of pathways that shape the response of the synapse to subsequent neurotransmitters. It increases the number and responsiveness of excitatory and inhibitory receptors, selectively increases or decreases the responsiveness of calcium, sodium, or potassium ion channels, and signals the nucleus of the neuron to increase gene transcription of other proteins that will help perpetuate the response up to a period of hours. It modulates the synapse, affecting how the postsynaptic neuron responds to whatever is coming at it. It causes a form of LTP, sculpting both the neuron's likelihood to fire and other actions it will take. In fact, recent discoveries in animal models show that BDNF is not just helpful, but required for LTP in the hippocampus.[23]

But there is also experimental evidence that doesn't support the neurotrophin hypothesis. In animal models, modifying the *bdnf* gene so that the substance cannot be produced in adult mice did not create the appearance of depression in those animals. Some studies of antidepressant action have failed to show any increase in BDNF from the medications. Researchers have also discovered that BDNF creates different effects in different brain regions, and increasing BDNF in some brain regions actually leads to depressive behavior. Some have suggested that a decrease in BDNF may be more involved with heightening and reinforcing the symptoms of depression than with their initial development.[24]

NEUROPLASTICITY

The neuroplasticity hypothesis wraps up all the components of the lifecycle of brain cells into the prevailing theory of depression. It paints a picture of the brain undergoing continual processes of abrasion and repair, and when abrasion overwhelms repair, you can get a neuropathic condition: depression. "Ultimately, what we and I think the field is focused on is the idea that stress and depression are associated with loss of synaptic connections in some key brain regions, like the prefrontal cortex and the hippocampus," Dr. Duman said. "One factor involved is BDNF, but there is also evidence that other factors involved in synaptic plasticity also are dysregulated and contribute to that loss of connectivity and function. Conversely, antidepressants, a chronic administration of traditional agents, and then more recent studies with ketamine, which works after a single dose, show that these drugs have the capability of producing the opposite effect. They increase BDNF and reverse the synaptic deficits, or the synaptic atrophy, that occurs with chronic stress."

But, a question this theory doesn't answer is with all those abrasive substances circulating throughout the entire brain, why do we see damage to the hippocampus, the subgenual cingulate cortex, and the prefrontal cortex, but not other areas? Why is it not just as likely that these forces would damage the motor cortex, for example, so that people under chronic stress have difficulty moving their arms or something? How do the factors underlying depression take aim at particular regions?

I asked Dr. Duman for his take on it. "There are a couple of ways to look at it," he responded. "One that we think about is how these key regions, particularly the prefrontal cortex, are involved in control of a lot of other processes in the brain, a lot of other brain regions. It's sort of looked at as a top-down region of regulation. So it has connections with the amygdala, which is involved in emotion, and mood and fear and anxiety; it has connections with brainstem regions that are involved in reactivity, like the locus coeruleus and the dorsal raphe; it also has connections with the nucleus accumbens, the dopamine system, which is involved with anhedonia and reward processes. These connectivity problems that occur in the prefrontal cortex could influence its ability to interact with these other regions in a fashion that allows their normal behavior and function." In this scenario, it really just takes a couple good hits to the prefrontal cortex to cause cascading effects on the areas it controls. Start with someone who has some level of existing vulnerability, from genetics or some other factor, and add in all the feedback loops that intensify an effect, and you are looking at a potent force against particular parts of the brain.

"The other possibility, and there's plenty of evidence for this too," he continued, "is that there are also imbalances within those other structures. Either the connectivity of those neurons, and/or the neurochemistry and signaling processes within those pathways. There's a lot of work that's been done on the mesolimbic dopamine system, again which controls reward, motivation, and anhedonia, and those senses also show a lot of dysregulation in chronic stress models as well as in postmortem tissue from depressed patients."

Antidepressants, ECT, and other therapies promote hippocampal neurogenesis and production of BDNF and other growth factors that help new neurons and glia survive and differentiate; function is gradually restored and symptoms diminish. Is it enough, though? I asked Dr. Duman if those lost structures can grow back... if the depressed brain can truly recover.

"Certainly it can be reversed in milder forms, or moderate cases," he responded. "This is based on our rodent models and also the brain imaging studies, where there's a reversal of the volume loss. The evidence is that in these processes, the atrophy can be reversed with time. I think the question becomes a little more difficult in cases that are very severe. You can even compare it to a physical problem, where somebody breaks their leg in ten places, it takes several years before they can recover, versus somebody who has a simple, single break, and is maybe out for a month or so. If you have really, really severe damage that has occurred over a long period of time, it's going to take much, much longer, and be much more difficult to reverse that problem and those deficits. It'd be possible, but it's going to be much harder. That's related to the complexity of the brain. Not only repairing connections in one brain region, but all of the underlying environmental, genetic, and other processes that have occurred over many years and gone into damage of these connections. So all of that has to be dealt with before you can really fully recover."

So the new model of depression centers on neuroplasticity. The chemicals matter because they either damage brain cells or help protect or restore them, but the bigger picture is how brain tissues and structures are being affected. It might seem as if this is trading one broad statement – depression results from a chemical imbalance – for another one – depression results from abrasive forces overwhelming repair capabilities. Moving beyond a chemical view of depression to focus on their impacts to specific tissues isn't just shifting the blame, though. By putting the spotlight on the brain's adaptation mechanisms – and how they are enabled or disabled – neuroplasticity illuminates the mechanisms that create or counter depression. It implies there may be a few ways to affect it directly: reduce abrasion, support repair. If we can get even a few of these tools under intentional control, shouldn't we be able to affect depression's course, from vulnerability through to recovery?

Chapter Six

GLUTAMATE AND GABA

"Ketamine may save my life," Carolyn texted me. It was the summer of 2019, and by then she was deeply depressed, expressing a calm certainty that her life would end, someday, by suicide. I had told her about ketamine's rapid antidepressant effect about a week earlier, and she researched the treatment and found a clinic nearby that could provide it. The U.S. Food and Drug Administration (FDA) had approved a form of ketamine administered as a nasal spray just a few months before, and though it was so new it hadn't reached her insurance schedule yet, the clinic could start her on intravenous (IV) ketamine right away.

I waited anxiously for her call after that first treatment, but when it came, the news was not good. Unlike traditional antidepressants, ketamine is effective within a couple of hours, for about 60 percent of patients treated... but apparently Carolyn was not in that 60 percent. She did not feel any better; in fact, she had experienced "depersonalization," a sense that her identity was gone. "I felt like dust," she said, and it had upset her considerably.

A clinical study in 2000 involving a very small number of patients first revealed ketamine's amazingly rapid antidepressant effect.[1] Further testing in the years that followed supported its short time to effect, but also highlighted limitations. It provided relief from the symptoms of depression in about 60 percent of patients within 24 hours of treatment, but the effects dissipated in a week to 10 days. In 2013, Dr. James Murrough of the Icahn School of Medicine at Mount Sinai[2] led a team that attempted to reach more people and extend the effects beyond a week through repeated doses of ketamine over a longer period. "We were building on the observation that a single infusion works so fast, but we know that people relapse after about a week," he said. "So we said, 'Let's give another one a few days later and see if we can keep them from relapsing.' And then another, and then another. We just borrowed an ECT model, which is Monday, Wednesday, Friday, for several weeks. So we started with 2 weeks, six treatments." It was successful. By the end of the 2 weeks, 70 percent of the 24 patients responded to treatment with reductions of depression scores of 50 percent or more, and a median duration of 18 days following the final treatment.[3] The protocol for ketamine when Carolyn started it was twice a week for 4 weeks, then once a week for 4 weeks, then once every 2 weeks, and so on until it was down to a dose every few months. So she duly went back a couple days later for her second dose, still without apparent benefit.

Her father died the next day. He had been in hospice care for more than 10 years with Alzheimer's disease. While not a surprise, his death was still a shock. She soldiered on, staying with the program and transitioning to the nasal spray. After the first few experiences with the drug – she said the session after her father's death actually made her feel more suicidal – it seemed to get better. She started taking a Xanax before treatment, and felt relaxed and peaceful during the session. Not necessarily in a good way; while the drug was administered she would imagine she was drifting off into death, which she found comforting.

Carolyn has been undergoing ketamine treatment for several months now, and I think it is working, but slowly. She has had some very serious down periods and doesn't seem resilient to setbacks, but overall she does seem to be on a more even keel. She now actively enjoys the treatments, dancing in her chair in the clinic as the drug takes hold. She seems to have lost her fascination and craving for suicide, but it is still too soon to know.

THE EVIDENCE

Just like the clinicians of the 1950s couldn't believe their eyes when they first saw the effects of imipramine and iproniazid, the rapid and lasting – for a week – effect of ketamine on major depression is disrupting existing models of the disorder. Ketamine seems to immediately turn on the forces of neuroplasticity with a surge of the growth factor BDNF, making synapses grow like weeds for a few hours.

The FDA announced approval of esketamine for treatment-resistant depression in March 2019. Sold under the brand name Spravato, esketamine is considered a breakthrough therapy: its rapid action (usually within 4 hours of administration) and lasting (for up to 10 days) effects make it a substantial improvement over anything else on the market. And as Dr. Murrough's team had shown, as doses are administered over the ensuing days, weeks, and months, the effect usually stretches out. Eventually a dose every few months should keep the sufferer in remission, and maybe, one hopes, even those will prove to be unnecessary someday. Compared to the weeks-long time lag for effectiveness of all the traditional antidepressants, and the fact that it reaches a bunch of people classified as treatment-resistant (they tried at least two different antidepressant medications for enough time at adequate dosage without responding), esketamine seems to be a gift from heaven.

Of course, esketamine is ketamine, a street drug known for its dissociative and psychotomimetic effects. I had to look up the word "psychotomimetic"; it means that it can give someone the symptoms of a psychotic break, similar to LSD. So that's a drawback. If abused (and the fact that it's a street drug means it *is* commonly abused), it can damage the brain, liver, and urinary tract. Low-dose ketamine administered through an IV or in the form of a nasal spray in a clinic under medical supervision, however, is uniquely effective for most – most – of the people who use it. The mind-altering side effects are gone before the clinic releases the patient, about 2 to 3 hours after starting treatment.[4]

What is unique about ketamine in any form, compared to other antidepressant medications, is that instead of targeting monoamine systems – the neurons that release or respond to serotonin, norepinephrine, or dopamine – it works on the system of neurons that use glutamate as a neurotransmitter. Early observations of the effects of the TCAs and MAOIs led to the monoamine deficiency hypothesis of depression, which later fell apart under greater scrutiny. As I spoke with Dr. Murrough, he reminded me of that bit of history, warning me not to assume that the rapid antidepressant effect of ketamine provides a simple explanation for depression, with glutamate at its core. I admit that is exactly what I would like to do. Putting ketamine aside, the glutamate system is still an attractive candidate for a root cause of depression. Glutamate is heavily involved in neuroplasticity, strengthening synapses to promote learning and memory, causing the transcription and release of the growth factor BDNF, and trimming off or remodeling neuronal processes. And when there is evidence that some areas of the brain are hyperactive and other areas hypoactive in major depression, it makes sense to look at the brain's primary excitatory neurotransmitter, glutamate, and its primary inhibitory neurotransmitter, gamma-aminobutyric acid (GABA).[5]

"The physiologic effect [of traditional antidepressants] is that they increase levels of serotonin in the synapse," Dr. Murrough reminded me. "So, that observation, going back decades, that serotonin increased from these medications, kind of gave rise to what might be the biology of depression. People started talking about the serotonin hypothesis of depression; they started talking about a serotonin deficiency syndrome, that maybe people were depressed because they had too little serotonin in the brain. But it turns out that that's probably not the case.

"The story with ketamine has been somewhat similar. There was a now-famous initial observation of an antidepressant effect of ketamine given to a very small handful of patients, published in 2000. That's the Berman paper, and it's considered the first report in the literature of a clinical effect of ketamine. There were a couple of remarkable things about it. One was that patients seemed to be showing reductions in their depression severity within a couple days or a day, and that had not been seen before with antidepressant [treatments], particularly medications. The time course was the biggest head-scratcher. And then later it was shown that it was effective in individuals that we might describe as treatment-resistant: they hadn't done well with standard serotonin-based antidepressants. Those people started responding. And after those observations happened, a lot of attention turned toward the glutamate system in depression."

Though he is not willing to jump on the "glutamate is the root of depression" train, Murrough does think that the rapid effect of ketamine tells us something important about antidepressant actions. "One way to think of it is that all those steps between changing serotonin transmission and ultimately changing neuroplasticity – probably the actual mechanism of why something would be antidepressant – involves the important in-between step of glutamate signaling. Somewhere in those steps from affecting serotonin to affecting neuroplasticity is affecting glutamate signaling, NMDA receptor-based signaling. If you gave something that jumped over a couple steps and went right after glutamate, that could explain, or at least be the beginning of a framework of explanation, for how there could be a drug that works that fast."

GLUTAMATE AND GABA

Though much of the focus of studies of depression has been monoamine neurotransmitters like serotonin and norepinephrine, those substances actually account for a small fraction of signaling in the brain. Glutamate is used in approximately 85 percent of the synapses in the human neocortex, and GABA in most of the rest.[6] As the most plentiful neurotransmitters in the brain, glutamate and GABA have been studied extensively. Glutamate is excitatory – it activates receptors to build up electrical potential to the point that the neuron fires an action potential through its axon on to the next neuron. More specifically, when glutamate attaches to its ionotropic receptor, it opens a channel that lets positively-charged ions into the neuron, so that the inside of the neuron becomes less negatively-charged. When the charge level reaches its threshold, more ion channels open to increase the electrical potential suddenly – in about a millisecond – and the neuron fires an action potential. It is the classic ionotropic receptor function.[7]

In contrast, as the primary *inhibitory* neurotransmitter, GABA activates receptors that increase the negative charge inside the neuron, holding it back from reaching its threshold. GABA receptor binding opens an ion channel allowing negatively-charged chloride ions into the cell, leaving it hyperpolarized. With a more negative charge, it will have to take in even more positively-charged ions to climb up to its threshold.[8] Eventually, more glutamate can still get it there, but it's going to take a lot more binding; it has a deeper hole to climb out of.

Glutamate can be converted into GABA and back, and both are synthesized from the amino acid glutamine. Glutamine is considered "conditionally-essential": though our bodies can manufacture it, sometimes we don't make enough and need to supplement it from the diet. The lifecycle of these substances starts with glutamine, transported across the blood-brain barrier. Neurons and some glia convert the glutamine to glutamate. After release by a neuron, most of the glutamate is swept up by an astrocyte, which can return the glutamate back to the neuron for reuse, hand it to a GABAergic neuron to convert it into GABA, or recycle it back to glutamine. Oligodendrocytes can also recycle glutamate back to glutamine. In this way, neurons, astrocytes, and oligodendrocytes all play roles in maintaining an equilibrium among glutamine, glutamate, and GABA.[9]

Glutamate acts through several types of receptors, of which the N-methyl-D-aspartate (NMDA) receptor is most often mentioned in studies of affective disorders. NMDA receptors are ionotropic – they directly manipulate ion channels – and they have some other interesting qualities. First, in addition to bringing in positively-charged sodium ions, NMDA receptors can allow calcium ions into the neuron, a process thought to be key to memory and cognition. NMDA receptor binding can cause the release of the growth factor BDNF and other substances that protect neurons. NMDA receptors are also unique in that they act as "AND" gates: they can only be activated when both presynaptic and postsynaptic cells are depolarized simultaneously. While it may not make immediate sense that these receptors only activate when both neurons are already depolarized – this should indicate a nerve impulse is already being generated – it is actually a really important quality in that it reinforces and amplifies neuronal connections. And when glutamate binds with NMDA receptors outside the synapses – on various other parts of neuronal processes – opening those calcium ion channels can result in excitotoxicity, killing off that portion of the neuron or glia. A high enough level of calcium ions entering the cell can trigger apoptosis, killing the entire cell.[10] So the binding of glutamate with NMDA receptors carries the powers of growth, synaptic strengthening, dendritic pruning, and cell death.

GABA has two types of receptors: GABA-A and GABA-B receptors. GABA-A receptors are ionotropic: when activated by GABA binding they open a channel to let in negatively-charged chloride ions, resulting in a deeper negative charge, or hyperpolarization. Benzodiazepines, commonly used to treat anxiety disorders, act through GABA-A receptors. GABA-B receptors are metabotropic: when activated they initiate a cascade of messengers that carry out other tasks. They are still inhibitory, though, and some of their actions prevent the neuron from firing for a long time.[11]

GABA also has unique attributes that make it interesting in studies of depression. In early brain development, GABA, acting through GABA-A receptors, is actually excitatory rather than inhibitory. Only after some amount of maturation does it switch roles to become inhibitory. In adults, GABA plays multiple nurturing roles essential to adult hippocampal neurogenesis. It again starts out as an excitatory neurotransmitter in the newborn neurons, with the effect of increasing release of the growth factor BDNF to protect the neurons as they mature. BDNF, in turn, prompts more GABA release[12] – a positive reinforcement loop, increasing neuroplasticity.

GLUTAMATE AND GABA IN DEPRESSION

Since the 1980s, studies have reported elevated glutamate and a decreased glutamine/glutamate ratio in blood plasma of depressed subjects compared to controls. That's outside the blood-brain barrier, though, and other studies looking for similar evidence of glutamate abnormalities in the brain through the cerebral spinal fluid and postmortem brain samples have had mixed results, with some showing increases and others not.[13]

Magnetic Resonance Spectroscopy (MRS) is a functional imaging technology that uses the slightly different radiofrequency signatures of various chemical substances to measure their concentration levels. The detectable substances include some amino acids, neurotransmitters, and metabolites of neurotransmitters, which means MRS can show some indications of neurotransmitter levels and binding activity in different areas of the brain.[14] Studies using MRS to measure concentrations of glutamate, glutamine, and GABA in different areas of the brain have also produced mixed results, but some meaningful trends. A large majority of studies show evidence of reduced glutamate metabolite levels in the frontal cortex and cingulate regions of patients when they are in a depressive episode, implying reduced excitatory neurotransmission. In contrast, the occipital and parietal/occipital regions show elevated glutamate metabolite measures in people in a major depressive episode.[15] GABA deficits have been seen in parts of the brain in people with major depressive disorder, particularly the occipital cortex, anterior cingulate cortex, and dorsolateral/dorsomedial cortices, implying reduced inhibition of neurotransmission. People with the melancholic subtype of depression or treatment-resistant depression show the most pronounced deficits in GABA in these areas, up to 50 percent below normal levels. Depressed patients not meeting those criteria still show up to a 20 percent reduction.[16]

The most compelling evidence of a role for glutamate in depression came from a 1994 study of how traditional antidepressants affect glutamate's NMDA receptors. Observing the effects of 17 antidepressant medications and electroconvulsive shock (ECS) in mice, researchers found that 16 of the 17 antidepressants and ECS all caused adaptive changes in the sensitivity of NMDA receptors. The one antidepressant that didn't affect NMDA receptors had been shown to be ineffective in humans in clinical trials, while the others were all clinically effective against depression. As a result of the study, the researchers proposed that perhaps glutamate, acting through NMDA receptors, is what makes traditional antidepressants effective – that it may be the final common pathway of antidepressant action.[17]

WHAT DOES KETAMINE TELL US ABOUT DEPRESSION?

So there is plenty of evidence that implicates glutamate and GABA in major depression and both glutamate and GABA hypotheses of depression have been proposed.[18] Ketamine's clinical success points a finger in the direction of glutamate, but adds some mystery as well. First, ketamine is an NMDA receptor *antagonist*: it blocks NMDA receptors. Since NMDA receptor binding promotes neuroplasticity, how could *blocking* those receptors have antidepressant effect? Second, ketamine has an antidepressant effect only when administered at a very low dose; a full, anesthetic dose of ketamine has no antidepressant effect. So a little ketamine is great, but a lot is ineffective?

And could it be that ketamine's efficacy doesn't come from blocking NMDA receptors at all? At least two other NMDA receptor antagonists have failed as antidepressants in clinical trials, and glutamate release inhibitors have shown no evidence of rapid effect, and inconsistent findings of antidepressant qualities even after sustained treatment, in two decades of human studies.[19] "It turns out there are several studies of other glutamate-based compounds that are not antidepressant, including ones that directly target the NMDA receptor," Dr. Murrough said. "So, in modern times, that's kind of the current vexing question in the field. We know that ketamine binds to the NMDA receptor several-fold more tightly than any other receptor in the brain. That being said, it interacts with many receptors, and actually that's the case with every drug. So when you see that a psychotropic drug affects 'X' receptor or protein, usually what they've done is a basic study where they generate binding constants that basically tell them how tightly this compound binds to an entire panel of receptors that we know about in the brain. And there could be hundreds. There's about 20 serotonin receptors, there's 20 acetylcholine receptors; you kind of just go through them all. And ketamine binds to lots of things. It binds to serotonin receptors, and now it's been found to bind to opiate receptors."

Bolstered by decades of animal studies, NMDA receptor activity is still the most likely source of ketamine's rapid antidepressant effect, though it does not seem to have a straightforward path there. Dr. Murrough commented, "One thing we know that happens when ketamine blocks the NMDA receptor – or we think happens; it has been shown in rodents – is that, somewhat paradoxically, you actually have an acute increase in release of glutamate in the synapse, a glutamate surge. So then the question becomes how does that happen? Because it's not immediately obvious why, if you gave a drug that blocks glutamate, why that would cause *more* glutamate signaling. It was an early paradox, but again there are classic rodent studies that showed you can inject a rat with ketamine, and there's definitely data that shows the glutamate goes up. At first, everyone was sort of 'Chalk that up to the list of things we're not sure how that happens.' But there's been a lot of work done over the last 5 to 10 years to unpack the basic mechanisms of how ketamine works."

Those theories about ketamine's mechanisms also address why only low doses of ketamine have antidepressant effect. One theory is that ketamine initially blocks NMDA receptors on GABA neurons only. By blocking those particular NMDA receptors, at least in the beginning, those neurons don't fire their GABA at other neurons, so they stop inhibiting glutamate neurons for a while. That pause in GABA gives those glutamate neurons a short time to run free – ketamine has a half-life of just a few hours, so it will be out of the system before long – spawning neuroplastic processes.[20]

Another theory involves the fact that even when at rest, glutamate neurons still do some trafficking in glutamate. That is, glutamate neurons keep releasing a very low level of glutamate into the synapse when they aren't depolarizing, binding to a few NMDA receptors on the postsynaptic neuron. This theory of ketamine action suggest that in the steady state, those NMDA receptors actually hold back forces of neuroplasticity. Blocking just those receptors involved in that steady-state activity through extremely low dose ketamine unleashes BDNF production and strengthens synapses.[21] Either way, the effective result of ketamine's blockage of NMDA receptors is a spurt of neuroplastic activity.

Back to Carolyn's predicament, though. If glutamate operating on NMDA receptors is the final link in the actions of antidepressants, then why does ketamine fail to provide relief for some, even after repeat doses? "That is one of the key questions that the field has been asking, was asking initially in the wave of the initial ketamine observations," Dr. Murrough replies. "The short answer is we don't know."

So, glutamate and GABA are disrupted in depression: too much here, too little there, with subsequent over- and under-activation of neurons supporting cognition, mood, and control of the brain's and body's systems. Their actions support the theory that increasing neuroplasticity is what gets someone out of a depressive episode, and may truly be the final common pathway of effective antidepressant response. We have to look beyond the tools, though, and find the handyman. What puts glutamate and GABA into a disrupted state?

Chapter Seven

GLIAL PATHOLOGY

It's the sort of immigrant story that, though not uncommon, just isn't told often enough. Girl comes to America with limited English, a suitcase, and a dream... and a post-doctorate fellowship at an Ivy League University.

For Grazyna Rajkowska, learning to speak a new language was challenging – she could write English well enough to get by, but not speak it – though it wasn't her first time overcoming such an obstacle. Born and raised in Poland, she attended college at the University of St. Petersburg where classes were taught in Russian. Learning the language along the way, she completed degrees in biology and zoology, and then picked up a PhD in neuroanatomy.

Moving to the U.S. for a post-doctorate fellowship in neurobiology at Yale, Rajkowska joined a team examining cells from the prefrontal cortices of schizophrenia patients, looking for a neuronal pathology that would explain the disorder. After 5 years there, she was offered an independent position in her own lab at the University of Mississippi Medical Center, where she switched her focus to studying cellular pathology in major depression, analyzing the cause and effects of that disorder at a cellular level. Rajkowska is credited with bringing to light glial involvement in mood disorders. It is glia, not neurons, that are so significantly reduced in number in depression. Though both neurons and glia show damage, the impact on glia – the long-disregarded "glue" of the brain – is much stronger than anyone had supposed. And given the role glia play in cycling and availability of glutamate and GABA, they become more interesting every day.

GLIA

The most numerous cells in the human brain, glia outnumber neurons 10 to 1, and have long been neglected in theories of how our brains work. Dismissively named after the Greek word for glue, glia were originally assessed to merely provide basic support for neurons, which carried out all higher order functions. Only in the last few decades has the rich and active functionality of glial cells come to light, along with their involvement in psychiatric disorders. It started at the end of the millennium when Rajkowska published the results of studies showing that much of the volumetric reductions seen in major depression come from reductions in the size of neurons and outright loss of glial cells.

I asked Dr. Rajkowska why, with everyone else staring at neurons, she had decided to investigate glia in major depression. "It was in the late 90s and there were some neuroimaging studies showing changes in glucose metabolism in the frontal cortex in depression," she replied. "I saw those studies and I thought that if there are changes in metabolism then it is very likely that neurons are involved. I searched the literature and I didn't find a single study that was counting and measuring neurons and glial cells in depression. And the reason that I looked at glia was very simple. There are two main types of cells in the brain: neurons and glia. I thought why look at neurons only? It could be glia that were involved. I didn't have any root scientific reason behind starting to look at glia, other than being very systematic. I was very surprised to see that glial changes were more prominent than neuronal changes in depression."

Throughout the 1990s many imaging studies of depression showed both structural and functional differences in the depressed brain compared to a healthy one. Rajkowska started with postmortem brain tissue, counting cells and measuring their size. When the brains of people who had died with depression were compared to matched, non-depressed controls, the ones with depression had fewer large neurons matched by more small neurons, and fewer total glia. These differences existed in some regions and not others, and indicated that neurons were either shrinking or not developing to maturity, while glia disappeared.[1]

Over the last two decades, researchers have continued digging into a glial role in major depression. Postmortem studies consistently report decreases in the number and packing density of glial cells in particular areas of the brain in deceased subjects who had major depressive disorder or bipolar disorder. In the depressed subjects, one early postmortem study of the subgenual cingulate cortex showed a 24 percent decrease in glial cells overall,[2] while other studies have shown lesser, but still very noticeable, glial cell reductions in the dorsolateral prefrontal cortex, supragenual cingulate cortex, orbitofrontal cortex, and amygdala. The change in packing densities in some areas points to a loss of glial processes rather than cell death, and some studies have shown a change in the size and shape of the nuclei of glial cells in mood disorders. Of course, there are also studies that fail to find significant differences, and evidence of glial pathology differs somewhat between major depressive disorder and bipolar disorder in degree and in the precise regions affected. Overall, though, depression shows a larger effect on glia than on neurons. Differences in neuronal cell size and packing density show up in fewer areas, and are of less magnitude than the changes seen in glia.[3]

There are three basic types of glial cells: microglia, astrocytes, and oligodendrocytes. In human embryonic development, microglia are born out of the same precursor cells that will give rise to the bone marrow, making them closely related to the immune cells of the body. The other two classes of glial cells come from the same precursor cell types as do neurons, and adult hippocampal neurogenesis gives birth to new neurons, astrocytes, and oligodendrocytes. New microglia, in contrast, come from existing microglia dividing and growing, often in response to some pathological condition like infection or injury.[4]

Microglia constitute the brain's primary immune system; they initiate and terminate inflammatory responses, consume invaders, and perform other protective functions. More recently, researchers have observed their important role in the brain's early development, where microglia participate in synapse development and pruning. Throughout adulthood, they are involved in synapse formation, neurogenesis, the actions of growth factors like BDNF, and in the pruning of synapses. Unlike astrocytes and oligodendrocytes, they don't seem to be reduced in number in depression. Animal studies do show an increase in microglial activation following chronic stress, and a decrease in indications of activation following antidepressant treatment.[5] So rather than being the victims of a pathology, they may themselves be pathological.

Astrocytes are the most plentiful and versatile types of cells in the brain. Among their many functions, astrocytes cooperate with the tightly-packed endothelial cells of the brain's blood vessels to build and maintain the blood-brain barrier. Wrapping their "end-feet" around the blood capillaries in the brain, they take up the glucose required for its energy needs and can also tighten to restrict blood flow to different areas. New technologies have shown that astrocytes express almost all of the receptor systems and ion channels that are found in neurons. They also have transporter systems that take up synaptic serotonin, norepinephrine, dopamine, and histamine, and they contain several enzymes, including monoamine oxidase, that together regulate the availability of neurotransmitters. Astrocytes are actively involved in the uptake, recycling, and metabolism of glutamate, making them an important part of neuroprotection and excitatory transmission. In fact, because they are so important to the balance of glutamine, glutamate, and GABA in the brain, they may be responsible for the glutamate and GABA dysfunctions seen in major depression. They synthesize and release growth factors that help the survival, growth, and differentiation of new neurons, including BDNF, glial-derived neurotrophic factor (GDNF), and others. They also play a role in inflammatory processes, synthesizing and releasing inflammatory cytokines.[6]

Astrocytes get the most attention in studies of depression because of the many postmortem and imaging studies that show significant losses of that type of cell in the key areas associated with depression's effects. This is especially prominent in younger patients, victims of early-onset depression. Studies that look for glial pathology in areas not involved with depression, such as the sensorimotor cortex, show normal glia in those areas.[7] In addition to outright loss of cells, there is evidence of deficits in the proteins that astrocytes make, including the glutamate transporters and enzymes so important in glutamate recycling and neurotransmission.[8] Animal studies also back up their importance, showing that chronic stress decreases astrocyte numbers while antidepressants have a positive effect on their function.[9]

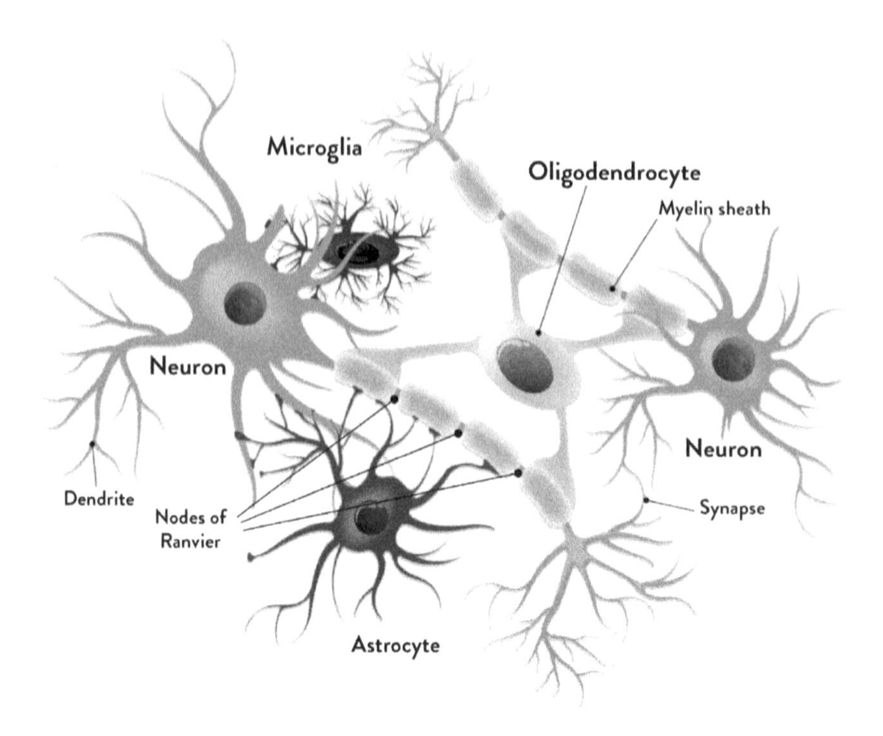

Figure 7-1: **Types of glial cells: astrocytes, oligodendrocytes, and microglia.**

That is not to say that oligodendrocytes are being ignored. While astrocytes show lower density and decreased protein expression in younger victims of major depressive disorder, its elderly victims tend to show more cell loss and pathology of oligodendrocytes instead.[10] Oligodendrocytes form the myelin sheaths that protect axons and speed a signal to its destination, wrapping their appendages around those axons. In fact, the pale appearance of those myelin sheaths is what gives "white matter" its name. White matter damage is frequently observed in late-life depression, called "white matter hyperintensities."[11] Some oligodendrocytes also participate in sweeping up extracellular glutamate and converting it to glutamine, and, just like neurons, they are sensitive to excitotoxic damage caused by excess glutamate.[12] Oligodendrocytes are specifically targeted for damage by glucocorticoids, so they would be especially vulnerable to a runaway HPA axis.[13]

DEVELOPMENT OF A PATHOLOGY

I asked the same question that I had asked Dr. Duman: why do we see this damage in some parts of the brain and not others in depression? How does depression take aim at particular areas?

Dr. Rajkowska's colleague at Mississippi and frequent research partner, Dr. Jose Miguel-Hidalgo, was happy to offer some thoughts. "It is possible that there is some specific, for example, genetic mutation that specifically targets glia in a particular area of the brain," he said. "I don't think this is the case, though, because depression is so widespread. There are so many people who suffer depression, that I think it's more functional: the responses of glial cells to functional and other neurochemical alterations in different brain areas."

As an alternative to an initially glia-based pathology, he suggests that perhaps the primary factor is a change in what the neurons are doing. "What has been seen in neuroimaging studies is that there are some areas where neuronal activity appears to be more affected than others. You can find some areas that are more activated than others, or some areas that are activated when others are depressed, in functional terms. Those differences can be very dramatic, by the way." The glia in those areas adapt, responding to the new neuronal behaviors. "So if the prefrontal cortex is depressed in some way, either functionally or structurally speaking, or some of the connections between cells or their responses to other brain areas are down, glutamate, GABA, even neurotrophic factors will be changed in that part of the brain, and that will impinge directly on the responses of glial cells. What they will do is to adapt to that situation in that particular brain area. So, over time, there will be a difference that you can detect using neurochemical methods or morphological, functional, postmortem, pre-mortem – they would show up eventually." These changes could be happening over 5 or 10 years before any sort of functional impact is seen, he added. Adaptation becomes maladaptation; glial cells create or maintain – or both – a vicious cycle. "So, neurons change, glial cells change, and after some time they adapt to this situation, and it makes it more difficult to go back to normal, until something happens, like some therapy. Because until something changes that situation, those glial cells are likely going to be either helping or maintaining the physiological changes that were caused by spending a long time in that state."

Dr. Miguel-Hidalgo talked about a recent study of oligodendrocytes in an animal model of chronic stress. "This researcher used mice, subjected them to isolation for 28 days and then compared them to their littermates who had not been isolated. At the end, they looked at the myelin thickness in various brain areas: the prefrontal cortex, the cerebellum, the nucleus accumbens. They were looking for problems with these glial cells – the oligodendrocytes, which produce the myelin – as a result of social isolation, which is one of the stressors associated with depression. If this is all over the place, myelin would change everywhere. They found that the isolation drove a dramatic change in the thickness of the myelin in the prefrontal cortex only; there was no effect in the nucleus accumbens or the cerebellum. This is a clear example that glial cells can be affected in some parts of the brain and not in others."

Drs. Rajkowska and Miguel-Hidalgo are working on a new study now, observing what goes on at the nodes of Ranvier: the gaps in the myelin sheath along an axon where one oligodendrocytic myelin-forming appendage meets up with the next. Or, nearly meets up; those small gaps are actually really important to neuronal signaling. As the electrical charge moves down the axon, it jumps from gap to gap, node to node. Astrocytes monitor those gaps, helping to maintain the myelin; it is a place where neurons, astrocytes, and oligodendrocytes all interact. "We think that maybe the nodes of Ranvier are changed in depression," Dr. Rajkowska says. "Maybe they are shorter, and maybe the molecules that are around them are expressed differently, and this is what we will be looking at in postmortem brain tissue of depressed subjects and tissue from homologous regions of rat brains, rats that were exposed to stress."

So, maybe glial cells, particularly astrocytes, significantly contribute to mechanisms through which an initially minor change in neuronal signaling – driven perhaps by the ravages of chronic stress or neuroinflammation – is maintained and reinforced over time. And with their essential role in the glutamate and GABA lifecycles, an alteration in their function could lead to an excess of glutamate outside the synapses, where it is poised to damage neurons and glia through excitotoxicity.[14] The damaged glia become even less capable of doing their jobs, and the effect grows. As Dr. Miguel-Hidalgo put it: "I think the whole system adapts in a way that is dysfunctional, eventually. We don't understand all the reasons, but maybe there are points of entry for treatments in the future."

Chapter Eight

BIOENERGETICS

Dr. Richard Boles owns a highly unusual medical practice. He built it on a concept he devised when, as a geneticist at Children's Hospital Los Angeles, he saw a large number of young people with certain illnesses: cyclic vomiting syndrome (thought to be a form of migraine), other types of migraine, autism spectrum disorder, and chronic fatigue syndrome. While that collection of illnesses seems like an odd grab-bag of serious neurological problems, they are actually closely related in that each is thought to have a strong mitochondrial component. Boles established his self-pay practice in Pasadena, California, to specialize in complex disorders associated with mitochondrial dysfunction. Patients find their way to him, he says, primarily through the internet or word of mouth, and he treats many of them through telemedicine. When Boles saw that many of his young patients had extremely high rates of depression – as high as 50 to 75 percent of those with cyclic vomiting syndrome, for example – he began to suspect that having a mitochondrial disorder may confer a higher risk of depression as well. Interestingly, not only did those young patients have much higher rates of depression, so did their mothers.

Mitochondria are incredible structures; these organelles provide the energy used in all the processes of the cell in the form of adenosine triphosphate (ATP). Science's best understanding of mitochondria says that about 2 billion years ago, an ancient one-celled organism enveloped an ATP-producing bacterium and kept it captive. The one-celled organism provided the protection of its internal environment while using the quantities of ATP churned out by the bacterium and its descendants to fuel its life processes. Reminiscent of their bacterial heritage, mitochondria replicate constantly in our cells, presumably replacing worn-out structures. These organelles are essential to cellular activities, and the busiest cells, including those in the brain, heart, and liver, host thousands of them.[1]

As Boles saw more and more patients with mitochondrial disorders and depression, and more of their mothers with depression as well, he began to think that mitochondrial dysfunction could be a heritable risk factor for depression. His suggestion was met with skepticism by his colleagues, though. "Even in my own institution, almost no one believed me," he said. "They said, 'Oh, it's just because you're looking at really sick patients,' and 'Oh, it's what happens when you have a kid who's affected; of course the mother's going to be depressed.' But that was not what was happening, because my other patients, with other metabolic disorders that were not mitochondrial, their mothers were not depressed any more than the average population."

There was clearly a genetic component to the depression Boles was seeing, and it looked like it may be a particular sort of inheritance pattern – that it came from the mitochondrial DNA, not the nuclear DNA. By convention, when we talk about someone's genetic material, we are talking about the DNA in the nucleus of the cell, contributed by both father and mother. The cell's mitochondria also hold a small but critical set of genes, inherited exclusively from the mother. The ovum, or egg, contains our first set of mitochondria along with all the other working parts of the cell. The father's sperm is much smaller in size than an egg; it really holds only a cell's nucleus for its DNA contribution and a tail for propulsion. So when a characteristic is consistently inherited from the mother only, mitochondrial transmission becomes a strong possibility.

Mitochondrial DNA contains the genes necessary to create some, but not all, of the protein subunits that comprise the respiratory chain – a key element of the cell's ATP-producing machinery. The rest of the genetic information necessary for that purpose resides in the cell's nuclear DNA; that portion can could be inherited from either the father or the mother. In all, 13 of the nearly 80 subunits necessary for the respiratory chain come from mitochondrial DNA and the remainder from nuclear DNA.[2] That minority contribution from the mitochondria is essential, though; cellular energetics cannot happen without it.

Boles went looking to test his hypothesis. His first study addressed only his own patients and their mothers: 15 mothers of children with maternally-inherited mitochondrial disorders and 17 mothers of children with different metabolic disorders known to be carried by recessive genes. This latter group was an excellent control for the study because the children had equally serious illnesses, but the condition must have been inherited from both mother and father to show up in the child. All the mothers completed depression, anxiety, and general mental health assessments. In the end, the scores showed that the mothers of the children with maternally-inherited mitochondrial disorders had depression and anxiety scores three times higher than the mothers of children who had inherited an illness through nuclear DNA.[3]

This was a small study, though, not enough to convince the skeptics. Boles led a follow-up study with more patients. "They were recruited from the United Mitochondrial Disease Foundation website," he said. "There were 55 families in the 'probable maternal inheritance' group, and 111 families in the 'probable non-maternal inheritance' group. Fifty-one percent of the mothers in the 'probable maternal inheritance' group had depression, while 12 percent of the mothers in the 'probable non-maternal inheritance' group had depression. As you know, 12 percent is basically the prevalence of depression in the general population." The prevalence of depression in the fathers was 9 to 12 percent also, but what was most telling was the effect on the aunts and uncles. Not only did the "probable maternal inheritance" mothers have much higher rates of depression, but so did their sisters and brothers – about three times the rate seen in the siblings of the fathers or of the "probable non-maternal inheritance" mothers.[4]

A strictly matrilineal inheritance – the mother, the mother's mother, etc. – would point to mitochondrial DNA as the culprit. Though Boles' data did not say depression was inherited from the mother in every case, it did show an excess of maternal inheritance, enough to say that, in many cases, it comes through mitochondrial DNA. Stepping aside from the carefully-selected mitochondrial disorder patients he used in previous studies, Boles moved on to a large-scale study of inheritance of recurrent early-onset major depressive disorder. Using multi-generational pedigrees of 672 individuals published in the Genetics of Recurrent Early-Onset Depression project, he found that matrilineal relatives – those with the same mitochondrial DNA as the individual studied – were significantly more likely to suffer from a mood disorder than were non-matrilineal relatives, with an odds ratio of 2 to 1.[5] "I was really shocked," he recounted. "I didn't think the odds ratio would be 2.0; I thought it would be a little lower than that. An odds ratio of 2.0 implies that half the inherited component is in the mitochondrial DNA with the other half in the nuclear DNA. And these are not mitochondrial disorder patients; these are just patients with early onset recurrent depression, which means that they had to have had at least two episodes of depression before their 31st birthday."

None of this suggests that you can only inherit depression, even early-onset depression, from your mother, or that mitochondrial DNA is necessarily involved. But, because manufacturing cellular energy is a vital function of mitochondria, the fact that mitochondrial disorders show such a robust relationship with depression points strongly toward cellular energetics in the development of the disorder. Other very important mitochondrial functions include buffering stores of calcium ions in the cell, synthesizing small molecules, and initiating apoptosis,[6] providing several opportunities for mitochondrial involvement. Studies are confirming that people with known mitochondrial disorders suffer psychiatric illness at rates higher than the general population. The largest sample size in a recent review involved 36 adults with mitochondrial disease, finding a 70 percent lifetime occurrence of psychiatric disorders, with some overlap: 54 percent with major depressive disorder, 17 percent with bipolar disorder, and 11 percent with panic disorder.[7]

In 2011, Boles and Dr. Ann Gardner of Karolinska Institutet in Stockholm proposed that major depression and other "affective spectrum disorders" could result from interactions of mitochondrial dysfunction with inflammation. These, in turn, interact with elements of the monoaminergic system to create and sustain depression.[8]

THE BRAIN'S ENERGY

Energy is not the same as electricity. Electrical charges are used for neuronal signaling, but energy is used in everything a cell does: growth, gene transcription and protein synthesis, shuttling molecules through a cellular membrane... everything. These functions are powered by the energy released by breaking the chemical bond holding the third phosphate group in a molecule of ATP. Mitochondria operate a process called aerobic oxidation, which consumes glucose and fatty acids to produce a stream of ATP molecules. In neurons, operating pumps and other signaling machinery is especially costly, requiring approximately 60 to 80 percent of the ATP consumed. In both neurons and glia, supporting transport and enzymatic activities, such as those involved with glutamate cycling, also pose a high price.[9] Additionally, as the power-producing component of the cell, mitochondria migrate to the areas of growth in a cell, making them essential to the processes of neuronal and glial growth and plasticity.[10]

Aerobic oxidation takes the breakdown products of glucose and fatty acids from food and oxygen from the bloodstream to produce energy, leaving carbon dioxide and water as waste. Through the machinery of this process – the respiratory chain – mitochondria take adenosine diphosphate (ADP) molecules and free phosphate groups and squish them together into new ATP molecules.[11] It is this consumption of glucose and oxygen that shows up as activity in functional imaging. PET directly measures glucose consumption and fMRI directly measures oxygen consumption, to which they attribute a level of ATP production. Assuming that production exactly matches demand – and evidence indicates it is a good assumption – higher or lower levels of glucose and oxygen consumption reflect higher or lower levels of cellular activity. Knowing how energetically expensive neuronal firing is, an upsurge of consumption is understood to represent neurons firing.[12]

ENERGY METABOLISM IN DEPRESSION

Even after Boles and his team published their findings, he saw little initial reaction from the broader community, and the mechanisms through which cellular energetics could be involved in depression remain under-studied. "As time went on and more and more papers came out," he said, "there was definitely an understanding that there is a correlation [between depression and cellular energy metabolism], at the very least. All these papers came out, and we hoped to move the needle, but it didn't move very fast."

There are several ways in which cellular energy metabolism could be impacted. The ATP-generating respiratory chain within mitochondria could be malformed because of genes inherited through mitochondrial DNA (13 of 80 subunits) or genes inherited through nuclear DNA (the rest). The respiratory chain could be perfectly formed, but under constant attack from free radicals. Finally, there are other substances that perform critical roles in the respiratory chain, including zinc, selenium, and co-enzyme Q10. Deficiencies in those substances could lead to ATP under-production, even in otherwise perfectly healthy mitochondria.[13]

The body of evidence that cellular energy metabolism could be involved in depression and other psychiatric disorders is growing. A couple of postmortem studies of people who died with depression, schizophrenia, or bipolar disorder turned up differences in comparison to psychiatrically-healthy controls. One such study found significantly decreased expression of 6 of the 13 respiratory chain subunits encoded in mitochondrial DNA in major depression, while another found some respiratory chain subunits altered in all three psychiatric disorders, in a manner specific to each disorder.[14] Functional imaging shows evidence of impaired cellular energy metabolism in the brains of living people with major depressive disorder. A recent MRS study found 26 percent lower mitochondrial energy production in glutamatergic neurons in the occipital cortex in patients with major depressive disorder compared to controls,[15] and a whole-brain MRS study found abnormalities in a key brain bioenergetics metabolite: higher in the gray matter and lower in the white matter of the depressed population versus controls.[16]

OXIDATIVE STRESS

Much like neurons, which rely on an electrical potential across their outer membrane to operate, mitochondria also maintain a differential electrical charge across a membrane, with positively-charged hydrogen ions (basically, free protons) on one side and free electrons on the other. As the protons are repelled from the positively-charged side toward the negatively-charged side, they flow out through pores in a protein complex straddling the membrane. It is a proton motor, using the flow of protons to generate mechanical movement, stamping out 30 ATP molecules for every glucose molecule consumed. When the membrane's charge differential is disrupted, there is nothing to make the protons flow and the respiratory chain ceases to function. With enough mitochondrial dysfunction, the cell has few options: use a much less efficient process that can only generate 2 ATP molecules for every molecule of glucose; or start the process of apoptosis, killing the cell because the environment is unfavorable for the functions needed.[17]

A primary source of damage to mitochondrial function comes from free radicals: molecules, atoms, or ions with one or more unpaired electrons. They are highly reactive, trying to push their free electron(s) on any willing or unwilling partner. In doing so, they damage DNA, proteins, and lipids. Some of the common free radicals come from superoxide (O2-), hydrogen peroxide (H2O2), nitric oxide (NO-), and peroxynitrite (ONOO-). Both internal metabolic processes and external influences such as exposure to air pollution, X-rays, ozone, and industrial chemicals produce free radicals; mitochondria themselves produce low levels of free radicals as they form ATP molecules.[18]

Our bodies have defenses against free radicals, though: antioxidants, molecules stable enough to absorb and neutralize rampaging free radicals. Some antioxidants are produced during normal metabolism in the body; others like vitamins E, C, and B-carotene are found in the diet. When we don't have enough antioxidants to balance the free radicals, the damage they cause to DNA and other cellular structures is called oxidative stress. With its high metabolic rate, the brain is especially susceptible to such damage.[19]

Oxygen is a very reactive element, so much so that the process of donating an electron is called oxidation, no matter what type of molecule donates it. The complementary process, accepting an electron, is called reduction, and the two together comprise a redox reaction (reduction of one molecule, atom, or ion paired with oxidation of another). Redox reactions are very common to biological processes; it only becomes a problem when a lot of ions are running around unable to find their redox match, pushing their extra electrons on or stealing them from other molecules. Because mitochondria host their own mitochondrial DNA, lack the repair mechanisms of nuclear DNA, and rely on a lipid membrane to function, they are especially vulnerable to oxidative stress. Damage to their inner membrane can open the way for leakage of even more free radicals within the mitochondrion, creating a cascade effect.[20]

Oxidative stress is deeply intertwined with inflammation, providing additional mechanisms for chronic inflammation to contribute to the development and maintenance of depression. For one thing, when free radicals attack lipid membranes, they change the chemical structure of the lipid and create new substances that the immune system may mark as invaders, prompting an autoimmune response. Oxidative agents also push tryptophan metabolism down the kynurenic pathway, reducing serotonin production. Inflammation itself increases the production of free radicals, potentially launching a vicious cycle of continuing damage to the cells of the brain and the body. Oxidative stress is now thought to make a significant contribution to all inflammatory diseases, heart disease, stroke, alcoholism, and neurological disorders such as Alzheimer's disease and Parkinson's disease.[21]

Depression also shows evidence of oxidative stress. Two recent meta-analyses both found significantly higher levels of oxidative stress and significantly lower levels of antioxidants in people with depression versus controls. The more recent one – involving the results of 273 studies – found that all of the markers of oxidative damage in the blood serum were significantly higher in depressed subjects than in controls, and most of those markers went down after treatment with antidepressants.[22]

TREATING DEPRESSION AS A MITOCHONDRIAL DISEASE

Once again, the final proof of a link from cellular bioenergetics to early-onset depression comes from the observation that supporting mitochondrial function is a strong factor in recovery for its young victims. Boles says that between a half and two-thirds of the patient population he sees in his medical practice have been diagnosed with clinical depression, so he treats their depression as well as their presenting illness. Most of his patients are on a psychotropic drug; he adds a formulation of dietary supplements of his own devising that he finds to be extremely effective.

I asked in how many of his patients he is able to successfully treat their depression. "We are able to manage the depression in all of them," he replied. "That's with the caveats that I have followed them, and I monitor their blood levels [of key metabolites] and I increase blood levels again when needed. I also have their DNA, and have figured out what the underlying problem is. I don't just diagnose, give them a prescription and say 'Here you go' and they get better. It takes time, and there are often obstacles in the way that impede progress and we have to figure it out. But eventually, all of them."

His secret formula – or not really a secret, because he has posted the ingredients on a website – includes 33 active ingredients plus extra coenzyme Q10, specifically in ubiquinol form. Coenzyme Q10 is an antioxidant, but more importantly, it is directly incorporated into the mitochondrial energy-producing machinery. Ubiquinol costs more than the other form of coenzyme Q10 (ubiquinone), he says, but it is more bio-available (you don't need to take as much for full effect). Each ingredient is available over the counter as a dietary supplement.

"It takes time," Boles says, "and you have to be careful that it's not just anything over the counter. You have to make sure that you have the right products at the right dose, and that they're compliant. You have to keep with it and monitor blood levels to make sure they're high enough, particularly the coQ10. I like coQ10 levels that are higher than 4, which is much higher than the normal range, and it works. The patients actually get better." I asked him about the timeline for improvement. "Sometimes they can get better faster, but it really takes 6 months before you see a big improvement. In 2 to 3 months you should start to see some improvement, but 6 months is really when you see it." He added that for research partner and fellow clinician Dr. Ann Gardner the antidepressant effect of a similar dietary supplement takes longer to see. "She has an older population, primarily [depressed] adults with hearing loss. She is saying it takes 6 to 12 months for her patients. These are different populations, of course; they may get better slower because their age. Mine are kids, mostly."

And of course he is not just prescribing a dietary supplement for his patients. "My day job is as a geneticist. I do whole genome sequencing – the whole 3 billion nucleotides – in all my patients, and usually the parents as well. I spend hours with them going over the nitty-gritty of everything to know about the patient, and order whole genome sequencing. I don't take the laboratory result from that; I go through each of the hundreds of variants flagged by the system as potentially impactful myself. I connect it to what the individual has, and we discuss everything that I found in their DNA, what that means for their health, and what they can do about it."

The recipe for the dietary supplement is listed on the website NeuroNeeds.com, as "SpectrumNeeds," with the utility of each ingredient spelled out. "I spent an entire month writing up all the stuff on there, so there's a huge amount of information," Boles said. "It's directed towards mitochondrial dysfunction, not specifically towards depression, but you can see all the ingredients on the left hand side. All you have to do is click on any one of them and you'll get more information than you'll ever want to know. That's not only what it is and what it does, but why I've put it in SpectrumNeeds." He recently added another formulation with 40 active ingredients in capsule form, named EnergyNeeds.

In the same way that low-level but chronic inflammation is a high risk factor for depression, mitochondrial dysfunction does not have to be severe to have significant impact. "I think that a lot of the problems in depression doesn't come primarily from a mitochondrial disorder, but individuals whose mitochondrial DNA or whose nuclear DNA making the mitochondrial proteins are just on the lower side of the average range. They don't have the resilience that someone else might have," Boles said. "Then when you get other mutations and environmental components, you go over the edge and don't have the resilience to fight back."

In the brain, resilience requires neuroplasticity, and neuroplasticity requires energy. Cellular growth is very demanding; forming new neurons and glia, or even new synaptic connections, takes a lot of cellular energy. Astrocytes consume a lot of energy cycling glutamine and glutamate. If a cell's energy production can't quite keep up with these demands, then neuroplasticity wouldn't quite keep up either. It is easy to see how a just slightly under-performing bioenergetics system would impact neurogenesis in the few locations where it takes place as well as adaptation and repair of neurons and glia throughout the brain. Whereas decades of chronic stress or chronic inflammation could build up enough abrasive factors in the brain to overcome a healthy growth and repair function, those factors would build up in a much shorter timeframe if growth and repair processes are disadvantaged.

But neuroplasticity does not appear to be disrupted throughout the whole brain in depression. Structural studies show losses in specific locations and not in others. The neuroplasticity model of depression needs something more. It needs a way to explain why relatively distant parts of the brain would be affected, with no apparent impact to the tissues between them. The developing concept of functional connectivity may start to provide an answer.

Chapter Nine

CONNECTIVITY

Imagine you just moved to another country. At first, everything seems strange, but eventually you learn your commute and where to get lunch, and make some friends. After a while, you notice something odd. On alternate Tuesdays, you see a lot of people wearing yellow hats. Just on alternate Tuesdays, no other days, and it's always the same people. They don't necessarily talk to each other or group together, but they do show that regular yellow-hat alignment. Watching more closely now, you notice another group of people who wear the same orange tie every Wednesday. Again, they may or may not group together or work in the same building, but they all wear that same tie every Wednesday. You begin to see other patterns among the rest of the people around you: red baseball caps every 10 days or purple T-shirts every 8 days; these people display a connection with one another, even though you could follow them all day long and never see them interact. Are they showing mutual support for a favorite sports team? Do they coordinate this somehow? You are seeing some sort of identity and alignment, but it is hard to pin down what it means.

Different parts of the brain, parts that may or may not directly connect with one another physically, indicate their identity and alignment too. The indications are so subtle that it took a long time for anyone to notice, and though it is hard to pin down where they come from and exactly what they mean, scientists now know that they reflect a higher-level organization of brain function. And it turns out that a depressed brain is organized differently than a healthy brain.

MAPPING THE BRAIN

The map of the brain used so far – the system of coordinates and clumps of gray matter – refers to its structural organization, not its functional organization. The functions in which many structures participate, like speech, vision, anger, or sadness, have been elaborated over the years, often through studies of stroke, epilepsy, or traumatic brain injury. How the brain coordinates and orchestrates these various functions has been elusive, though. I see a shiny object on the floor, I pick it up, and I become happy because I like shiny objects. There are several different functions involved in that chain of events, including visual processing, movement, and emotional control, and those functions need to interact in a coordinated manner.

Thanks to modern technology, we can look into the brain and see what is going on, or approximate what is going on, at least, through such a chain of events. Functional imaging measures changes in blood flow, oxygen utilization, and other such indicators as evidence of increased or decreased metabolism in specific areas of the brain. A neuron consumes substantial resources to manage polarization of its membrane and conduct the activities that result from receptor binding: gene expression, opening and closing ion channels, moving pools of neurotransmitters around, and so on. So when researchers see increased activity immediately upon having a subject initiate a task, it is reasonable to infer that they are seeing activation of the parts of the brain responsible for that task.[1] When I look at the shiny object, there is heightened metabolic activity in my visual cortex and in other parts of my brain involved with control and image processing. Picking it up correlates with increased activity in parts of the motor cortex. With such information, researchers have identified networks of correlated brain activity; a visual processing network, for example, or sensorimotor network. These are task-based activations, and they stand out in functional imaging with increased use of resources. One task follows another, and then the brain goes on to the next state as appropriate to the situation.

More recently, however, researchers have identified additional networks involved in attention, emotional processing, and cognitive control. Importantly, these networks are different in people with depression (and other psychiatric disorders) than in healthy people.

THE BRAIN'S INTRINSIC FUNCTIONAL NETWORKS

Even when we aren't engaged in a task, the brain is always active; it remains active while we sleep. Dr. Marcus Raichle of Washington University describes the brain as "a very expensive piece of real estate" in terms of energy consumption. Though it accounts for only 2 percent of the body's mass, even at rest it uses 20 percent of the energy consumed by the whole body.[2] "You can do a metabolic scan of the entire body with PET, and measure how much glucose all the organs of the body are using, and the brain is just like a light bulb in the head compared to anything else," he commented. Plus, it is always 'on', he adds. "If you take a walk or go out for a run, your muscles go from pretty much zero to 100 percent. The brain lives in a world where it starts at 90 to 95 percent and goes to 98 percent. There's no rest period for your brain. It goes down a little bit in slow-wave sleep, but it doesn't go off."

In 1997, Washington University researchers reviewing the results of nine PET studies of human visual information processing noticed something interesting. When the study subjects started their task, the areas of the brain associated with visual processing increased activity, as expected. What was not expected, though, were that other areas of the brain consistently showed blood flow *decreases* – evidence of diminished activity – at exactly that same time. So, a study subject would be in the PET scanner awaiting a task (passive viewing) while the scanner recorded blood flow indications. Then the task was presented and the subject did what was asked of him. Later, when the passive-viewing images were compared to the same subject's task-active images, some areas of the brain had actually *reduced* their activity level as the task started. Subjects finished the task, went back to passive viewing, and the changes went the other direction. It was as if intentional cognitive activity took place at the expense of some other, subconsciously-directed, brain activity.[3] What are we thinking about when we're not thinking about anything? Random thoughts, day dreams, scouring the lab for food or predators?

Dr. Raichle led the research that exposed what is going on in the brain when nothing in particular is going on. In 2000, exploring the hypothesis that perhaps, rather than signifying random thoughts popping into one's head, this resting state activity may be part of an organized mode of brain function, his team gathered PET images from three groups of healthy volunteers, totaling 49 people. The first two groups had been used as a control condition for other studies, and had been scanned resting quietly, awake with eyes closed. The last group were resting images of 11 people, with images captured first resting awake with eyes closed, and then again with eyes open and fixed on a cross-hairs target. Across all the subjects, some areas showed consistent activity in the quiet, resting state, eyes opened or closed. Comparing these resting state results with PET studies involving a variety of tasks, they identified areas that were consistently and significantly more active in the resting state than in a task state. This indicated "the existence of an organized, baseline default mode of brain function that is suspended during specific goal-directed behaviors." They called it the brain's default mode.[4]

Initially, Dr. Raichle assessed that the brain's default mode probably served to monitor the internal and external environment when awake and alert, constantly evaluating the emotional salience of that environment. Then, when an attention-demanding task arose, the brain would be ready to switch on whatever was needed to deal with it.[5] There was a great deal that was not known about how this default mode operated, though, or the overall organization of what could be several different operating modes of the brain. A breakthrough came from Stanford University, leveraging a new way of looking at the brain's functional connectivity that had been discovered by a PhD student just a few years earlier.

In 1995, Bharat Biswal was working on his doctoral thesis at the Medical College of Wisconsin when he was tasked by his mentor to take a look at a wiggly, noisy signal seen in fMRI studies, Dr. Raichle recounts. "The magnetic resonance scanner used in fMRI is very sensitive to a signal that has to do with the brain's metabolism. If we put you in an MR scanner and look at this particular signal, and we ask you to move your hand or talk or whatever, we'll see the parts of the brain change their activity. But in the course of doing that, the signal that was being used here was kind of wiggling along and looking noisy, even when you weren't doing anything. The strategy at the time was to average that noise out; just get rid of it." Biswal looked deeper into that "noise" and found evidence of functional connectivity among various parts of the brain. Basically, he saw that related areas across the brain showed some very slight energy fluctuations – just tiny little spikes in the fMRI signal – in common, even when not actively engaged together in a task.[6]

Biswal's subjects were given a simple finger tapping task, touching each finger to the thumb of one hand for 20 seconds, then resting for 20 seconds, then finger tapping again. Beginning before the finger tapping task was started and continuing through and between tasks, Biswal took a series of fMRI images at short intervals. Awaiting their task, each subject had been told not to think about anything in particular. Some were told that their eventual task would involve finger tapping, but most of them were told, falsely, that it would be an auditory task, in order to keep them from anticipating their movements.[7]

As expected, the images taken while the subjects were tapping their fingers all showed activations in areas of the brain that control hand movement. Biswal then searched the resting state images of each subject to see if there were activations in those same hand-movement areas when the subject was *not* performing the task. And he found some – periodic activations in those same areas of the brain, but taking place outside of any attempt to actually move a hand. They were very small activations; really just a fluctuation of the scanner signal. And not only were those fluctuations seen in the areas that were directly involved in the finger-tapping task, they also showed up in two additional brain areas: one that is thought to control intention to initiate movement, and another that is associated with moving the legs or feet. So these areas were all movement-related, but not necessarily related to hand movement, and some of them hadn't been actively engaged in the task state. The fluctuations took place at the same time in all those areas at a really slow pace. A few pixels would light up on an fMRI image, then those pixels would be dark again for the next 10 seconds, then they would light up again, and so on, in that slow cycle. In brain time (where individual neurons fire on the order of a few milliseconds) that's slower than slow; and just like on a light spectrum the area beyond red is called infra-red, this beyond-slow frequency is called "infra-slow." This was all happening in the resting state – before and between tasks – including in the subjects who had no reason to think they were about to be asked to move their hand. Biswal used a principle that had been laid out in a 1993 paper to propose that such consistently time-correlated signals represent functional connectivity: different, and often distant, brain areas showing unity towards performing a particular function, whether or not they were actively performing that function. The neurons in those areas didn't need to be actively firing and the areas didn't need to be directly, physically connected to one another through axons and synapses. Through those fluctuations they were demonstrating coordination, coherence, and common intent.[8]

This new idea of functional connectivity among neurons in distant areas of the brain was not an easy sell. "It was, of course, like everything in science," Dr. Raichle remembers. "People said 'Obviously, this can't be; it's a mistake; it's an artifact,' or whatever." A few years later, however, when Raichle had identified the brain's default mode, Biswal's work provided a method of analysis that could reveal more about it. "After we published the default mode papers in 2001," Dr. Raichle said, "a friend of mine at Stanford, Mike Greicius, played the same trick with the default mode network. There's a part at the back of the brain, and he just looked at the same funny signal in the brain, and said, 'Who is it correlated with?' Well bloody well, there was the whole default mode network, front to back, the whole thing was right there. And this opened up this whole area of resting state functional connectivity. By 'connectivity' in this sense, it's related to this time-varying signal that exists across the brain, but is highly correlated within the brain systems. So each system can be separated out... I could put you in the scanner and we could map out all the major systems of your brain, while you just lie there." The scanner would detect the periodic fluctuations from the nodes of the default mode network; separately it would detect periodic fluctuations from the nodes of the movement network, and other fluctuations denoting other networks. The "map" of the brain's systems that resulted from such a scan would show which parts of the brain expect to work together on different pursuits.

Over the next few years the baton would pass between Washington University and Stanford, as they mapped out individual networks and how they interact. "All of a sudden, we could map out the entire cortex of the brain, which is the upper surface, and all their connections down below as well," Dr. Raichle said. "And so you began to see a very large scale layout of how the brain was working." About 20 or so large-scale intrinsic networks have been specified, some of them – a visual processing network, for example – already known to exist, but others showing up unexpectedly and bringing a new perspective to basic functioning of the brain.

The default mode network includes the structures that are most active in the resting state, and then fade to the background when a task starts. It is anchored in the posterior cingulate cortex, precuneus, and ventromedial prefrontal cortex, and includes a few additional areas. Those structures indicate its function: it is "an integrated system for self-related cognitive activity, including autobiographical, self-monitoring, and social functions."[9] Connectivity doesn't necessarily mean that these areas are talking to each other. "They may be talking with each other, or talking to themselves," Dr. Raichle comments. It is a collection of activities correlated in time. Some of the areas are directly linked together structurally through tracts of axons; others have only indirect structural connections.[10]

The functions of the posterior cingulate cortex and the precuneus were not well known when research into the default mode network started; not surprising if they are most active when a person is resting and take a back seat when he or she starts to do something. Both of those areas seem to be very important to a sense of self, conscious awareness, and of agency: where the concept of "I" resides. The posterior cingulate cortex has a central role in supporting internally-directed thoughts and shows increased activity when people retrieve autobiographical memories or plan for the future.[11] The ventromedial prefrontal cortex is a main part of the "emotion circuit," connecting to structures of the limbic system including the amygdala, hypothalamus, and brainstem. So what are we thinking about when we're not thinking about anything in particular? Ourselves. Our experiences, our feelings, our plans... our place in a self-centered world.

Two more of these large scale intrinsic networks are known to be important in depression. The central executive network is anchored in the dorsolateral prefrontal cortex and the posterior parietal cortex. It is the task execution monitoring system. At its most active when there is goal-directed behavior, it maintains and manipulates information in working memory and serves judgment and decision-making functions.[12] The salience network, with primary nodes in the right fronto-insular cortex and the anterior cingulate cortex, seems to provide the switching function necessary to tamp down the default mode network and engage the central executive network.[13] So, the idea is that while we are at rest, the default mode network is spinning along, actively running through self-centered thinking. The central executive network and salience network are sitting in the background, with comparatively low energy usage. When something happens that needs attention and a response, the salience network catches it, tamps down the default mode network, and spins up the central executive network to deal with the task.

These mode switches were demonstrated in 2013, using transcranial magnetic stimulation (TMS). Applying a series of TMS pulses interleaved with fMRI scans on 24 healthy volunteers, researchers showed that high frequency (excitatory) pulses to sites in the central executive network or the salience network caused a decrease in default mode network activity, while a low frequency (inhibitory) pulse to those sites caused an increase in default mode network activity. By using the salience network and the central executive network to ramp up or down the default mode network, they proved that the networks interact with each other, rather than simply co-existing.[14] Other experiments showed that the salience network comes up first, then awakens the central executive network for action.[15] The salience network appears to orchestrate the interactions of those two different networks: the "task-positive" central executive network, and the "task-negative" default mode network. Dr. Raichle likens it to a symphony orchestra. "You have the first and second violins, the violas, the cellos, basses, woodwinds and percussion, and throw in the brass if you want. But when it's operating, they're all coordinated together in space and time... it turns out there's this beautiful coordination of all this."

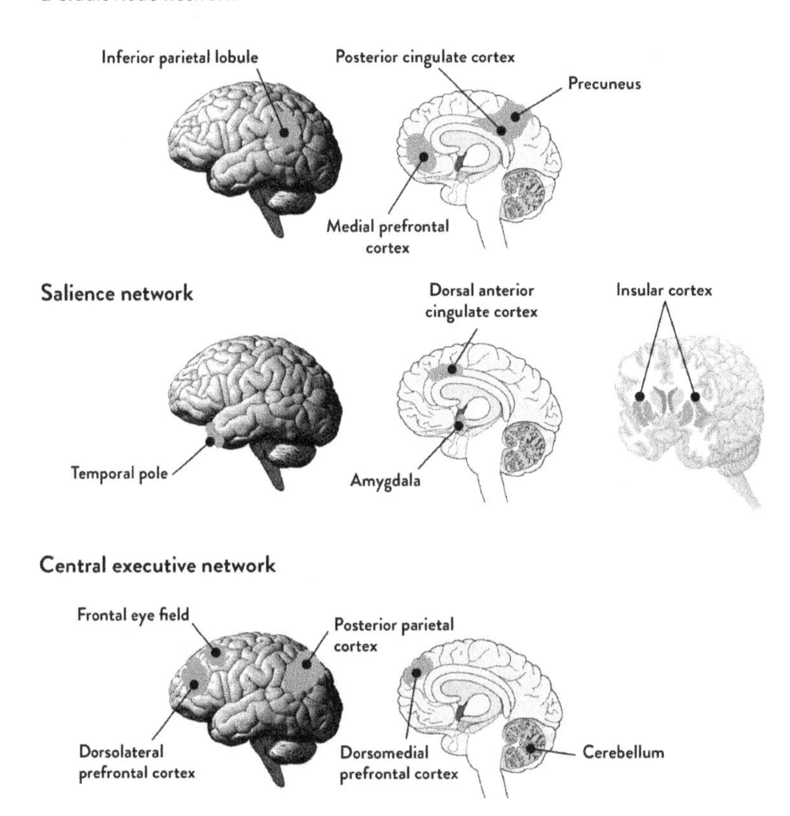

Figure 9-1: **Large-scale intrinsic networks strongly implicated in depression, showing primary nodes.** Based on Van Oort, J., Tendolkar, I., Hermans, E. J., Mulders, P. C., Beckmann, C. F., Schene, A. H., ... & van Eijndhoven, P. F. (2017). How the brain connects in response to acute stress: A review at the human brain systems level. *Neuroscience & Biobehavioral Reviews, 83*, 281-297.

INTRINSIC NETWORKS IN DEPRESSION

When researchers extended studies of functional networks from healthy volunteers to people with psychiatric disorders, the importance of these neural networks to psychopathology quickly became apparent. Dysfunction across one or more of those networks has been identified in autism, schizophrenia, depression, anxiety, Alzheimer's disease, and frontotemporal dementia. Depression has its own unique fingerprint in the interactions among and within neural networks, distinct from any of the other psychiatric disorders studied.[16] The composition of the networks change. The default mode network incorporates additional brain structures, and some nodes of other key networks have apparently lost synchronization. The brain's symphony orchestra is no longer playing the music it was assigned. Different instruments pipe in when they are not supposed to, or miss their entry altogether.

By 2015, there were enough resting state functional imaging studies of neural networks in depression to support a meta-analysis of the differences between depressed and non-depressed groups. Results of 25 papers, involving a combined total of 556 subjects with major depressive disorder and 518 healthy controls, showed hyper-connectivity* within the default mode network in depressed patients, with the hippocampus and the medial prefrontal cortex more tightly connected – displaying a more coherent "I'm here" blip, in a tighter time range – than seen in healthy controls. They also showed that network to be hyper-connected with some nodes of the central executive network, and hypo-connected with the nucleus accumbens. The central executive network seemed almost split apart, with the dorsolateral prefrontal cortex and the cerebellum hypo-connected to the posterior parietal cortex. Several other connectivity differences within and between networks distinguished the depressed patients from healthy

---

* Hyper-connectivity refers to more or stronger positive connections either within a network or encompassing additional areas; hypo-connectivity is fewer or weaker connections.

controls as well. It is a complex reorganization of networks, supporting the frequent observations that people with depression have a bias toward internal thought at the expense of external engagement, and difficulties regulating their mood.[17]

Connectivity matters. In depression, hyper-connectivity of the default mode network has been matched to the destructive practice of rumination.

Rumination is a train of thought narrowly focused on one's distress, its causes and outcomes; for many with depression, it seems inescapable. Moshe Bar, who proposed a cognitive neuroscience hypothesis of mood and depression, offers an example: dwelling on "the context of my bad comment over dinner last night; what I said, what I should have said, the resulting facial expressions and verbal responses to it, possible future implications..." This is in contrast to the wandering train of thought of a healthier brain: "In spite of that miserable comment, dinner was really tasty; never tried the combination of figs with prosciutto before; need to go grocery shopping tomorrow..."[18] Rumination involves repetitively and passively focusing on symptoms of distress and on the possible causes and consequences of these symptoms; a core tendency of depression.[19]

Changes in network connectivity in depression open a window into how rumination becomes so pervasive and uncontrollable in depression. Among the hyper-connectivities in depression is the inclusion of the subgenual cingulate cortex into the default mode network, coupling its activities to that of the posterior cingulate cortex. The subgenual cingulate cortex is highly activated in the memory of negative events and during tasks that involve generating feelings of sadness. Though it is not normally part of the default mode network, a 2007 study showed that it becomes a very significant node of that network in depression.[20] A later study showed that this increased connectivity between the posterior cingulate cortex and the subgenual cingulate cortex exists during rest periods in depression, when the default mode network is most active, and correlates with rumination.[21] So now, the resting state of a person with depression has the usual self-centered focus of the posterior cingulate cortex, but with it comes the negativity and sadness focus of the subgenual cingulate cortex.

In 2010, Dr. Yvette Sheline, then of Washington University, identified a "dorsal nexus," an area in which several networks seemed to be "hot-wired" together in depression, and not in healthy controls. This location in the dorsomedial prefrontal cortex appears as a part of the default mode network, the affective network (which includes the subgenual cingulate cortex), and the central executive network, again, in depressed subjects only. As she noted in her paper, "The discovery that these regions are linked together through the dorsal nexus provides a potential mechanism that could explain how symptoms of major depression thought to arise in distinct networks – decreased ability to focus on cognitive tasks, rumination, excessive self-focus, increased vigilance, and emotional, visceral, and autonomic dysregulation –could occur concurrently and behave synergistically."[22] Such a re-wired brain would incorporate all those symptoms of depression as part of its basic functioning, at rest and during tasks, as spillover from one network activates another one inappropriately. It draws a picture of someone with depression trying to do a mental task – a math problem, for example – but these intrusive thoughts of herself and her misery keep coming up. She can't focus, can't concentrate, and can't carry out normal mental tasks because her expanded default mode network just won't shut up.

Dr. Sheline's recent work focuses on the relationship of the brain's networks with major depressive disorder, and what happens to those networks when someone receives effective treatment. Among the projects she's worked on in the last few years are studies that show the results of antidepressants and psychotherapy. "These regions need to be in balance with each other, the prefrontal cortex and the limbic system, and they become out of alignment during depression," she says, "but by treating them with antidepressants, or we also showed you could do it with CBT, that balance is restored." But is balance truly restored? "The thing that hasn't been well studied is to what extent people actually do completely return to normal," she responds. "They return in the direction of normal, but I wouldn't say it necessarily is all the way there." And with the recurrent nature of major depression, it matters. Research shows that a lingering over-connectivity in remitted major depression predicts recurrence.[23]

A hyper-connectivity model of depression is starting to take shape, but with many questions still to answer. How and why do functional networks become reconfigured? Does it reflect structural changes, and is there any relationship with the left-side hypo-activity and right-side hyper-activity often documented in major depressive disorder?

Various cellular and neurobiological abnormalities that could cause or contribute to hypo- or hyper-connectivity within and between networks are being investigated in the brain's serotonin and dopamine systems, and in glutamate and GABA. Evidence shows that dopamine and serotonin neurotransmission modulate activity of the default mode network, with dopamine transporters, serotonin transporters, and 5-HT1A auto-receptors all implicated.[24] And with glutamate as the primary excitatory neurotransmitter and GABA as the major inhibitory one, it makes sense that they might be involved in the activation of structures within neural networks. Looking to see if glutamate and GABA individually or together control the activation level of the default mode network, one study performed MRS and fMRI on 20 healthy male volunteers. They found that higher levels of glutamate in the area they examined seemed to keep the default mode network amped up, while higher levels of GABA tuned it down.[25]

There is a long way to go to understand what these changes in functional connectivity tell us about depression or other psychiatric disorders, but the view of the higher-order organization of the brain is opening the door to understanding how we experience the world, in ways unforeseen just a few decades ago. And I wonder if discovery of these networks answers another question about depression. If depression results from the abrasive forces of excess glucocorticoid and glutamate signaling and other toxic substances that overwhelm regrowth and repair capabilities, then how does it take aim at particular brain regions? Why do we see tissue loss in the prefrontal cortex and hippocampus, but not the motor cortex? This view of brain organization showing distant parts working together may hold clues to the answer.

As a conjecture, maybe damage from chronic stress or inflammation strikes everywhere, but grows into a destructive, self-reinforcing loop specifically in some element of the default mode network... simply because we leave it running all the time. The default mode network consumes a great deal of the energy used in the brain; like the refrigerator, it is responsible for a lot of the power bill each month because it is always on. With the most activity, the nodes of this network are first to see a build-up of damage. The brain tries to repair itself but can't quite keep up, and the damage builds to a point that it impairs emotional control. As the network of "me" – my plans, my sense of self, my hopes and dreams – the default mode network is well positioned to become a core dysfunctional element of an illness that is so personal in its pain. Perhaps the disease process crosses a threshold when glia are destroyed or neurogenesis interrupted. Interconnected circuits of the brain spread the dysfunction to other areas, reverberating through the brain's plentiful positive and negative feedback loops. The system of the brain crashes and drags the sufferer into a major depressive episode.

That idea is certainly speculative, but I think it supports the argument that the model of depression needs another layer, one that brings in the organization of the brain and how it orchestrates its functions. Neuroplasticity provides a view of growth, repair, adaptation, and degeneration, but doesn't indicate why we would see losses in some areas and not others. An organizational view, focusing on how activity levels interact across the brain, on top of neuroplasticity's tissue-level view could lead to conclusions about how point damages eventually turns into depression.

The discovery of large-scale intrinsic networks and the interactions among them is opening doors to understanding how the brain steers someone's existence from one moment to the next. The differences in that orchestration between people with depression and those without indicate that it is important in the disorder. There is an even larger organizing and orchestrating scheme at work in our lives, though, one that steers our existence around the clock and orients us for life on earth. And it turns out that this scheme is also organized differently in people with depression than in those without.

Chapter Ten

CIRCADIAN RHYTHMS

Pausing at the top of the last hill, the patient dismounted a bit stiffly and set his bicycle against a tree. It was midday already; he'd stayed out much later than he'd intended. It had just felt so good to think he didn't have to go back – he wasn't a prisoner there – and maybe he'd never go back, never go anywhere again. It suddenly struck him as odd; not that he had been considering never going back, but that all throughout the night, the idea of riding on forever, to the end of the earth or at least the end of himself, had seemed completely natural, completely normal. He had a pretty good memory of the full course of his illness, going back years; but when had the constant thought of his own death become "normal" to him? Now, exhausted but uplifted from his all-night ride, he finally felt some distance from it.

Though in theory he was free to come and go from the institution at will, in his many months there he had never stayed out all night. They would be worried; he really should go back. He turned to retrieve his bike.

It was at that moment – a sudden fresh mountain breeze, a flash of sunlight through the trees, a birdsong abruptly cut off – that he realized what had been nagging at him for almost an hour now. It was a sense of freedom, of release. Not only release from those constant thoughts of death, but also from hopelessness, confusion, and sadness. It was as if a weight had been lifted from his soul. Though tired, he had energy, clarity... he felt fine. After years of intractable depression, he felt fine.

It was an experience he would repeat many times over the next few months, as would other patients and even a few of the orderlies. The bad part was that it didn't last. After one night's sleep, the depression came back. But by staying up through the night, with or without bicycling, he would be free of depression for the next day. Tired, of course, but free. It worked for almost all the other patients as well, though the orderlies were simply grumpy after being up all night. It was unexpected and unexplainable, but it was real.

SLEEP AND DEPRESSION

It is still unexplained, and still real. Sleep deprivation therapy – now called "wake therapy" because intentionally depriving depressed patients of their hard-won sleep is counterintuitive – has been practiced since the 1960s, its effectiveness demonstrated over and over again.

Sleep has a special relationship with depression. Profound sleep disturbance is one of the most reliable characteristics of depression, reported by more than 90 percent of its sufferers. In fact, abnormal timing and distribution of rapid eye movement (REM) and non-REM sleep during the sleep cycle is regarded as a primary characteristic of depression. About two-thirds of people with depression experience insomnia, with problems initiating sleep, maintaining sleep, early-morning awakenings, or combinations of those. Hypersomnia – over-sleeping – is less common, and tends to be a feature of atypical depression. And some patients experience both insomnia and hypersomnia during the same depressive episode.[1]

Sleep is an essential process for maintaining homeostasis. The metabolic and energy demands of processing sensory inputs, learning, forming memories, and planning and executing motor functions are intense; these daytime activities must be followed by a period of reset, of clearing out the metabolic byproducts and restoring energy resources. Over 100 years of research has shown that sleep and circadian rhythms play an important role in effective learning and memory formation. It seems that sleep provides the window of opportunity (and possibly active mechanisms) for synaptic plasticity to re-normalize, for some of its artifacts to fade away, restoring attentional capacity for the oncoming day.[2]

Sleep is controlled by a two-part process. The "C-process" (for circadian) represents the 24-hour cycle of rising and falling melatonin production and core body temperature, entrained to the local day-night cycle. It is a skewed sine wave, reaching its high in the late afternoon or early evening and its lowest point in the early hours of the morning. The "S-process," sometimes called "sleep debt," is a homeostatic process in which the need for sleep builds up until relieved by sleep. The two processes interact with one another; we sleep when they are farthest apart and awaken when they are closest. Some researchers theorize that depression is a result of, or at least involves, a deficiency of the S-process: depressed people don't build up sleep debt as they should, remaining constantly out of phase with their circadian rhythms. They suggest that sleep deprivation therapy works by resynchronizing the C- and S-processes; that keeping someone awake for approximately 36 hours allows enough sleep debt to build up that their bodies' homeostatic processes temporarily re-align with their circadian rhythms.[3]

Measures of electrical activity, eye movement, respiratory activity, and muscle tone show a strong organization in the progression of sleep. Sleep consists of 4 stages; the first three are numbered and the fourth stage is REM sleep. Stage 1 is the transition from wakefulness to sleep: sleep is imminent but not yet developed. Stage 2 is where sleep truly begins. An electroencephalogram (EEG) shows the start of "spindle waves," oscillations of a second or two that show the slowing synaptic and neuronal activity of the thalamus and cerebral cortex. In this stage, breathing becomes slower and deeper, muscle tone relaxes, the eyes roll slowly back and forth, and the body temperature begins to fall. In stage 3 (formerly considered two stages, 3 and 4) there is a gradual synchronization of the thalamus and cerebral cortex. Body temperature continues to fall along with heart rate; muscle tone relaxes even further. This stage is known as slow wave sleep due to the presence of delta waves, which oscillate at a frequency between 0.5 and 4 Hertz. In the last stage, REM sleep, almost all the skeletal muscles of the body are inhibited (except eye movement) and overall muscle tone is much lower. REM sleep involves vivid dreams, rarely remembered except for the ones taking place late in the sleep cycle. REM dreams are usually long, visual, often emotional, and may not be obviously connected to anything going on in one's life. Dreams take place in non-REM sleep as well, but these are shorter, less emotional, less visual, and more of a recap of the day's events: what Freud called "the day's residue."[4]

A usual night's sleep cycles through the stages of sleep five to six times, taking about 90 minutes per cycle. From waking, we drift off through stage 1 to stage 2 and 3, then back up through stages 2 and 1. The first descent all the way from stage 1 to 3 usually takes 30 minutes or less, and then we spend another 30 minutes in stage 3 before climbing back up. At the top, however, we don't wake; we go into REM sleep. Then back down into the depths of sleep and back up to another session of REM. As the night progresses, we don't climb down as far – maybe just stage 2 for the last few cycles – and the time spent in REM sleep increases. All in all, we spend about a quarter of the sleep time in REM; approximately 2 hours. The process, from waking to the cycles of sleep and eventually back to waking, is highly organized.[5]

In depression, the pattern of sleep – the "sleep architecture" – tends to be different. People with depression have reduced slow wave sleep and significant differences in REM sleep: it takes them less time to get to REM sleep – zipping through stages 1 to 3, less time in stage 3, then back up the ladder to REM – and then they spend more time there compared to the other stages.[6] So, even when a depressed person does sleep, they tend toward REM sleep at the expense of slow-wave sleep.

Some scientists classify sleep disruption as "prodromal" in depression: it often occurs first, sometimes weeks before the onset of other symptoms. A survey of almost 15,000 Europeans found that in those diagnosed with mood disorders, insomnia was the first symptom to appear in more than 40 percent of cases, leading all other symptoms.[7] Longitudinal studies show people suffering from insomnia are twice as likely as the general population to eventually get depression.[8]

And, functional imaging studies have shown that several brain areas implicated in depression, including the prefrontal cortex, cingulate cortex, hippocampus, striatum, amygdala, and thalamus, are affected during the experience of sleep disorders as well.[9]

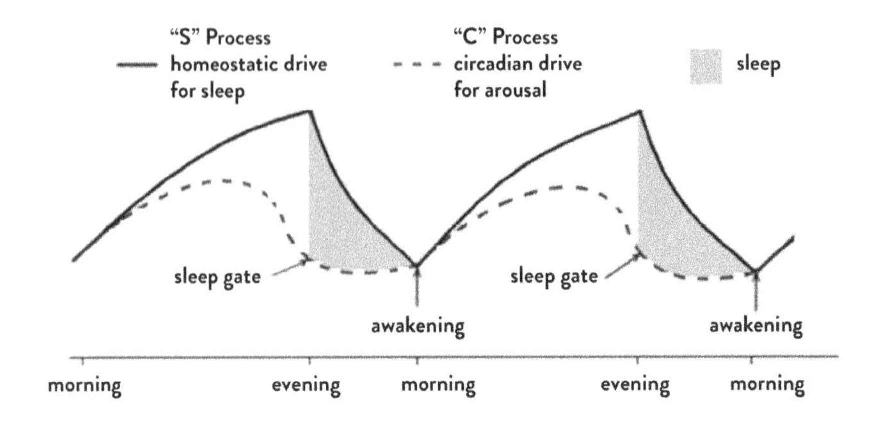

Figure 10-1: **Two-process model of sleep. Based on Borbély, A. A., Daan, S., Wirz-Justice, A., & Deboer, T. (2016).** The two-process model of sleep regulation: a reappraisal. *Journal of sleep research*, 25(2), 131-143.

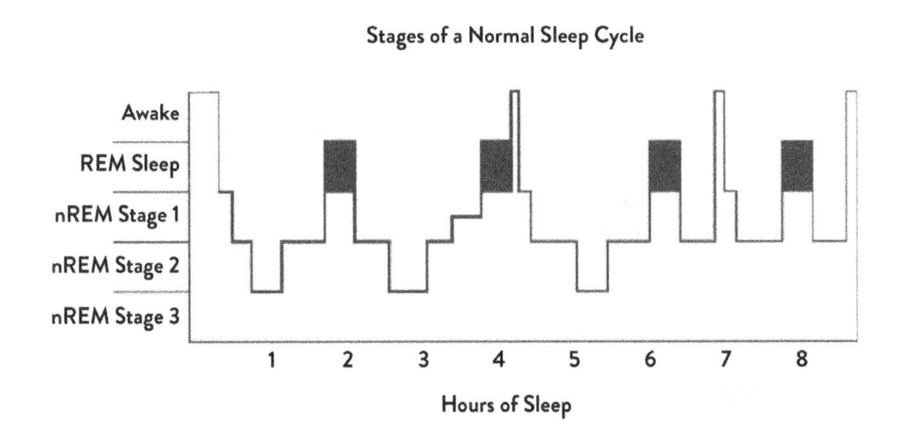

Figure 10-2: **Model sleep architecture.** Courtesy of www.utzy.com.

Of course, the most startling evidence of this special relationship is the observation that sleep deprivation creates a fast-acting relief of the symptoms of depression in about 60 percent of the people who try it. It was first described in the late 1960s when a depressed patient saw remarkable improvement in his symptoms after biking through the night. This led to a series of studies replicating the rapid antidepressant effect of a night without sleep, not necessarily including biking or other exercise. However, it is difficult, to say the least, to conduct double-blind, placebo-controlled trials of sleep deprivation – patients generally know if they've been kept awake all night – so the "gold standard" of clinical proof may never be available. Despite this, a very significant load of evidence supports the effectiveness of this therapy. It is not obvious that people with a disorder that includes very poor sleep should be helped by even more disruption, yet that is exactly what has been observed.[10]

CIRCADIAN RHYTHMS

Sleep and the day-night cycle are the most prominent circadian rhythms we experience, but they are not the only ones, not even the only ones closely associated with depression. Cortisol production follows a cyclic pattern over the 24-hour day, reaching its lowest value in the very early morning hours and then rising to its peak just before awakening. Production of serotonin, dopamine, norepinephrine, the sex hormones, and other substances implicated in depression are affected by circadian rhythms.[11] For a long time, scientists looked among these substances for insights into depression, and for biomarkers: something measurable to distinguish healthy individuals from those with depression, and hopefully, cast a light on the nature of the disorder.

Dr. Anna Wirz-Justice was on the search for biomarkers in the 1970s as a biochemist in the University Psychiatric Clinic in Basel, Switzerland. "In those days," she remembers, "one was looking, perhaps foolishly, perhaps ideally, for a marker for depression, in blood or urine, thinking it was as easy as diabetes. I had my little blood markers, and I was testing the values in healthy subjects. I wanted to get very nice normal values with which to compare my patients.

And then I started studying men versus women, different times of day, the menstrual cycle, and even got around to looking at the same people throughout an entire year." But the data refused to display a meaningful profile. "Just the search for a nice, tight, normal value for a couple of assays led me to realize that the changes over the day and over the year were greater than any I could find between a so-called control group and depressed patients. So, how can you do statistics if maybe the difference between your groups is just a phase shift: they're living in another circadian time? And if they have lower values, is it because they're later, and not because their values, *per se*, give an indication that low serotonin, for example, is the reason that they're depressed?"

The patterns themselves were intriguing, though, and clearly relevant to depression, if not in the way she sought. The antidepressant effect of sleep deprivation was also making its way into journals, and Wirz-Justice began to work on understanding how circadian rhythms affect mood. In the early 1980s, she joined the research staff at NIMH in Bethesda. "This whole concept of 'rhythms and blues' was just beginning, and I was there with this group that began the idea of light therapy. We all worked with Tom Wehr and Fred Goodwin, who was head of the psychobiology branch at that time. Tom was interested in the phenomena of rhythms in bipolar patients, where they are the most strongly visible. He studied the cyclicity of the disorder over weeks and months, and the cyclicity of mood across the day."

Shortly after the work with bipolar depression began, another researcher, Norm Rosenthal, joined the NIMH staff with a related quest. As Wirz-Justice recounts, "He was inspired by a patient who had actually come to NIH and said, 'I'm really affected by the seasons. I get depressed every October and I get better every May.'" It was the first case they studied of what would come to be known as Seasonal Affective Disorder (SAD). "They followed his mood cycles for years, and he was the first patient who was treated with light. It's really classical clinical research inspired by a single patient, who provides the evidence for a conceptual background. In this case it was that rhythms really are so crucial; that seasonality is the trigger; that changing day length is the mechanism, and that you can reverse the depression by simulating a spring day with light. So there were these two things, really basic rhythm disturbances in bipolar disorder, with Tom Wehr being the leading person, and Norm Rosenthal going out to look at light as therapy."

Several years later, Wirz-Justice brought her experience back to Switzerland. Basel was not necessarily the best town to bring the concept of light therapy. It was at that time, and still is, a pharmaceutical town. Geigy, the firm that originated the first antidepressant medication, had been absorbed into Ciba-Geigy, headquartered there. They have since been further absorbed into Novartis, still headquartered in Basel, along with Hoffman-La Roche and dozens of biotechnology companies. Plus, it was not necessarily a good time to look for interest in light therapies anywhere. The SSRIs were coming out, promising greater effectiveness and a better side effect profile than the older antidepressants; the monoamine deficiency hypothesis of depression reigned supreme.

Despite these drawbacks, Wirz-Justice was able to get funding from the Swiss National Science Foundation to pursue the idea of light therapy in Switzerland. "That's what I started with when I returned. I said, 'Well, if there's winter depression in Washington, then there's winter depression in Basel as well.' And lo and behold, Europeans turned out to be seasonally depressed, too. Other people in Northern Europe, of course in the Scandinavian countries, were also convinced that there was something in this 'winter depression.'" But her initial and enduring interest went beyond using light as therapy for mood disorders. Light is essential to maintaining our circadian rhythms, but the bigger story, for her, was in those rhythms and their effects on sleep and mood disorders.

Wirz-Justice became a "rhythmicist," joining a field that had started in the 1950s. "It was a very fascinating community to be in at the beginning, say 40 years ago, because it was a bit of a way-out field, not quite taken seriously. You know, 'biorhythms,' that sort of unserious stuff. But the discovery of its genetic basis, of clock genes, suddenly made our field entirely respectable." The role of the suprachiasmatic nucleus (SCN) in driving circadian rhythms had been identified in 1972 by two groups simultaneously, just as Wirz-Justice began her studies of circadian aspects of mood disorders. "The discovery of the SCN provided a morphological substrate for this idea of an internal clock in the brain. And that was really a fantastic discovery. And then 20 years later, the discovery of similar clock genes in flies and humans and plants documented the ubiquity of these processes as being necessary in evolution for predicting survival. Because the point of biological clocks is to have timed behavior starting before it's needed. So you are ready to wake up when you wake up, to face the stresses of whatever the day brings. And you're ready in the evening to fall asleep. This is a predictive behavior. And since flies and humans and mice all have similar clock genes, it must be important."

CIRCADIAN MECHANISMS

Life on the earth's surface is attuned to the 24-hour day, to light and dark periods in particular. In fact, circadian mechanisms have been identified in so many forms of life, from bread mold and plants to fruit flies, mice, and on up the evolutionary tree that they appear to be a basic requirement for life on earth.[12] The sleep-wake cycle is our most obvious circadian rhythm, but they also control or influence many other aspects of life, with eating, mood, cognitive abilities, body temperature, and hormone release among them. Our bodies synchronize with the day-night cycle through the SCN, a region in the hypothalamus that acts as the master clock. By itself, the human master clock doesn't keep perfect time. When isolated from all environmental time cues, it will internally maintain a period of just over 24 hours. This means it must be reset every day. To do so, there are a small number of specialized cells in the retina that respond to light and project directly to the SCN. The SCN also receives input from eating, social interactions, physical exercise, and other behavioral habits; this is called "non-photic" information. The SCN integrates the behavioral and environmental cues to generate a rhythm synchronized both to daytime and to the individual's unique daily routine. Serotonin is very influential in this function: serotonin signaling during the daytime advances the circadian clock, while serotonin signaling during the nighttime delays it.[13]

Once set, the SCN uses a collection of tools to communicate timing signals throughout the body. The primary tool is melatonin, a downstream product of serotonin. The SCN projects indirectly to the pineal gland, and approximately 2 hours before a person's normal bedtime, the pineal gland starts to synthesize and release melatonin. Melatonin builds up at night, reaching its peak during the darkest hours. As it builds, it becomes active in a negative feedback loop, working through melatonin receptors in the SCN to inhibit or phase-shift their electrical activity.[14] High levels of melatonin in the night begin to inhibit its own further production from the pineal gland until it reaches its lowest level in the middle of the day. Through its direct reach into the hypothalamus, the SCN uses other tools provided by the sympathetic nervous system, such as raising or lowering the core body temperature, to convey a timing signal throughout the body.[15] And it is a timing signal with a great deal of impact: researchers have discovered that a lot of genes are transcribed into their proteins based on a circadian schedule.[16]

At the cellular level, a core set of clock genes and their proteins interact to oscillate on a 24-hour basis, generating virtually all circadian rhythms throughout the body. *BMAL1* and *CLOCK* genes produce self-named proteins that join together into a single unit. This unit activates production of PER and CRY proteins, which also unite to form a single unit that travels back into the nucleus of the cell to inhibit the BMAL1-CLOCK activities. It is like a pendulum swinging; levels of each protein product rises and falls as the two teams compete with one another. As a result, the amounts of BMAL1 and CLOCK are at their maximums 12 hours off from the peak amounts of PER and CRY, giving the cell its 24-hour oscillation property.[17]

CIRCADIAN RHYTHMS IN DEPRESSION

Back on the faculty of the University of Basel with secure funding for light's effect on mood, Wirz-Justice was finally able to add in studies of sleep deprivation. She had been involved with sleep deprivation therapy before leaving for America – it was a European discovery, after all – and it had remained a compelling puzzle for her. "For me as a biochemist, it was an irresistible model," Wirz-Justice recalls. "If a deeply depressed patient stays up all night, and the next day has no symptoms of his melancholy anymore, there's something very strange going on in the brain of this person. And it's a fantastic short-term model to look at what the underlying mechanisms would be. So for me, sleep deprivation became a model to look at depression versus recovery in a short time frame without interaction with drugs." It certainly eliminates many sources of variability that can interfere with experimental results. The short time duration to get an antidepressant effect from sleep deprivation means that there can be minimal impact of changes in diet, exercise, menstrual cycle, response to traditional antidepressants, or even life's challenges, on the study observations. There simply isn't time for such potential confounding variables to assert themselves.

By the early 2000s, studies of circadian rhythms were showing a deep impact on many of the brain processes associated with depression. Such studies showed that exposure to light and darkness can trigger neurotransmitter switching, in which a neuron that was signaling through somatostatin switches to dopamine, or from dopamine to GABA, or back again.[18] They showed that CLOCK proteins interact directly with glucocorticoid receptors, leading to decreased sensitivity to glucocorticoids in the morning and increased sensitivity at night. Researchers saw that serotonin, norepinephrine, and dopamine all have circadian rhythms in their levels, release, synthesis-related enzymes, and receptors.[19] They observed the rate of serotonin synthesis in the brain to be directly related to the prevailing duration of bright sunlight: lowest in the short days of winter and highest in the long days of summer.[20] Day length and amount of sunlight also affect how readily serotonin binds with its various receptors, such that shorter days result in lower available serotonin and longer days in its higher availability.[21]

Of all the effects of circadian rhythms on the brain and body, though, possibly the most impactful are the changes in gene expression. A study in 2014 showed that a lot of genes have a circadian rhythm in the degree to which they produce their protein. Using blood from 22 healthy volunteers, researchers found evidence of gene transcription with consistent timing of peaks and valleys throughout the 24-hour day corresponding to 6.4 percent of the human genome. When those volunteers were put through a delayed sleep protocol (4 hours delay for 3 consecutive nights), only 1 percent of the genome showed the usual circadian rhythm of transcription.[22]

Many of the genes that respond to circadian rhythms lose synchrony in people with depression, just like they did in the delayed sleep study. A postmortem study of multiple brain regions in 34 people who died with depression compared to 55 controls analyzed the 24-hour cyclic patterns of gene expression in several areas of the brain. Subjects and controls were matched for the time of day of their deaths so that differences in daily patterns of gene transcription could be seen. More than a hundred gene transcripts showed a consistent rhythmicity and phase synchrony across those regions, evidence that those transcripts respond to circadian clocks. In subjects who had died with depression, the cyclic patterns evident in the controls were much weaker, with shifts in the timing of peak gene expression and disrupted phase relationships between individual circadian genes. Focusing in on the anterior cingulate cortex because of its association with depression, the top twelve circadian genes (in terms of consistency for time of peak expression) showed a distinct pattern of expression through the day in the controls. In the patients who had died with depression, though, only one of those twelve genes showed evidence it adhered to that pattern; the other eleven did not match the level of expression they should have had based on the time of death of the subject.[23]

The effect of circadian rhythms on gene expression links sleep deprivation therapy to the other rapid-acting antidepressant: low-dose ketamine. A recent study comparing gene expression in the anterior cingulate cortex of mice following sleep deprivation or ketamine infusion found that either intervention resulted in changes in circadian gene expression, with 64 of those genes in common.[24] A human study of ketamine treatment found that the depressed patients who responded rapidly – shedding their symptoms within 24 hours – became phase-advanced in their circadian day by about 50 minutes (judging by peaks of daily activity as measured by a wearable device).[25] This change in circadian rhythms echoes findings from a study of response to sleep deprivation and light therapy in bipolar I patients. In this study, the rapid responders showed a 57-minute phase advance, again judging by activity levels throughout the day.[26]

Both low-dose ketamine and sleep deprivation therapy can have antidepressant effects in hours, and both are effective for approximately 60 percent of patients – an interesting coincidence. From what science knows now about the course of depression, it seems even more incredible that *anything* could work that quickly. Depression involves loss of glial cells and synaptic connections in several areas of the brain, and interruption to the production and survival of new brain cells in the hippocampus. The impacts of these deficits can be seen in the many structural and functional imaging studies of the disorder. Antidepressants and ECT are believed to be effective because they repair these deficits; that's why they take weeks to work – they have serious structural issues to overcome. So how could ketamine or sleep deprivation pull someone out of depression in just a few hours?

Perhaps the answer is that unlike the structural differences associated with depression, the functional differences – the hyperactivity in some brain areas and hypoactivity in others – can change relatively quickly, certainly in a matter of hours. Brain activity results from the actions of neurotransmitters and depends on the availability of those substances and their receptors. Those factors, in turn, result from timely expression of the genes coding for the various receptors and enzymes involved in synthesizing neurotransmitters. So maybe the rapid antidepressant effects of sleep deprivation and ketamine come from restoring "normal" brain activity levels through expressing specific genes. That would imply that the distressing mood and somatic features of major depression come from the changes in activity levels of particular brain areas, like the subgenual ACC and dorsolateral prefrontal cortex, while the structural changes associated with depression simply lock those activity levels in place. The changes in gene expression invoked by either of those modes of therapy would provide relief from depression for as long as those protein products linger, and the antidepressant effects of ketamine therapy may last even longer because of its effect on NMDA receptors, growing synaptic connections. This view of depression is certainly very speculative, but it is fair to say that by controlling even some gene expression, circadian rhythms exert a powerful influence over our lives and our mental health.

Sleep deprivation (wake) therapy, bright light therapy, combinations of the two… these are done successfully at several clinics. So why not more? Why are millions of people taking antidepressant medications, enduring the side effects, and undergoing ECT and other therapies?

"I've spent 40 years trying to convince psychiatrists to use this non-pharmacological treatment of keeping insomniac depressed patients up all night," Wirz-Justice told me. "As you can imagine, it's really a joy to try to convince them that taking away their patients' sleep is going to make them better. Ha ha! But that's just beginning to happen. I think the most important work in the last years has been done by my friend and colleague Francesco Benedetti in Milan, because he's been using sleep deprivation, which we now call 'wake therapy,' for the last 20 years in his seriously depressed bipolar patients. And combining it with various other modalities such as light to prevent the partial relapse after recovery sleep. Because, I guess the reason sleep deprivation didn't catch on as a treatment was why should we go to all that effort if they relapse into depression after recovery sleep? And so the development over the last years has been to use light, shifted sleep timing, SSRIs, or lithium in bipolar patients, combined with sleep deprivation to get the quick response that is then maintained. That's what we want. And I think these non-pharmacological therapies: light therapy, wake therapy, shifted and stabilized sleep – in other words, sleep hygiene – they are very simple methods that can get people out of severe depressions much more quickly than the 5 or 6 weeks of an antidepressant. But I'm not saying either/or; I'm saying they can be combined."

Dr. Wirz-Justice is a Professor emeritus now, retired from her position at the University of Basel. Hers is an active retirement, though, in which she promotes both chronotherapeutics and light-oriented, healthful behaviors. "My effort in the last few years has been to try to implement the research into the clinic and into public awareness.

I talk about the importance of light-oriented behavior for health in interviews with all the newspapers I can reach. I'm interested in how we can change or improve architecture to have more daylight in buildings, to inform people that their behavior should be more light-oriented: go outside for half an hour a day. So I do a lot of teaching and workshops." Her activities promote awareness of the biological activity of light and its non-visual and emotional effects. The higher daylight availability experienced by hospitalized patients in east-facing rooms, with bigger windows, or with hospital beds nearer to windows, consistently speeds up recovery from depression, encouraging architects to design psychiatric wards and retirement homes using chronotherapeutic principles. There is also a growing awareness of the potential influence of LED lighting on mental illness, particularly in windowless industrial environments.[27]

Wirz-Justice also works to get psychiatrists interested and active in setting up chronotherapy units in their clinics. "The nice thing is that slowly but surely, there is a growing interest in implementing chronotherapy in the clinic, which is fabulous. This began about 10 years ago in some private clinics in Switzerland, because they could market it as 'We're treating people with something natural: light.' No drugs; there is an anti-drug tendency, for using 'natural' remedies. Yet 'natural' light has extremely powerful effects on the brain! And now, university clinics are one by one wanting workshops: 'How can we set up chronotherapy in our university clinic?' And when it enters the university it might eventually enter treatment guidelines, and thus future practices. Even though we don't have all the answers, even though the research is not complete. When people say 'You don't know the mechanism of how light therapy works,' I always argue, 'Hey, we've been using antidepressants for 50 years but we don't know the mechanisms.' I just want controlled trials and good studies that show in different sleep disorders, or different psychiatric or neurological disorders, that light can help certain aspects. Maybe not treat the entire disorder, but for example stabilize the sleep-wake cycle. In many, many illnesses that's a key factor in going towards health."

The earliest models of depression advanced by Schildkraut, Coppen, and their contemporaries in the 1960s were incredibly useful to science and to patients alike. By shifting the view of the disorder from a morbidly moody state to that of an illness, they pushed the medical world to look for physical causes and treatments to help sufferers. These were models built around chemical substances – single molecules that create effects in the brain. Other substances asserted a role in depression: glucocorticoids, pro-inflammatory cytokines, glutamate, and more. Each of these chemicals exists for a reason, though. In some quantities, locations, and contexts they perform a necessary and even beneficial purpose, while under other conditions they may be damaging or toxic. None of them are good or bad or protective or toxic on their own; that assessment depends on what the brain is doing with them.

The neuroplasticity hypothesis shifted the focus from the chemicals to the functional and structural components of the brain that use or are impacted by them. It brought in processes of growth, repair, and adaptation and showed how a brain can change physically with depression. The energy-producing machinery of the cell needs to work for those growth and adaptation mechanisms to function. Likewise, the support provided by glia plays an essential role in the functioning of a healthy brain; when they are damaged or underfunctioning, that health suffers. Science has made great strides in understanding these aspects of the biology of depression, and patients benefit from that.

Progress has been impressive, but there are still unanswered questions. Why does damage strike some areas so hard, and leave other areas alone? Why is neuroplasticity interrupted in some areas and not in others? New discoveries about large-scale brain networks show a moment-to-moment orchestration of activities across regions and structures, while discoveries about circadian rhythms show orchestration of activities across day and night. All these advances are coming together to shed light on the biology of depression.

There has been incredible progress in answering the question "what is depression?" But the real point of all these efforts is to answer the more important question: how do we predict it, prevent it, and cure it?

Part Two

ORIGINS

Chapter Eleven

THE ROAD TAKEN

A white dog sat in the road, a little hard to see through the snow. It watched Meredith, apparently waiting for her. Meredith closed her eyes, took a deep breath, and opened them. The dog was gone and it was a beautiful spring day once more; not a snowflake in sight.

It was another hallucination, a vivid one, though at least not as upsetting as when she saw the shadow people. A glimpse from the corner of her eye; a cough or a voice behind her; the feeling someone was following her. Or the time she had stopped on the stairs heading to the basement to do laundry. It was the early hours of the morning, but just for a minute, the 19th century house echoed with the sounds of a party. People ran on the floor above her, stomping and dancing. She heard a string quartet, glasses clinking, and people talking in a language that wasn't quite Norwegian and wasn't quite German. It was probably Dutch; the homeowner had been in the Dutch fur trade in the early 1800s. And then it just went away.

Meredith has bipolar II disorder. In college she had cycled into hypomania often, but for the last 3 years, she has had a constant depression. She first saw a psychologist as a child, aged 6, because she had a plan to kill herself. Treatment seemed to work; after a while her psychologist determined she had been cured, and she went on her way. It didn't last for long, though. In middle school, her mother and two aunts were diagnosed with cancer; the aunts died soon after. Her mother survived and is a survivor still, but Meredith remembers a sense of being shut down, of mourning and grieving all through high school and into college. In her second year of college the hallucinations became so oppressive that she realized she needed help.

If Meredith were a case study for how someone develops depression, it is hard to find a box she doesn't check. Her mother, RuthAnn, has suffered and is currently suffering from unipolar major depressive disorder. It started a few months before Meredith was born, when RuthAnn was told her brother had AIDS. RuthAnn remembers being so depressed after his death that she would lie on the couch, containing toddler Meredith within the crook of her knees to play without coming to harm. RuthAnn had few time periods as Meredith was growing up that she was not depressed. Her brother's death was followed by her father's a few years later. It was a hard childhood for Meredith. And now RuthAnn wonders if she "gave" her child a psychiatric disorder, through genetics, early life adversity, or being depressed while pregnant. It is something that she lives with, despite the fact there is nothing she could have done about any of those factors.

How did Meredith end up with bipolar II disorder? Was her illness an inevitable outcome of genetic and early life factors? And is it guaranteed to continue, or can she truly recover?

In the early 2000s, a team of researchers took on the task of identifying the pathways through which someone develops depression. Using data from an on-going study of Caucasian twin-pairs from the Virginia Twin Registry, they started with genetic risk and then added other potential risk factors, from childhood through adulthood, with the goal of building a mathematical model that would predict the onset of a major depressive episode in the past year.

Starting with the female twin pairs, the researchers plotted their experiences against suspected risk factors and influencers throughout the stages of life. In childhood, those factors were genetics, a disturbed family environment, childhood sexual abuse, and childhood parental loss. For early adolescence, the influencers were neuroticism, low self-esteem, and early-onset anxiety or conduct disorder. In late adolescence, they included educational attainment, lifetime traumas, social support, and substance misuse. Adulthood could include history of divorce and past history of major depression, and in the last 12 months, the subjects may have seen marital problems, other difficulties, and stressful life events.[1]

As one would expect for such a complex disorder, the resulting model is also complicated, with many variables and pathways to depression, but the data best fit three main routes: internalizing symptoms, externalizing symptoms, and psychosocial adversity. The internalizing pathway starts with genetic risk and flows through neuroticism, low-self esteem, and early-onset anxiety in early adolescence. These factors lead to a higher than normal chance of having suffered an episode of depression in the past as an adult, and from there to suffering an episode of depression in the last 12 months. The externalizing pathway begins with conduct disorder in early adolescence, followed by substance abuse in late adolescence, leading to a higher than normal chance of a major depressive episode as an adult, and to an episode within the last year. The adversity pathway is more extensive, beginning with any of the childhood risk factors of disturbed family environment, childhood sexual abuse, or parental loss. This flows through low educational attainment, lifetime trauma, and low social support in late adolescence, with higher chances in adulthood of getting divorced and/or suffering a major depressive episode. From there, any of the last year risk factors (marital problems, difficulties, or stressful life events) produce a significant risk of going into a major depressive episode during that year.[2]

The model scored 52 percent success in predicting the onset of a major depressive episode in the most recent year for the women studied,[3] so it was correct half the time. A survey during that same timeframe determined that just under 7 percent of women in the U.S. had suffered an episode of major depression in the last 12 months,[4] so the model had much more accuracy than they could expect from just guessing. A similar study on the pathways to depression in men scored 49 percent success in predicting the onset of a major depressive episode in the last year. The pathways were the same as those for women, except that childhood parental loss and low self-esteem weighed more heavily in the men, and genetic risk affected more pathways to a higher degree in the men studied than in the women.[5]

Beyond the weighting of specific factors, there are a couple additional messages in these results. First, there are a lot of different pathways to depression. That's to be expected; depression is such a heterogeneous disorder that it would be surprising to find only one way there. There's another message in the flipside of those 52 percent and 49 percent predictive accuracy scores, however. Though the statistics are complicated, it is fair to say that of the people who hit all the marks and were predicted to have a major depressive episode in that last 12-month period, fully half of them didn't. There is no inevitability here.

Further, even though major depressive disorder can be chronic or recurrent, it doesn't have to be. A study of 3,481 residents in Baltimore, Maryland, observed that about half of the 92 people with a first-onset major depressive episode recovered and had no further episodes during the 23 years of the study. The disorder was unrelenting in 15 percent of those with depression – they didn't see a year without at least one episode – and recurrent but not as dire in another 35 percent.[6] The Lundby Study cohort, in which 344 people from a community sample who had been diagnosed with their first episode of depression were followed for up to 49 years, found a 60 percent probability of remaining free of recurrence across the whole sample.[7]

Why do some people get depressed, suffering one major depressive episode after another or trapped in a chronic, unremitting depression, while others, who seem to live the same sort of lives, don't? Is vulnerability to depression a genetic trait, or driven by something in the environment, or both? A genetic inheritance is set and unchanging, but if environmental factors are avoidable, maybe a vulnerable individual can take action to tip the balance away from depression. Science has identified many factors that contribute to the development of depression, from genetics through a variety of personal and environmental exposures. The chapters that follow show what we have learned about the development of depression, laying the groundwork for resistance and recovery.

Meredith is an artist, though it has been a long time since she has been able to enjoy her skills. With numerous awards and scholarships out of high school, she earned her way to a prestigious art program in college. Even when her symptoms became so severe that she started seeing a psychiatrist again, she still managed to excel, focusing the darkness into her art. This was also the time when she had hypomanic episodes: periods of intense creativity where the ideas came so fast she couldn't even write them all down. She won a fellowship to study art in Rome, and graduated in the top three students in her class.

By this time Meredith's disorder was getting harder to handle, though. She was suffering the whole panoply of symptoms for a mood disorder, including disturbed sleep and appetite, and a constant struggle with her memory and cognition. She was terrified that she was forgetting everything. After 5 months in Rome she returned home and took jobs in a pet store, a coffee shop, and, for a blissful period, in a private art museum. She was hospitalized three times, and did another stint as a day patient, battling her symptoms. Through this time her health insurance was frequently interrupted, and she could not always afford her medications. The effect was very destructive on her work life: she suffered anxiety, panic attacks, and feelings that people at work were conspiring against her. She left each job when she felt she couldn't go out and face them anymore; better to quit than be fired.

Now, untrusted to drive or to manage her own medications, Meredith lives with her mother. The state declared her disabled, bringing some financial relief, and she is back on her medications and in regular therapy. She has a long way to go, but there are some encouraging signs. She is determined to get out in the world again. She is still young, and, I hope, too defiant to give in.

Chapter Twelve

GENETICS

RuthAnn often thought of killing herself. Everything was so difficult; living was just too hard. Walking around the house, driving somewhere to eat; it was an effort simply to walk and to breathe. How do you do the dishes, how do you collect them, how do you clean them, when just walking across the room is difficult? Easy things were hard, like when she went to vacuum and couldn't remember how to turn the vacuum cleaner on. But she trudged through. Anytime the thought of killing herself arose, she realized she couldn't bear to leave her child, Meredith, behind. There is so much pain in this world; you can't leave a child here alone.

When RuthAnn was growing up, people didn't talk about mental disorders, especially not in the family. Looking back now, she could see serious mental issues in her family across generations. Her mother was never diagnosed, but was treated for generalized anxiety disorder and probably had ADHD too. There was a cousin with bipolar disorder. Her father's flash temper was always a factor growing up. One time at the dinner table, he got mad at her mother and upended the table, flinging all the plates and the food on her lap. When the anger was gone, he was distant, hard to reach. Her mother would call the police, but he was a police officer himself. They would come, and be like, "Dan, what's going on?" He'd be sitting there reading a newspaper or something and say, "Nothing. Nothing's going on." And you could see smashed stuff everywhere.

So there was violence and abuse in RuthAnn's life growing up. It usually wasn't directed at her, but rather at her brothers. But they hadn't shown up with depression. It was like the orchid and the dandelion. Her brothers were dandelions; they could thrive in the crack of a sidewalk. Not RuthAnn. She was an orchid.

The worst part was wondering if she was somehow responsible for her daughter's bipolar II disorder. Whether it came through a direct genetic inheritance, or was a bit more roundabout – the way RuthAnn's depression and so many family illnesses affected Meredith's early life – felt important to her. She had shielded her child from family violence, but couldn't shield her from grief, from fear when RuthAnn had cancer, or from her own lack of availability when she was depressed, when she slept all day and couldn't stir up the energy to do anything. Was she responsible for Meredith's disabling illness?

Is depression a fate written into our genes? Almost all of the victims of depression with whom I spoke feel it is. They point toward mother, father, child, or siblings who also have depression, or an anxiety disorder, or something. But if 15 percent of the U.S. population experiences depression in their lifetime – approximately one in every seven individuals – would you not expect that anyone with an extended family size of seven or more could point to at least one blood relative with depression, just from the odds?

Environmental influences must be considered too. In genetic studies, the "environment" is anything that is not genetic: exposure to a disease in one's youth, toxic mold in the walls of a house, parenting style, Cesarean birth, exposure to sunlight, diet… anything. Family members are more likely than strangers to experience the same environmental factors over a long period of time, so a shared family environment could make a disorder look genetically-based when it isn't. And we cannot say with certainty that major depression is a single disorder. Melancholic and atypical depression manifest with such different symptoms that we must hold open the possibility that depression may be a collection of disorders that share mood symptoms.

Several studies have tackled the question of whether major depression comes from the genome* or the environment, or, most probably, a combination of the two. In 2000 a team analyzed the results of all the valid studies they could find to come to an overall conclusion about the heritability of major depression. They looked for family studies, twin studies, and adoption studies that drew a clear distinction between major depressive disorder and bipolar disorder, showed a systematic recruitment of subjects with depression and their relatives, used direct collection of diagnostic data from all or nearly all the subjects, used diagnostic criteria proven through active use in clinics, and used assessors who were unaware of the diagnoses of other family members ("blinded") when they made a diagnosis of a subject. In their sweep of published reports, they ended up with five family studies and six twin studies that met all their criteria.[1]

In a family study, a proband (a person with depression) is usually matched with a comparison subject who has no history or indications of depression. If depression runs in families, there should be higher incidence of relatives with depression in the proband group than in the comparison group. The combined results of the five family studies showed an approximately 3-to-1 odds ratio: a person was significantly more likely to have major depressive disorder if a first-degree relative had it. In all, the team concluded that major depression does run in families. Family studies alone cannot distinguish genetic causes from environmental causes that also affect whole families, however. For that sort of distinction, they needed twin and adoption studies.[2]

None of the three available adoption studies met the criteria and were only considered in terms of the overall result: a yes/no conclusion that major depressive disorder was heritable. In an adoption study,

* Note that all the studies covered here address nuclear DNA, not mitochondrial DNA.

children are raised by parents with whom they have no blood relation, highlighting environmental factors in contrast to genetic ones. Two of the studies did conclude there was some level of heritability of depression, while the other did not.[3]

Twin studies compare outcomes for identical twins versus fraternal twins. Identical twins come from the same egg and sperm; each twin has the same genetic information as his or her co-twin. Fraternal twins come from two separate eggs and two different sperm; as full siblings, they share half their genetic information on average. Typically, all twins share the same environmental influences from conception and usually through childhood, so genetic factors are more readily differentiated from environmental factors. If major depression is entirely inherited through the genome, then we would expect that either both or none of the twins in identical twin pairs would have the disorder. In fraternal twin pairs, if one twin had it, then, on average, 50 percent of the time their co-twin would have it too. If even just part of the liability for depression is genetic, then there should still be higher concordance among identical twins compared to fraternal twins, just not a 100 percent match. The six twin studies gathered by the meta-analysis team involved a total of more than 21,000 individuals. The broad conclusion from the meta-analysis was that major depression is a complex disorder that does not result from either genetic or environmental influences alone but rather from both. Of those environmental factors, the contribution from a shared family environment was negligible; what mattered was environmental influences on an individual. Their estimate of 31 to 42 percent genetic heritability, with 37 percent as the common mid-point, is often cited as the heritability of depression.[4] A heritability of 37 percent translates to increased risk – but nowhere near a guarantee – that a child of someone with depression will themselves suffer from it someday. It means that if about 15 percent of the population suffers from major depression at some point in their lifetime, then about 22 percent of the people with depression in their immediate family will get it someday – they have 1.5 times the likelihood of getting depression compared to someone without a first-degree relative with depression.[5]

OUR GENES

Research has identified many candidate genes for depression, though some enjoy more support than others. Studies have delved into specific pathways implicated in depression, like serotonin transporters and the stress system. Now, massive genome-wide association studies compare millions of points across whole genomes looking for commonalities among people with depression.

Each of us has approximately 25,000 genes on 46 chromosomes; two copies each of chromosomes 1 through 22, plus either an "X" and a "Y" or two "X"s. The genetic information for an individual is his or her "genotype," and the physical characteristics that result from it is the "phenotype." Each variation of a gene is called an "allele" and can result in a different phenotypic characteristic. For example, one can inherit an allele for blue eyes from her mother and an allele for brown eyes from her father, and in that case the dominant allele – brown eyes – prevails in the phenotype. With two copies of every gene, we potentially have two different versions of each characteristic.[6]

There is a principle that is so fundamental to life that it is called the central dogma of biology: DNA is transcribed into ribonucleic acid (RNA) which is then translated into a protein. (Of course, scientists then discovered retroviruses, which transcribe RNA to DNA, making their heads spin.) Each of those 46 chromosomes is one long DNA molecule wrapped around some structural elements. Information is encoded in DNA as a sequence of nucleotides: adenine (A), guanine (G), cytosine (C), and thymine (T). The DNA molecule forms a double helix, two spirals wound around each other connected throughout with A pairing with T and C with G, like a twisted ladder. Amazingly, only about 1 percent of that DNA specifies sequences of proteins; the rest is non-coding. Non-coding DNA does have purpose, though not all of them are known at this time. For example, glucocorticoid response elements are non-coding DNA; they provide a landing site for an activated glucocorticoid receptor, telling it where transcription of the target gene should start. Even within genes there are chunks of DNA that specify parts of the resulting protein separated by chunks of non-coding DNA.[7]

Gene expression starts when a particular enzyme binds to the gene's promoter region. After binding, the enzyme separates the DNA strands – unzips the double helix – to make the bases in the template available for pairing with the complementary RNA base. The RNA bases are A, G, C, and uracil (U) substituted in for thymine (T). The enzyme moves along the DNA strand like a bubble, unzipping the strands to read the next base for the growing RNA chain, and zipping it back together as it leaves, continuing until it reaches a termination sequence. The result is a strand of pre-messenger RNA that will be spliced to get rid of the non-coding information and become messenger RNA (mRNA). This mRNA is a single-stranded nucleotide chain that is translated into a protein by transfer RNA. Every three nucleotides in the mRNA chain specifies an amino acid. Transfer RNA reads three nucleotides, grabs the appropriate amino acid and adds it to the growing protein chain, then reads the next three, and so on. As the protein grows, whether it is a receptor, hormone, enzyme, or anything else, it folds into its characteristic shape as the chain forms. Meanwhile, before they are eventually broken down and the parts reused, lingering strands of mRNA can tell scientists which genes were recently expressed and in what quantity.[8]

THE CANDIDATES

The most vigorous debate in the depression genetics community swirls around the significance of the promoter region of the serotonin transporter gene, *SLC6A4*. Part of the non-coding DNA for that gene, the promoter region controls how often that gene is expressed; i.e., how many copies of the serotonin transporter protein are produced. The serotonin transporter protein has been a target of antidepressant medications since the beginning: TCAs and SSRIs both block it. It sits on the terminal of a presynaptic neuron, and when that cell releases serotonin some of it binds to the protein and is pulled back into the originating neuron, to be made available for reuse. SSRIs in particular were developed to block the serotonin transporters so that serotonin lingers longer in the synaptic cleft, prolonging serotonin signaling.[9]

By sequencing the *SLC6A4* gene in 52 unrelated individuals, a team from the University of Wurzburg found two versions of the gene. One version, the "long" allele, contained a specific sequence of 44 base-pairs that was absent in the "short" allele. This variable region of the gene is called the serotonin transporter linked polymorphic region, or 5-HTTLPR. By analyzing the amounts of serotonin transporter mRNA in cell cultures, researchers could see that cells with two long alleles produced 1.4 to 1.7 times as much serotonin transporter protein mRNA than a cell with one long and one short allele or a cell with two short alleles, leading to about twice the level of serotonin uptake. This could be hugely important in depression: a gene that could, by itself, result in having fewer serotonin transporters available would seem to perform an SSRI function on its own. Researchers immediately began exploring the possibility that the different alleles formed a biological basis for behavior,[10] and further, for depression.

Individual studies produced conflicting evidence, though; some showed a significant link to antidepressant effectiveness,[11] while others failed to find such a link. Only four of sixteen studies looking for a linkage between the 5-HTTLPR variation and mood disorders in a variety of unipolar and bipolar samples found such an association; three of the four showed the short-short genotype to be associated with depression while the other found the long-long genotype to be associated with depression. A postmortem analysis of the prefrontal cortices of suicide and non-suicide victims with and without depression showed no link between the 5-HTTLPR variant and binding of serotonin in the brain.[12] Other studies found positive signs of such links.[13]

The long versus short allele of the serotonin transporter was heading to be considered a difference that makes no difference, until a major breakthrough came in the form of a 2003 study by Caspi et al indicating that subjects with one or more 5-HTTLPR short alleles were more susceptible to developing depression than were subjects with two long alleles, but only subsequent to stressful life events. In their longitudinal study in New Zealand, subjects were enrolled at birth and assessed ten times between ages 3 and 26. A total of 846 Caucasian members were divided into groups according to their allelic variation: 265 individuals with two long alleles, 435 individuals with one long and one short allele, and 147 individuals with two short alleles. The number of stressful life events involving employment, housing, health, relationships, and more, between ages 21 and 26, were assessed. Thirty percent of the subjects had experienced no stressful life events; 25 percent had experienced one such event; 20 percent had experienced two events; 11 percent had experienced three events; and 15 percent had experienced four or more events.[14]

The study members were then assessed for past-year depression. Seventeen percent of the study members had met the criteria for a major depressive episode in the past year. When sorted by genotype and number of stressful life events, the results indicated that 10 to 15 percent of the subjects with the long-long genotype had shown a past-year major depressive episode, no matter how many stressful life events they had suffered since age 21. That means approximately 15 percent of the long-longs with one stressful life event suffered from depression, as did approximately 15 percent of the long-longs with four or more stressful life events. Subjects with either one or two short alleles, however, showed a progression in incidence of past-year major depressive episode by number of stressful life events, starting at approximately 10 percent (statistically no different than the long-longs) with zero stressful life events, 12 percent with one event, 15 percent for two events, 26 percent for three events, and 33 percent for four or more events. This apparent dose-response relationship indicated a particular gene-environment interaction, with the existence of the short allele interacting with environmental exposure to stressful life events to create a different outcome: higher incidence of depression.[15]

The researchers then looked to see if interaction of the 5-HTTLPR and maltreatment specifically in childhood predicted adult depression. Sure enough, they found that childhood maltreatment predicted adult depression among the subjects with at least one short allele, but not among study members with the long-long genotype. That is, study members with the long-long genotype who were maltreated as children had the same probability of developing depression as an adult as did long-long genotypes who had not been maltreated. In contrast, a study member with one short and one long allele who had been severely maltreated as a child had 50 percent greater chance of developing depression as an adult than a short-long with no childhood maltreatment, and a short-short who had been maltreated as a child had twice the probability of developing depression in adulthood than did a short-short with no childhood maltreatment.[16]

These results spurred a series of studies of the interaction between the 5-HTTLPR genotype, life stresses, and depression, with some supporting Caspi's results and others not supporting it. Reviews and meta-analyses involving larger and larger numbers of subjects also followed; again, some supported the hypothesis that the 5-HTTLPR moderated lifetime probability of depression in response to life stresses, especially early life adversity, while others failed to find a meaningful connection. A 2011 meta-analysis of 54 studies, totaling 40,749 subjects, came out in strong support. However, that same year, a separate study that tried to replicate the Caspi results on a very similar cohort (a longitudinal study of 893 individuals also in New Zealand, who had been followed from birth to age 30) found no conclusion that there is a stable gene-environment interaction involving 5-HTTLPR, life stress, and mental disorders.[17]

Of course, the careful reader will have noticed that the results Caspi saw – more depression in short-shorts and short-longs following stressful life experiences – is the opposite effect you would expect. The short variants produce fewer serotonin transporters, so it should be as if he or she was on a natural SSRI. This apparent contradiction had been noted, but it is always possible that people who live with this variant from birth have a compensating mechanism at work to balance – or over-balance – serotonin re-uptake.

One of the effects of the Caspi article was to prompt a similar search for other genes that may cause depression through environmental interaction. Attention focused on the mechanisms that are implicated in depression, particularly those involved with the stress response: HPA axis function and glucocorticoid receptors (GR). The GR, when not activated, sits in the cell with a set of chaperone proteins, which keep it sequestered and inactive. When glucocorticoids bind to the GR, the GR sheds its protein chaperones and relocates to the nucleus of the cell, where it binds to glucocorticoid response element and initiates gene transcription. The FKBP5 protein is one of those chaperones. It inhibits GR function, making the GR less likely to bind with cortisol and also making relocation of the receptor to the nucleus less efficient. When the GR does eventually make it into the nucleus to initiate gene transcription, one of the genes transcribed is the *FKBP5* gene – creating more FKBP5 to inhibit the functioning of GRs. It is a very short negative feedback loop: activation of the GR results in a protein that makes the next GR activation a little harder. With glucocorticoid resistance a characteristic of depression, attention focused on *FKBP5*.[18]

The *FKBP5* gene has several different alleles, affecting how much new FKBP5 is produced by the activation of the GR, differences in GR sensitivity, and stress hormone system regulation. Each of these variants is formed by a single nucleotide polymorphism (SNP) of FKBP5. (In an SNP, one of the bases in a sequence of nucleotides – A, T, C, or G – is swapped for another.) When studied on their own, the various alleles of *FKBP5* sometimes showed an effect on depression and sometimes didn't. When, similar to Caspi's research, the different alleles were studied in association with childhood adversity, however, a strong gene-environment interaction was revealed. A meta-analysis of 7 independent studies, involving a total of 7,135 individuals with depression, found that having the "T" allele of a particular *FKBP5* SNP was significantly associated with increased risk of developing depression after childhood adversity.[19]

There are about a hundred candidate genes for depression, but the last one I want to mention is a polymorphism of the *BDNF* gene. The growth factor BDNF is very important in neuroplasticity, so a difference that results in more or less BDNF in various parts of the brain would impact development of and recovery from depression. A common *BDNF* polymorphism that has the eventual effect of building a BDNF protein containing either the amino acid valine or methionine ("Val" or "Met") at a particular spot is known as the "Val66Met" polymorphism. With inheritance from each parent, a person can end up with two "Vals", two "Mets", or one "Val" and one "Met." A 2009 study of the interaction of the *BDNF* gene with early life adversity found that carriers of a "Met" allele, when exposed to greater early life stress, have smaller hippocampal and amygdala volumes, elevated heart rate, and a decline in working memory, predicting higher rates of depression.[20]

GENOME-WIDE ASSOCIATION STUDIES

In the modern day we have big data and the technology to handle it. Rather than crawl through every possible variation of genes in the (known) pathways to depression, researchers can analyze millions of points in the genomes of hundreds or thousands of people, and see what stands out, using a genome-wide association study (GWAS). This approach addresses the genome objectively, without the influence of our understanding of candidate pathways for depression, by comparing the genomes of a large number of unrelated people with depression and a large number of unrelated controls to a database of a million or more SNPs, looking for those that rise to a level of significance in the people with depression versus the controls. The first GWAS of depression was published in 2009 and included 1,738 cases and 1,802 controls. No SNPs reached genome-wide significance. A 2011 GWAS found an SNP that showed genome-wide significance for depression, but other GWASs did not replicate that finding.[21]

In 2007, an international Psychiatric Genomics Consortium was established to define the spectrum of risk variants across psychiatric disorders, including major depression, schizophrenia, bipolar disorder, autism, and attention deficit hyperactivity disorder. Six years later they published the results of a GWAS mega-analysis of depression involving 9 primary samples that included 9,240 cases of depression and 9,519 controls. Although this was the largest sample to date, no SNP reached genome-wide significance.[22] By 2015, fifteen GWAS studies for depression had been conducted, but with few significant findings, and those findings were not replicated in other studies. Sample sizes grew: a 2013 GWAS included 34,549 individuals with depression, yet no SNP reached genome-wide significance. And, significantly, none of the nearly one hundred popular candidate genes like *SLC6A4*, *FKBP5*, or *BDNF* showed evidence of significant association with depression in the results.[23]

As depression is a complex disorder, any one SNP might have a small effect, driving the need for considerably larger sample sizes to get meaningful results. It is also very heterogeneous in its expression;

it is possible to meet DSM-IV or DSM-5 diagnostic criteria for a major depressive episode through at least 227 different combinations of symptoms, with atypical features, melancholic features, a seasonal pattern, peripartum onset, and so on; the different subtypes of depression could each result from many different combinations of genes, each weakly expressed. Still, one would think that if the influence of a genetic variant is strong enough to show up in focused studies involving about 100 people – even though it has to be crossed with adverse childhood experiences – it would be strong enough to show up in nearly 35,000 cases of depression.

The increased availability of genetic testing has made larger sample sizes available. In 2018, the Consortium published results that included 135,458 cases of major depression and 344,901 controls, run against a database of almost 10 million SNPs. That mega-analysis identified 44 independent significant loci (locations), that they mapped to 153 genes. Though they found several SNP hits in each of those 44 loci, none of those locations was on the coding part of a gene. Instead, the hits clustered on regulatory areas or other non-coding portions of the chromosomes, potentially impacting many genes downstream of each location.[24]

The GWAS also identified some significant pathways: higher order groupings of genes that work together toward a functional result. Among the pathways implicated were developmental processes, through which neurons differentiate, project to their targets, and develop synapses. Their results connected to genes expressed in neurons but not in glia. Among the genes of significance were some encoding voltage-gated calcium channels – important to neuroplasticity – and others involved in cytokine and immune responses, including neuroinflammation. One of the genes identified has a functional role that may be consistent with chronic HPA axis hyperactivity; two are associated with high body mass index. The results set also included the gene that codes for dopamine D2 receptors (*DRD2*), one for an estrogen receptor, two types of glutamate receptor, and a retinoid X (a derivative of vitamin A) receptor. Notably, once again, even with these huge sample sizes, the studies did not find a significant association with *FKBP5*, *SLC6A4*, *BDNF*, or other popular candidate genes.[25]

I asked one of the scientists who conducted the study why the candidate genes didn't show up in the results. The 5-HTTLPR allele is an insertion/deletion polymorphism, after all, not an SNP. Maybe their study wasn't set up to find it? Nope. "If they're common," she responded, "then they will be tagged in some way by the other common variants. It's really things that are not common that we're not picking up. On the whole, we're not finding that DNA variants of those candidate genes are associated with disease. And so, all those candidate genes, well, generally they're discredited. The candidate gene studies tended to be small and it is generally accepted that there was a lot of publication bias. That's why the genetic research community moved to this large-scale whole genome approach, which lets the data tell us what is associated. However, one of genes from the candidate gene era, the *DRD2* gene is coming up as associated with psychiatric disorders, and *DRD2* is a target for antipsychotics." But not the others.

So back to the first question: is depression a fate written into our genes? No, but some level of vulnerability is. Everyone has some degree of vulnerability to depression based on their genetic inheritance; some people have a lot. Even on the high end of heritability estimates, though, there is plenty of room for resistance to the condition.

Chapter Thirteen

EARLY LIFE ADVERSITY

Joshua had become used to his suicidal thoughts, but when he started feeling homicidal, it scared him enough to have himself admitted to a hospital.

"Growing up," he said, "I was bullied a lot in school. I had an afro; people looked at me and thought I was adopted." Joshua's mother was Caucasian and Jewish. His father, he told me shortly, was mixed African-American and Native American, and a Baptist. Joshua remembers always having been gender dysphoric. He came out as gay while still a small child, and later as transgender. His mother was very accepting; Dad the Baptist was not. "I'd come out as gay; I was seen as this weak kid. I was bullied; people fought with me all the time. That didn't help."

Joshua struggled through school and his parents separated early on, but he had the loving support of his mother and maternal grandparents, and he got by. In 2012, however, a triple tragedy started to take hold as all three of those relatives became terminally ill. Joshua, by then in his late 20s, was their only caregiver.

"From 2012 to 2015 I was taking care of my mom and my grandparents," he said. "During those 3 years I didn't care for myself at all. My main priority was my mom and my grandparents and their wellbeing. I wasn't sleeping at all; I was afraid to sleep because I thought I'd miss a call. So I'd literally get by on 2 hours and a candy bar from the vending machine at the hospital. And then they all went, almost all together in 2015. My grandfather at the end of May, my mom in the end of June, and then my grandmother was the last person to pass away in August."

With the deaths of his family members, Joshua was evicted from his apartment and ended up in a homeless shelter, where his trauma continued. "I was always on guard, always depressed," he recalled. He suffered anxiety, post-traumatic stress disorder (PTSD), and panic disorder as well. "There, I didn't sleep at all because then I'd started having nightmares. Sirens would trigger flashbacks of when my mom was in an ambulance or something, and flashbacks whenever I would hear someone screaming, because my mom had stage 4 cancer and in the end all that she could do was scream because of the pain. And I couldn't do anything; all I could do was hold her hand. I went through a very, very depressed stage during that point in time and I became extremely suicidal. I started self-harming. It was really bad. And then I realized I needed to go to grief support. From there they got me in touch with an actual therapist, where I was diagnosed with major depressive disorder."

In 2017, still in the homeless shelter, Joshua's disorder just got worse. "I woke up extremely depressed. I was suicidal, and for the first time in my life I was homicidal. And that really scared me. It was like, everyone else, they had their mom; I didn't feel like it was fair that I didn't have mine, and it was also around the anniversary of her death as well. I was put in inpatient care in the hospital for a week."

In the mid-1980s, Dr. Charles Nemeroff, co-director of Duke University's mood disorders unit, was growing more frustrated by the day. "I had patients who came into the hospital with depression," he recalls. "During those years the mean length of stay was quite long; we'd have patients in the hospital for up to 30 days. And in the course of evaluating those patients I was really thunderstruck by the prevalence of childhood maltreatment in that population. It was absolutely extraordinary." But his clinical observations, and applications for research funding to follow up on them, fell on deaf ears.

Today, it is widely accepted that early life adversity has clinical and neurobiological consequences later in life, but it was a slow march toward acknowledgment of the problem. Around 1980, the publication of several very influential books broke a long silence about child sexual victimization. Clinicians and caseworkers started developing a knowledge base of long-term outcomes associated with childhood molestation, documenting "a variety of psychological and social difficulties in adulthood, including depression, guilt and low self-esteem, interpersonal problems, law breaking and substance abuse, suicidality, sexual problems, and an increased likelihood of being re-victimized in the future."[1] Nemeroff and a few others kept pushing for funding to investigate their clinical observations in animal models, to explore what was really going on in a developing brain under adversity. "The fundamental issue was simply this disbelief that events that could happen early in life could have long-lasting biological consequences, or medical consequences," he said. "There were the hard-core sort of rigid, orthodox scientists who just couldn't conceptualize, with what we knew about neurobiology then, how a series of traumatic events at age 5 could have a long-lasting effect on the brain or on the body throughout adult life."

It took the support of what was considered almost a rival community to turn the tide. "Ironically, one of the biggest helpers here turned out to be the psychoanalytic community," Nemeroff commented, "because for years, they've been talking about the importance of early life. And so, this remarkable, unwritten partnership between psychoanalysts, largely based on Freud's work, who believed that early life adversity is really important for adult psychopathology, sort of merged with the biologists who were doing this work, and that held together to convert the psychiatrists."

Of all the harmful things that can happen to children, certainly the worst must be abuse or neglect from the people they look to for love, shelter, and support. The U.S. Department of Health and Human Services reported that child protective services responded to 3,358,000 incidents of child abuse or neglect in the U.S. in 2015 – accounting for roughly 4.5 percent of U.S. children. There were 683,000 confirmed victims: 75 percent were neglected, 17 percent were physically abused, and 8 percent were sexually abused. An estimated 1,670 children died of abuse and neglect that year.[2] And authorities agree that the vast majority of cases of child abuse and neglect go unreported.

Research funds started to flow, and by the late 1990s studies showed that childhood adversity was associated with later-life medical and psychiatric illness, and also revealed the sort of dose-dependent relationship that implies causation. In 1998 the U.S. Centers for Disease Control and Prevention (CDC) and Kaiser Permanente reported results of the first wave of the Adverse Childhood Experiences (ACE) study, involving approximately 9,500 Kaiser Health Plan members who completed standardized medical evaluations and responded to a questionnaire on childhood abuse and exposure to forms of household dysfunction. Questions included experiences of emotional/psychological, physical, or sexual abuse; violence towards the mother or stepmother; living with a problem drinker, alcoholic, or street-drug user; and a member of the household going to prison during the respondent's childhood. The results showed a strong, graded relationship between the breadth of exposure to abuse or household dysfunction during childhood and multiple risk factors for several of the leading causes of death in adults, including alcoholism, drug abuse, depression, and suicide attempt.[3] In that first wave group, 40 percent of the respondents reported zero ACEs and 9 percent reported four or more. Lifetime prevalence of a depressive disorder grew from 6.5 percent of the 3,807 participants reporting zero ACEs, to 22 percent of those with one ACE, 29 percent with two ACEs, 37 percent with three ACEs, 44 percent with four ACEs, to 54 percent of the 332 people who reported five or more ACEs.[4]

EARLY LIFE ADVERSITY AND DEPRESSION

It is important to emphasize that none of these data imply that someone who develops depression must have experienced childhood abuse, or that the fact of having experienced childhood abuse or other forms of early life adversity doom an individual to depression. After all, while 54 percent of the people who suffered five or more types of adverse childhood events developed depression at some point leading up to the study, 46 percent did not. With an average age of 56.6 years,[5] they had enough time for such an outcome to show up. Early life adversity is a strong factor, but not a mandate, for depression.

The ACE study and others also show that the type of adversity matters to the outcome, and not always in an intuitive manner. Childhood emotional abuse posed the greatest risk of any of the adverse childhood events, conferring almost three times the likelihood of developing a depressive disorder compared to someone who had not suffered such abuse.[6] Neglect or childhood sexual abuse gave someone about twice the likelihood, and childhood physical abuse gave someone about 50 percent higher likelihood of developing depression as an adult compared to someone without such experiences.[7] An Israeli study addressing the relationship of early parental loss (before age 17), through death or permanent separation, found that someone who had lost a parent as a child was about three times as likely to have major depression as an adult than someone who had not. Unsurprisingly, loss of a parent before age 9 was more impactful on development of major depression than was such a loss at age 9 or after; what was surprising was that loss due to separation was more impactful than loss due to death.[8]

In vivo imaging had shown volumetric reductions in certain brain regions of people with depression, but couldn't determine if those differences were a consequence of depression or a previously-existing risk factor. The results of an influential imaging study published in 2002 involving 21 women with current depression who had suffered physical and/or sexual abuse before puberty, 11 women with current depression with no history of abuse before puberty, and 14 healthy, non-abused female controls helped resolve the issue. The depressed, abused women averaged an 18 percent smaller left hippocampal volume than the depressed, non-abused women, who were similar to the controls. Right hippocampal volume was similar across all three groups. The implication was that childhood abuse, not depression itself, is associated with reduced left hippocampal volume.[9]

FROM EARLY LIFE ADVERSITY TO DEPRESSION

The human brain takes a long time to fully develop and mature. An infant's brain has about 100 billion neurons at birth, but is only one-quarter to one-third of its future adult volume. It will continue to grow and specialize according to a precise genetic program, modified by environmental influences. In a "bloom-and-prune" process throughout childhood, neurons first greatly increase their dendritic branches and synaptic connections then prune many away, leading to a more efficient set of connections that will be continuously remodeled throughout life. Different regions of the brain mature at various times. For example, synaptic overproduction in the visual cortex reaches its highest level at about the fourth month following birth, then synapse elimination takes place until preschool age. In the medial prefrontal cortex, peak synapse production occurs at 3 to 4 years of age, but substantial pruning waits until mid-to-late adolescence.[10] The prefrontal cortex is the last area of the brain to fully develop, reaching full maturity in a person's mid-20s.

The brains of girls and boys develop differently, with males typically over-producing, then pruning, synapses and signaling mechanisms to a greater degree than females. The sex hormones have an organizing effect starting as a fetus and later during puberty. Over time, brain regions become more efficiently connected as glia form myelin sheaths around axons, speeding neuronal signals to their destinations. These processes of dendritic pruning and myelination change the composition of the brain over childhood and adolescence, so that the brain of a young adult actually has less gray matter, but more white matter, than that of a child.[11]

These regional differences in the phases of synaptic development and connectivity potentially open region-specific windows of vulnerability to harm at different ages, and there is a growing amount of evidence that the timing of early life adversity matters. When MRI scans of 26 women who had experienced repeated episodes of childhood sexual abuse were compared to those of 17 healthy female controls without childhood abuse, the results showed a strong developmentally-based outcome: the young adults who had experienced sexual abuse between either ages 3 to 5 or 11 to 13 years had the most reductions in hippocampal volume, while those who had experienced sexual abuse between the ages of 14 and 16 showed reduced frontal cortex volume.[12] Puberty in girls usually occurs between the ages of 10 and 14 years. The hormonal changes that lead to the onset of puberty can start many years earlier, however, and the increasing levels of sex steroids may have important implications for neuronal plasticity and stress responses throughout that time. It is possible that when you mix the timing and type of the abuse with the phase of synaptic development of each brain structure, and add in the actions of sex hormones, you get a result specifically patterned for that formula.[13]

EARLY LIFE ADVERSITY AND NEURAL NETWORKS

With the discovery of the default mode and other intrinsic brain networks, several teams have been looking for differences in network composition and connectivity in people with current major depression versus those with no psychiatric disorder. One team doing such a search found, to their surprise, a distinct pattern that showed up in depressed patients who had a history of early life adversity, but not in depressed patients without such a history. This study involved resting state functional imaging of 189 patients with current major depressive disorder and 39 healthy controls. As the team reported:

> *The primary finding in this study was the dramatic primary association of brain resting-state network connectivity abnormalities with a history of childhood trauma in major depressive disorder (MDD). Even though participants in this study were not selected for a history of trauma and the brain imaging took place decades after trauma occurrence, the scar of prior trauma was evident in functional dysconnectivity. Among patients with MDD, a history of childhood trauma and current symptoms quantified by clinical assessments were associated with a multivariate pattern of seven different within- and between-network connectivities involving the dorsal attention network, frontoparietal network, cingulo-opercular network,*

> *subcortical regions, ventral attention network, auditory network, sensorimotor network, and visual network. Overall, our study showed that traumatic childhood experiences and dimensional symptoms are linked to abnormal network architecture in MDD.*[14]

These differences in networks were in addition the network connectivity disruptions of major depressive disorder involving the default mode network, central executive network, and salience network. Further, the type of trauma suffered – physical abuse, emotional abuse/neglect, or sexual abuse – each left different traces in terms of network connectivity.[15]

In 30 years, the scientific and clinical communities have amply shown that early life adversity conveys a significantly greater risk of depression later in life. But, just like a genetic inheritance, it's a risk factor, not a mandate. And, like a genetic inheritance, it has to be taken as a fact: it happened; it is unchangeable now.

Joshua is fighting his way back from adversity to health, using medications, psychotherapy, and supportive connections to do so. "The first thing I was ever on was Zoloft," he said. "That helped for a time, because I was in a partial hospitalization program; that was a day program. I was in there for 2 or 3 months. I was put on Zoloft, and Klonopin for the anxiety, then found out that the Klonopin was not working; it just made me dizzy. So now I'm on Propanolol, and I'm now on Prozac [an SSRI]. And it's working well. I love my meds!"

He also feels that medication alone is not sufficient, and has been actively engaged in dialectic behavior therapy, or DBT. "I've always thought that medication is only one piece of the puzzle," he said. "You can't just pop a Prozac and say 'I'm happy now; it'll cure me.' Medication is not an actual cure. It's to help the symptoms. That's something that I've had to come to understand, because a big part of it is the DBT, making an effort and finding ways to cope: having solid coping skills. The medication is there to help it, but you have to remember to keep yourself accountable. You can't just rely on the medications."

Spending parts of his days in a center that he describes as "kind of a drop-in place for homeless and for low income people," Joshua is turning the skills he's gained toward putting his life back on track. "It's been incredible," he said. "In DBT, we learn how to try to find that other piece... the silver lining in a situation. I find myself trying to use those skills and trying to teach other people. Because I live with some people with negativity, I try to help, but you can only help someone so much. I personally think you're responsible for your recovery. You have to hold yourself accountable, and not blame any outside sources, like this person died, or this person was mean to me so I have a right to be depressed."

He'd like to pay it forward someday, working with other people on mental health. "When you have mental health issues or you have any challenge in your life, trying to figure out some kind of silver lining or trying to look at changing your perspective can make a huge difference. Because a lot of people, we look at our depression and our mental health, and my PTSD, and it's like it's all negative, I can't do anything; I can't function. But then it could be a gift. It could show you a new way of living. I do have a right to have days when I feel blah. It's been difficult, and it has its negative things, but there's also an aspect of it where I've met some amazing people. In that respect, I'll say, thank you!"

Chapter Fourteen

EPIGENETICS

I have to get this right, Tyler thought, looking down at his newborn son. He was pink and wrinkly, but so animated, so alive, and so small and vulnerable. *He's not going to grow up like I did; it wouldn't be fair. I need to buckle down and get my mental health figured out.*

Tyler had suffered from mood swings for as long as he could remember. He didn't know it was depression; he didn't talk to anyone about how he felt. He didn't trust anyone enough to talk to them. Tyler's step-father had beat him starting when he was 3 years old, and though that particular relationship hadn't lasted beyond a few years, his mother was not a reliable person. She lied to him, a lot. It was as if she didn't even do it on purpose; she just couldn't help herself. So, aside from a few friends in school, he didn't talk to anyone about how bad he felt. He couldn't talk to his family about it. His parents couldn't be trusted, and as the oldest of three kids, he was the one everyone leaned on. It was a hard life for a kid. He remembered that one time as a teenager when he had even flirted with suicide. His baby sister had saved him then, though she didn't even know it. It was more or less a cry for help and not a legitimate plan to kill himself; he knew that even early on. He and his father had been skeet-shooting all day. Dad and his stepmom had gone out for a little while, and it was just Tyler and his baby sister at the house. He was cleaning the guns and stuff; Dad had recently shown him how. And, then, without really thinking about it, Tyler finished cleaning the handgun, put it back together, and put it under his pillow. He went into his little sister's room to love on her; to say goodbye. But when he saw her, he felt so selfish. How could he leave her there by herself? He had never even remotely thought about suicide ever again.

As an adult, Tyler had a tendency to blow up when he got upset. He didn't like it; in fact, it really bothered him to feel he wasn't in control of his own actions. There had been a couple of times during his military service when they wanted him to go talk with someone, just to make sure everything was cool, and he had obligingly done so. But he knew he hadn't really participated; he hadn't been truthful with them or with himself about what was going on. He had just been ignoring things, trying to avoid them until they didn't matter. *That doesn't work*, he thought. *The problem with a circle is that it comes back around.*

Now, looking down at his child, he knew with certainty that his son was not going to grow up in the same sort of environment that he had. He was not going to grow up unable to trust; he was not going to grow up thinking that it's OK to just explode at any time just because you're upset. *I'm going to get this worked out,* he thought.

Depression is understood to result from a gene-environment interaction, when a certain level of genetic vulnerability is exposed to environmental factors – with early life stress weighing in as a very potent environmental factor. That raises a big question, though. *How does an environmental exposure carve itself into the brain such that decades later it can lead to depression?*

The field of epigenetics has started to provide answers. "Epigenetics" is a reference to something that happens "on top of" the genome. It is about changes to how our genetic information is read and processed – differences in gene expression based on something other than changes in the DNA sequence. Even though our genetic information is the same from cell to cell and from conception to decomposition, how often a gene is expressed, or the silencing of a gene so it won't be expressed, is a process of epigenetics.

The term epigenetics has been in use since the 1940s, even before the structure of DNA was discovered. It is attributed to a developmental biologist, Conrad H. Waddington, who used it to describe "the causal interactions between genes and their products which bring the phenotype into being."[1] If all our cells have the same genetic information – the same DNA – then how does a cell know that it is a liver cell, and needs to produce liver enzymes, rather than a neuron that needs to form dendrites and forward a signal? Somehow, information is conveyed over and above the genome to differentiate and operate tissues.

EPIGENETIC PROCESSES

In the winter of 1944 to 1945, Nazi Germany banned food transports to the western part of The Netherlands as punishment for the Dutch government-in-exile's support of the Allies, causing a famine that killed more than 20,000 people. Between November 1944 and April 1945 – a period known as the *Hongerwinter* (Hunger Winter) – the average person's daily food consumption dropped first to 1,000 kcals and then to 500 kcals in that region. Their children who were conceived or born during this time still showed the effects of that prenatal experience decades later, with higher average body weight as adults, and higher levels of triglycerides and LDL cholesterol in their middle age. These children also experienced higher rates of obesity, diabetes, and schizophrenia; more than their parents or siblings born under more favorable circumstances. With careful record-keeping before, during, and after the Hunger Winter, this tragic episode became a natural experiment in which to observe the life-long impacts of starvation in the womb. With modern technology, it also became an opportunity to see how some prenatal exposures become a lasting part of someone's life.[2]

There are many mechanisms of epigenetics; some that have been observed in the processes of depression include DNA methylation, histone methylation, and histone acetylation. Methylation typically prevents the transcription of a gene into mRNA, while acetylation facilitates it; they operate like the brakes and the accelerator of gene transcription. The chromosomes that comprise our DNA spend a lot of their existence condensed into a very small space, becoming uncondensed in places as needed to support transcription or replication. If all chromosomes in a single human cell were stretched out and laid end-to-end, its DNA would be about 2 meters in length; all that is stuffed into the nucleus of a cell. For perspective, if the cell were blown up to be the size of a pea, the relative length of the DNA would be over a mile.[3]

Specialized proteins organize and compact DNA so that it fits into the nucleus and can still be accessed when needed. DNA reaches this compacted state by wrapping around proteins called histones, like thread around a spool. The combination of a histone with DNA strands wrapped around it is called chromatin. Tiny fibers protrude from these histones, with areas for methyl groups and acetyl groups to bind. If an acetyl group binds to the histone, the chromatin relaxes so that the DNA can be reached for transcription. If a methyl group binds instead, it makes the chromatin condense even more, blocking access for transcription factors. DNA itself can be directly methylated, blocking transcription. The methyl or acetyl group is stuck on its target by a specialized protein, and will stay in place unless and until a different specialized protein removes it. If the cell divides, then daughter cells inherit the methyl or acetyl group along with the DNA strand.[4] There is also evidence that when the cells that form eggs or sperm undergo epigenetic modification, those modifications may be carried forward to future generations.[5]

And so these epigenetic factors can have a long term impact. In 2009, researchers studied volunteers from the Hunger Winter cohort: 60 individuals who were exposed to famine early in their embryonic development and 62 individuals who were exposed to famine shortly before they were born, i.e., late in their embryonic development. For each group, the study also involved their same-sex siblings conceived and born outside the famine period as controls. The purpose of the study was to see if some of the differences those subjects showed as adults could be traced to epigenetic changes in the gene that codes for insulin-like growth factor II (IGF2) during their embryonic development. IGF2 is a key factor in human growth and development, particularly fetal development. By comparing the degree of methylation of *IGF2* genes between the individuals either conceived or born during the famine with that of their same-sex sibling, the researchers found that the individuals exposed to famine early, but not late, in embryonic development had less methylation of that gene compared to their siblings. It was a stable epigenetic difference, detectable even 60 years after the exposure. Because methylation represses gene expression, reduced methylation in those people would have meant more IGF2 exposure as they developed as embryos, presumably contributing to a lifetime of adverse health outcomes.[6]

EPIGENETIC TRACES IN STRESS

At the end of the 1990s, landmark studies concerning maternal care in rats showed epigenetic mechanisms for transmitting stress reactivity across generations. Female rats show a range of maternal behaviors in the degree to which they lick and groom their pups, and do "arched back" nursing. These behaviors are stable: an individual mother rat always acts the same way with her pups. The offspring of mothers who do a lot of licking, grooming, and arched-back nursing ("high-LG-ABN") are less fearful with a more modest HPA response to stress as adults than are offspring of "low-LG-ABN" mothers.[7]

In 1999, Francis et al from McGill University showed that such behavior not only affected the pups those rats raised, but was also transmitted through generations. Taking a few rat pups from litters of low-LG-ABN mothers and swapping them for a few pups from high-LG-ABN mothers, they ensured the mother rats each fostered several pups that were not genetically her own. When those pups reached adulthood they were bred and their maternal behaviors observed. Not only did the pups carry forward the behavior they had experienced with their foster-mother to their own litters, but so did their pups, and their grand-pups, and so on in succeeding generations. Likewise, no matter what their biological parentage, the pups raised by high-LG-ABN mothers were less fearful and showed a restrained HPA response to stress than did pups raised by low-LG-ABN mothers. The foster pups had inherited a characteristic behavior through mechanisms outside genetics.[8]

A few years later, Weaver et al from McGill found the mechanism for that change in behavior. Their study showed that maternal care altered DNA methylation and histone acetylation in the promoter region of the *nr3c1* gene – the gene for the glucocorticoid receptor (GR) – leading to a stable change in GR expression and HPA response to stress. The rats raised by low-LG-ABN mothers had extensive methylation in the promoter region of the *nr3c1* gene in the hippocampus, while the rats raised by high-LG-ABN mothers had increased histone acetylation in that promoter region instead. Because methylation inhibits transcriptional activity and acetylation facilitates it, the low-LG-ABN rats expressed fewer GRs than the high-LG-ABN rats. The GR is a key component of the negative feedback loop for the HPA axis, and so the reduced-GR rats had a sharper and longer-lasting stress response. These epigenetic markers appeared during the first week of life and persisted through adulthood. The researchers also showed that the condition could be reversed: when the low-LG-ABN-raised rats were infused with a histone acetylator, it reversed their epigenetic markers and both GR expression and the rats' stress response went back to normal – just as if they had been raised by high-LG-ABN mothers.[9]

In 2009, these studies moved from rats to humans when McGowan et al, also of McGill University, examined postmortem hippocampal tissue from male subjects organized into three groups: 12 suicide victims, all of whom had suffered childhood abuse or neglect; 12 suicide victims who had not suffered abuse or neglect as a child; and 12 controls who had died suddenly of unrelated causes and had not been abused or neglected as a child. The study showed that the total GR mRNA in suicide victims with a history of childhood abuse was significantly reduced relative to the non-abused suicide victims or controls. Because mRNA is a lingering template that shows a gene had recently been transcribed, this finding indicated that victims of early abuse had fewer GRs in their hippocampi when they died. As the mechanism for that difference, the McGowan team observed that the abused cohort had much higher DNA methylation in the promoter region of the *NR3C1* gene than did suicide victims without childhood abuse or controls. This methylation meant that the *NR3C1* gene was less likely to be transcribed into a GR in the brain of someone who had been abused as a child than in someone who had not. For the non-abused individuals, the levels of GR mRNA didn't differ whether they died by suicide or some other way, indicating the low GR mRNA result was a product of early abuse, not of suicide.[10]

Many more studies followed. By 2016 there had been 13 animal and 27 human studies specifically looking at methylation status of *NR3C1* in relationship to early life stress, parental stress, and the development of disorders later in life. Despite significant variations in the type of stress involved in these studies, 70 percent of the animal studies and 90 percent of the human studies found a significant increase in methylation status of the *NR3C1* promoter region after early life adversity. Studies of the impact of maternal stress (women who had experienced anxiety and mood disorders, pregnancy-related anxiety, violence, or war stress during pregnancy) all revealed increased DNA methylation in the *NR3C1* promoter regions in their children. Overall, the results have been very consistent, showing epigenetic mechanisms translating childhood or prenatal stresses into a long-lasting, stress-related trait.[11]

Epigenetic changes have also been detected in other genes that control pathways implicated in depression. A series of studies showed the "T" allele of *FKBP5* to be associated with an increased risk of developing depression following childhood adversity (though that SNP did not show significance in the 2018 GWAS results). The *FKBP5* gene codes for a protein that makes the GR less efficient, so when there is a lot of that protein active in the cells of the brain, the GRs don't function as well in their role of shutting off the HPA stress response. In a 2013 study, a multinational team looked for indications of DNA methylation in the gene as a potential means by which early life adversity is recorded and maintained to affect later-life stress responses. The results indicated that DNA methylation was reduced, but only after childhood, not adult, trauma.[12]

Since the McGill studies, epigenetic modifications associated with different types and degrees of stress have been characterized in genes coding for stress-related proteins CRH and vasopressin, growth factors GDNF and BDNF, and more. DNA methylation and histone modifications are the most studied, and patterns of these epigenetic changes after acute and chronic stress have been documented. These epigenetic changes tend to be specific to different tissues; for example, researchers may observe histone acetylation in one part of the hippocampus while other brain tissues are unaffected. The actions of various antidepressant treatments are also being studied for their epigenetic impact. One animal study found epigenetic modifications to over 2,000 genes resulting from chronic stress, and also found that chronic imipramine treatment reversed most of those epigenetic changes. Overall, these studies show that epigenetic mechanisms are strongly involved in the stress response in the central nervous system.[13]

There is growing evidence of the strength of epigenetics, emphasizing some of the difficulties in identifying how and if depression is inherited. Considering the effect of maternal behaviors on succeeding generations of rats, our responses to stress may be passed on through epigenetic modification of a gene, rather than or in addition to genetic variations. No genome-wide association study would find such a trait; it's not a difference in alleles. And maybe we also passed down epigenetic modifications of neuroplasticity factors, or potentially any pathway related to depression, on top of the genome, rather than through it. There's a difference between a trait passed down through epigenetics rather than genetics, though. An epigenetic change can be undone.

It has been 3 years since Tyler began pursuing treatment for his depression, and though it's been an unsteady path, he's maintained his dedication to getting better. He and his wife are still happily married. Tyler works from their house in a Virginia suburb most of the time – he's not yet willing to trust someone else to look after his son. It took him over a year to find a therapist that he trusts and works well with, but he's found that now. "She must have a line into someone upstairs or something," he says. "She really knows how to talk to me." He still has his low points, which he says are generally triggered by stress. "If I get into any kind of overwhelming situation, basically when my stress goes really high, I pretty much shoot straight down."

After resisting medications for years, he is finally taking them and seeing success. "I was super-opposed to them in the beginning," he says. "I've seen people that weren't bad people, but due to the situation they were put in they ended up becoming addicted. There's a lot of alcoholism in my family and I've always shied away from anything addictive. I mean, anything can become cocaine." He added, "We keep trying to give people medications for things when we are barely scratching the surface of what we understand. It's ridiculous." Tyler takes Zoloft and n-acetyl-cysteine (NAC), per his therapist's advice. "It's basically like a brain vitamin," he says. "The woman that I'm seeing, she said it's not necessary but it helps, and I have seen a difference when I don't have the NAC with my medication."

Tyler looks back now at his young adulthood, at the time he spent in the Army and the early years of his marriage, and sees the opportunities lost because he would never let himself acknowledge he needed help. He had powered through, being tough, but that toughness cost him. Now, when asked if he has advice for the reader, he takes a clearer view. "The biggest thing I could say is, if you think, ever, maybe I should talk to someone, go do it. Go do it. If it's a thought, ever, that you need it, do it. It doesn't have to become a thing that you do all the time, but if you need help, get it. Ask for it. A closed mouth eats no food. If you don't ask for help, no one's going to know that you need it."

Chapter Fifteen

DIET

When Jerome Sarris began graduate studies at the University of Queensland in Brisbane, Australia, his field didn't exist. In his clinical practice as a naturopath, acupuncturist, and nutritionist, Sarris had embraced a natural approach to helping his clients achieve and maintain good health, including good mental health. After years in practice, though, he found his interests coalescing around a more rigorous understanding of the interventions he was working with. Entering the University of Queensland, he was the first person in the country to pursue a doctorate in the integrative mental health realm. "I didn't have anybody to draw on," he said. "I had no idea where to go, what to do and I was very lucky that I had a wonderful supervisor, Professor David Kavanagh." Kavanagh was a psychologist in the Department of Psychiatry, and Sarris earned his PhD in the field of psychiatry.

From as early as his teens, Sarris had been drawn to learning about the body, the mind, and the spirit. While in high school, he took courses in aroma therapy and massage therapy at a local college, moving to England after graduation to study yoga instruction, exercise therapy, and plant-based medicines. He returned to Australia a few years later, earned a bachelor's degree and started his clinical practice. Diet and nutrition had always claimed a prominent place in his interests. "I was always fascinated that people could consume things and it could alter their state of mind... it could worsen their mental health or improve it," he said. "But I look back certainly in my late teens, early twenties, nobody had any idea in terms of the diet affecting mental health. Certainly it wasn't discussed. You'd just eat a normal, average western diet; you wouldn't even think about it."

Now, after a post-doctoral fellowship and years in academia, Dr. Sarris is Professor of Integrative Mental Health at Western Sydney University, where he pursues research into integrative medicine, nutraceuticals and plant medicines, and lifestyle medicine, particularly in anxiety and mood disorders. Just a few years ago, he helped establish the International Society for Nutritional Psychiatry Research (ISNPR) and currently serves on its executive committee. "It was the brain child primarily of Professor Felice Jacka [Director of the Food & Mood Centre at Deakin University and President of the ISNPR], and I was pleased enough to be one of the founding members of the organization. It's an international collaborative effort, looking at advancing our understanding of the impact of diet, nutrition, but also of the potential for select supplementation of nutrients to improve mental health and to treat various psychiatric disorders," he said. "If you look at areas such as cardiology or gastroenterology there's a good understanding of the impact of the diet. When it comes to the brain, traditionally, psychiatry didn't even think about diet. Which is crazy considering the various processes in terms of inflammation, in terms of the building blocks for neurochemicals, and obviously a range of processes and deficiencies." Other contributors to depression, including oxidative stress, neuroplasticity, and mitochondrial function are also affected by diet.[1] "There is absolutely a very strong link between brain health and what we eat, and that's what we're trying to advance through our research."

And research shows that diet does matter in depression. In 2014, a meta-analysis of 13 studies found that overall, a "healthy" diet characterized by high intakes of fruit, vegetables, fish, and whole grains was associated with about 15 percent reduced odds of having depression, all else being equal.[2] However, because most of the studies used in that analysis were cross-sectional (they looked at a snapshot in time rather than following the same population over many years) their results may partially reflect dietary choices people make *because* they are depressed.

Prospective studies eliminate that source of error – they follow an at-risk population for years before they have depression, to see what happens. The Seguimiento Universidad de Navarra longitudinal cohort study is an ongoing, prospective study looking for the influence of diet in the development of depression. By June 2014 it had enrolled 15,093 healthy Spanish university graduates and followed them for a median duration of 8.5 years. By that 2014 follow-up, 1,051 new cases of depression had developed among the cohort. Comparing the participants' level of adherence to particular dietary patterns to their outcome, the researchers were able to show that some diets reduced the risk of developing depression. The dietary patterns they assessed at that time were the Mediterranean diet (characterized by an abundance of fruits, vegetables, grains, and monounsaturated or omega-3 polyunsaturated fats), a "pro-vegetarian" diet, and the "Alternative Healthy Eating Index (AHEI)-2010." In the pro-vegetarian diet, people preferentially consume plant-derived instead of animal-derived foods, but don't follow a strict vegetarian diet. It includes seven food groups from plant origin (fruits, vegetables, nuts, cereals, legumes, olive oil, and potatoes) and five food groups from animal origin (added animal fats, eggs, fish, dairy products, and meats and meat products). The AHEI-2010 scores eleven groups of foods or nutrients (vegetables, fruits, whole-grain bread, sugar-sweetened beverages and fruit juice, nuts and legumes, red/processed meat, trans fatty acids, long-chain omega-3 fatty acids, polyunsaturated fatty acids, sodium, and alcohol intake) from 0 to 10 (worst to best).[3]

After adjusting for potential confounding factors (age, sex, body mass index, smoking, physical activity, vitamin supplementation, total energy intake, and presence of several diseases at baseline), each of those dietary patterns showed a protective benefit against the development of depression. For the Mediterranean diet, even moderate adherence conferred a protective advantage, reducing the risk of developing depression by about 20 percent. Moderate or better adherence to the pro-vegetarian dietary pattern or the AHEI-2010 dietary pattern reduced the risk of developing depression by about 30 to 35 percent. For all the dietary patterns, some of the protective benefit attenuated at the highest levels of adherence, in a sort of U-shaped curve. This sort of result is often observed in dietary studies, implying any diet or dietary factor is healthier in moderation than at either extreme.[4]

In whole-diet studies, the "bad" diet the others are compared to is usually the western dietary pattern, but though analyses report a trend connecting that dietary pattern to greater incidence of depression, there just haven't been enough studies to decisively confirm the association.[5] Within the western diet, it seems to be its excesses – the high-fat, processed foods and refined sugar – that tend to lead to poor mental health outcomes. A 3-year longitudinal study involving almost 70,000 postmenopausal women found that those whose diet included the highest dietary glycemic index or highest consumption of dietary added sugars had about 20 percent higher risk of developing depression than those with the lowest values in those areas.[6]

Even if a good diet provides a relatively small degree of protection, it is one of the few tools widely available to affect the course of depression. Scientists have also delved into the impacts of components of various diets, from macronutrients (food groups) to micronutrients (vitamins and minerals) to assess their individual impacts. "At the end of the day, we always advocate for whole food diets, a culturally appropriate whole food, unprocessed diet," Sarris commented. "That's the starting point. But that being said, there's work which has shown that even beyond a good diet there are certain supplements that can affect people's mental health."

MACRONUTRIENTS

One of the observations that ignited interest in the association of nutrition with mental health was the apparent effect of fish consumption on the incidence of depression. Comparing annual fish intake to prevalence of depression in 9 countries around the world, researchers found an inverse relationship: more fish, less depression.[7]

Fish, along with some nuts and seeds, contain high levels of polyunsaturated fatty acids (PUFAs). Part of the building blocks of cells, PUFAs are used to create membranes and generate cellular energy. The best-known PUFAs are omega-3 fatty acids (including alpha-linolenic acid or ALA, docosahexaenoic acid or DHA, and eicosapentaenoic acid or EPA) and omega-6 fatty acids (including linoleic acid or LNA, and arachidonic acid or ARA). The omega-3 fatty acid ALA and the omega-6 fatty acid LNA are both essential fatty acids: they must come from the diet. Each is also the beginning of its omega-3 or -6 family. In the omega-3 family, ALA can be converted to EPA, which can be converted to DHA. In the omega-6 family, LNA can be converted to ARA. These conversions are very inefficient, though, so EPA, DHA, and ARA are considered conditionally-essential nutrients.[8]

PUFAs are of great interest in studies of depression because they are known to regulate both the structure and function of neurons, glia, and endothelial cells. They alter membrane lipid composition, cellular metabolism, signaling, and gene expression. The omega-3 PUFAs in particular affect the expression of genes involved in synaptic plasticity, formation and function of cellular membranes, energy metabolism, and regulatory proteins. PUFAs and their metabolic products can activate neuronal receptors and cell signaling pathways; they can also be converted into endocannabinoids, with a host of functions that includes suppressing the release of glutamate, GABA, and monoamine neurotransmitters. DHA has a role in learning and memory, though the mechanism is unknown, and in the creation and survival of neurons.[9]

Though both are PUFAs, the balance of omega-3 to omega-6 fatty acids affects inflammatory processes and mitochondrial function and survival. Increasing omega-3 PUFAs at the expense of omega-6 PUFAs makes mitochondria more efficient and helps protect them from some stresses.[10] That ratio also drives pro- and anti-inflammatory processes. Inflammatory cells of the immune system contain large amounts of the omega-6 fatty acid ARA, which they use to generate mostly pro-inflammatory cytokines. When ARA is replaced in those cells by either of the omega-3 fatty acids EPA or DHA, anti-inflammatory cytokines are produced instead. Plus, EPA and DHA are used in the production of multiple substances that help resolve an inflammatory state back to normal.[11]

An analysis involving more than 6,500 Spanish men and women found that those who consumed a moderate amount (about 10 to 25 grams) of fatty fish per day were about 30 percent less likely to have depression than someone with low or no consumption. Fatty fish like mackerel or salmon are very high in omega-3 fatty acids and also high in omega-6 fatty acids. More is not always better, though. The results also reflected a U-shape, with the maximum fatty fish-eaters (more than 50 grams per day) seeing less benefit than the moderates.[12]

Studies of food groups have teased out a small but significant effect for many other major components within the neuroprotective diets as well. In terms of average daily consumption, the sweet spots for each of the types of food measured in various prospective studies was: at least two pieces of non-juice fruit, between two and four servings of vegetables,[13] 25 grams of legumes (such as beans, peas, and lentils),[14] and more than 20 grams of fiber.[15] Even alcohol has a place in a protective diet, with 5 to 15 grams of alcohol (which equates to two to seven drinks per week) showing protective effects against depression.[16]

MICRONUTRIENTS

As with essential amino acids and essential fatty acids, we get most of our vitamin and all of our mineral needs from the diet. All vitamins are required for normal brain function, but some are intimately involved with the basic functioning of neurons and other brain cells. Several vitamins and minerals have been directly associated with depression, with varying degrees of consistency and effect. It is difficult, after all, to separate out the contribution of one integral part of someone's diet from all the other factors that ride with it.

Vitamin A has a complicated relationship with depression: either too much or too little of its main derivatives (called retinoids) is associated with depression. In fact, some patients being treated for acne with isotretinoin, a type of retinoid, became depressed and even committed suicide. Retinoids are essential to the development and proliferation of brain cells in the fetus, and animal studies indicate that they are required for synaptic plasticity in the adult hippocampus too, where they regulate neurogenesis and neuron growth. They can enter the brain and directly modulate dopamine pathways and, to a lesser extent, serotonin and norepinephrine pathways.[17] The 2018 depression GWAS identified a retinoid-X receptor gene to be significantly associated with major depression,[18] so interest can be expected to grow.

Several elements in the vitamin B complex (specifically B9 and B12) have been shown to have a neuroprotective function. When many patients with depression and other psychiatric disorders showed low levels of vitamin B9 (folate) and sometimes vitamin B12 (cobalamin) in their blood serum, investigators began looking for a role in mood disorders. B9 works in conjunction with B12 to convert a toxic substance – homocysteine – into a useful one: methionine. Excess homocysteine stimulates glutamate's NMDA receptors to the point of excitotoxicity and also damages DNA in neurons, while methionine is used in DNA methylation, impeding transcription of some genes. People with low B9 levels are about 40 percent more likely to be depressed than those with normal levels, and lower folate and vitamin B12 levels and higher homocysteine levels predict about a 30 percent greater chance of developing depression.[19]

Vitamin D (specifically, D3) is a latecomer to the family of nutrients believed to be involved in psychiatric disorders. Vitamin D is involved in regulation of the brain's immune system and growth factors and in neuroplasticity and brain development. The results of three longitudinal cohort studies indicated that the participants with the lowest levels of vitamin D were about twice as likely to develop depression in the years that followed than the participants with the highest levels.[20] On its own, vitamin D supplementation as a treatment option has not been very successful, though. Randomized controlled trials have failed to show a significant effect of such supplements in reducing depression; however, most of those trials involved subjects with mild depression and a sufficient level of vitamin D at baseline, and had a lot of variation in the amount and duration of vitamin D administration.[21]

Zinc is the most abundant metal in the body, after iron. It has a structural role in several proteins used in genetic processes and is involved in neurotransmission in some neurons. It is crucial for immune responses and affects the balance of pro- versus anti-inflammatory cytokines. Zinc protects cells from oxidation damage by free radicals, and has an important role in apoptosis. It is known to be involved in the regulation of more than 300 enzymes that control numerous cell processes including DNA synthesis, normal growth, brain development, behavioral response, reproduction, fetal development, membrane stability, bone formation, and wound healing.[22] In the early 2000s, a role for zinc in depression was suggested from observations of low zinc levels in the plasma serum of people with depression. Animal studies showed that experimental zinc deficiency could induce depressive-like behaviors, which could be effectively reversed by zinc supplementation. Human studies have shown significantly lower zinc concentrations in depressed subjects than in controls, with greater depression severity associated with greater relative zinc deficiency.[23] The few randomized controlled trials of zinc by itself on depression produced mixed results, but as a supplement to antidepressant drugs it significantly lowered depressive symptom scores.[24]

The impact of diet on the course of depression is becoming clearer, thanks to focus from ISNPR and a growing body of evidence from studies and clinical trials. With small protective effect sizes in individual elements or across whole diets, the impact on the individual may not be enough to prevent an episode altogether, but it may extend remission, reduce the severity of a depressive episode, or improve treatment response. In 2017, a 12-week, randomized controlled trial of dietary intervention as an addition to psychotherapy or pharmacotherapy reported solid success, with 10 of 31 patients in the dietary intervention group achieving remission, compared to 2 of 25 in the control group.[25] "People's diet absolutely does affect not only their physical health but their mental health," Sarris commented. "And when people eat well and feel physically well, then it's going to encourage them to exercise because they've got more energy. And this in turn helps with weight regulation, with effects on self-esteem and self-mastery, as well as if you feel physically better, of course you feel mentally better." And,

Sarris notes, unlike many risk factors for depression, diet is a daily, intimate, and *modifiable* environmental exposure. "When we look at depression, obviously there's genetic predisposition involved; there are environmental factors, chronic stressors, but also there's a lot of lifestyle elements that impact the trajectory of a depressive episode. Dietary quality is certainly one of those."

Chapter Sixteen

THE GUT MICROBIOME

Professor Graham Rook of the University College London swept into the pub I had selected from the Internet, a bit breathless rushing from another meeting. Professor Rook is best known for his work in evolutionary medicine, particularly his formulation of the "Old Friends Hypothesis" as a way of capturing how our interaction with microbes shapes the immune system to resist allergic and chronic inflammatory disorders. This work has surged in the last decade as more studies are showing how microbes in the human gut shape immune system and brain development and play a role in psychiatric disorders, with depression front and center.

The idea that microbial organisms have some relationship with mental health is not new. As Dr. Harry Oken had commented when I met with him in Baltimore, it's been known for more than a century that syphilis can cause "madness," including symptoms of "melancholia, and less frequently, mania."[1] Antidepressants have shown antibiotic effects dating from the time of their first discovery. Iproniazid, the first MAOI, was developed to be an anti-tuberculosis drug, and its sister drug, isoniazid, was shown to have antidepressant qualities alongside its anti-microbial effects. Isoniazid was the more effective against tuberculosis with a weaker "euphoric" effect, and iproniazid was the more effective as an antidepressant and less so as an anti-tubercular agent, so the drugs went their separate ways.[2]

For Rook, an immunologist and microbiologist, a conversation with a fellow research professor, Dr. Christopher Lowry, lured him into the topic of the effects of microbes on the brain. "Chris had been trying to 'wake up' certain neurons in the brain, serotonergic neurons, in a part of the brain that is involved in damping down anxiety." Rook explained. "He was having terrific difficulty activating these particular neurons, because if you just give someone a serotonin reuptake inhibitor, obviously, you increase the activity of all the serotonergic neurons. But he knew that in neighboring parts of the brain, there are some that make you more anxious and some that make you less anxious, and he just wanted to turn on the ones that make you less anxious. And I think we must have been in the pub when we had the idea, because it was a crazy idea, that a particular microorganism from the environment that we'd been looking at [*Mycobacterium vaccae*] might be doing something in the brain. It had been in a clinical trial of lung cancer, and it had made a difference in the survival of a small cohort of lung cancer patients, but not overall. It had been noticed that the people who'd received this immunotherapy just seemed happier. It was rather unofficially done, but there was a sort of scoring system used which suggested it was actually affecting their mood, and they were less upset by the fact that they were dying.[3]

"That was why we thought, 'Well, let's stick it in these mice'," Rook continued. "Talk about a longshot! It was crazy, but it actually did activate this particular subset of serotonergic neurons that he was interested in. And so he's been studying that particular organism ever since. In the most recent experiment that he's done with it, he had a strain of mice, which, if you upset them by putting a dominant male in the cage, they get a change in the gut microbiota. They get a nasty gut inflammation, colitis, and they get striking behavioral changes: they sit moping in a corner, and respond in a similar way to various other tests. So you stress these mice and they get microbiota changes, colitis, and they become depressed.[4] But what he found was that if you immunized them with this particular microorganism, none of these things happen."

THE HUMAN MICROBIOME

The human gut harbors a complex ecosystem called the microbiome, consisting mostly of bacteria but also including viruses and bacteriophages, protozoa, archaea, and fungi. It is estimated that the gut is inhabited by about 100 trillion microorganisms, meaning our bodies hold several times more microbial cells than human cells. The total weight of these gut microbiota is roughly equivalent to that of the adult human brain. There are more than 1,000 species of microorganisms that can inhabit the human gut, together possessing about 150 times as many genes as do humans.[5]

Mammals co-evolved with microbes and maintain a fundamental dependence on them for survival, just as those microbes depend on us. For the most part, our relationship with our microbes is mutualistic: both parties benefit. The normal composition of gut microbiota benefits our health, and a disruption of that normal balance makes us more susceptible to disease. Two phylum of bacteria predominate in the human microbiome: *Bacteroidetes* and *Firmicutes*, with *Proteobacteria*, *Actinobacteria*, *Fusobacteria* and *Verrucomicrobia* also present in lower numbers. The gut of an infant is colonized starting at birth, when delivery through the birth canal exposes it to its mother's microbiota, establishing an initial maternal signature to the microbiota in the child. Children born by Cesarean section get a different microbial composition that resembles the microbial populations of the mother's skin instead; a signature that has been shown to persist into adulthood.[6] Studies show that infants delivered by Cesarean section are more likely to suffer from childhood obesity, allergies, asthma, and diabetes later in life.[7] Entering old age, we generally show reduced *Firmicutes* and *Actinobacteria*, and increased *Bacteroidetes* and *Proteobacteria*, accompanied by increased frailty and chronic low-level inflammation.[8] Throughout our lives, we affect our gut microbiota through mode of birth, whether we were breastfed, use of antibiotics or probiotics, diet, environmental exposures, stress responses, sleep disruption, and more.[9]

MICROBIOTA AND IMMUNE SYSTEM DEVELOPMENT

In turn, researchers are uncovering the variety of ways in which our gut microbiota affect us. Studies using germ-free animals show startling differences in brain development caused by an absence of microbiota. Though there is still so much unknown about how those microorganisms affect development and how the two-way communication in the microbiome-gut-brain axis works, some of the answers lie in the immune system and its development.

The immune system turned out to be a mediator in the *M. vaccae* experiments Professor Rook described. "The critical thing in this study was that if you blocked the development of immune cells called regulatory T-lymphocytes, T-reg for short; the injection didn't work. In other words, the ability of an injection of *M. vaccae to* stop the change in the microbiota, the colitis, and the change in behavior – all three things – was dependent upon the production of T-reg following the injection. And of course, that was extraordinary, really, because that implied that by regulating the immune system better, you could block all of these phenomena. And that took one back to the increasing amount of epidemiology suggesting that depression is more likely to occur in people who have raised levels of background inflammation."

Rook attributes the failure to switch off this background inflammation as a failure of immunoregulation. "In the modern world, we are suffering from two kinds of immunoregulatory disorder. We see a lot more allergies, autoimmune disorders, and inflammatory bowel disease, which are basically diseases where the immune system is attacking things it should not be attacking. I call these the 'forbidden target' diseases. And then there are all these other disorders in which your risk is increased if you have raised levels of background inflammation. And that's not just depression; it's also metabolic disorders like diabetes, heart disease, and so on. But all of these things look like failures of regulation of the immune system. Which all fits with Lowry's finding that you needed the T-reg."

He spoke of a study that associated exposure to a varied microbiota throughout childhood with psychosocial stress and inflammatory markers later in life. "A paper had been published in *Nature*, 2 or 3 years ago, by a German group claiming that if you compared children brought up in a rural background for the first 15 years of their life with children brought up in an urban background for the first 15 years of their life, and put them in a fMRI scanner, and then I think they made them do difficult mental arithmetic. But what they found was that the reactions of the brains, of the rural upbringing and the urban upbringing, were different. So they wrote a paper saying 'It just shows the stress of an urban upbringing is doing something to the development of their brain.' But I looked at this paper and thought, excuse me, I know some of the towns where they were recruiting these people, and if you know small German towns, they are just about the most peaceful, organized, wonderful, leafy, pleasant places on the planet! Why, it's a pure prejudice to say that life in such a place is more stressful!"

A team in Germany led by Stefan Reber (a collaborator of Professors Rook and Lowry) redid the experiment, with results published in 2018 in the *Proceedings of the National Academy of Sciences*. "We were saying to ourselves that what's actually happening is the ones with the urban upbringing are getting an inflammatory response to the stressor, because their T-reg haven't been turned on. Whereas the ones with the rural upbringing, they're not getting an inflammatory response. And so, Reber did another experiment. It's almost exactly like the previous *Nature* paper, except that what was measured this time was the inflammatory response, which was done by measuring both *ex vivo*, the levels of the inflammatory mediators over time, the number of white blood cells. And in order to make the urban and the rural subjects as different as possible, the urban ones didn't have pets in the home, and the rural ones came from houses that did have exposure to animals. And what they found was in the ones with the urban upbringing, you got inflammatory mediators produced, and a deficit of stuff called IL-10, which is an anti-inflammatory mediator. Which was quite different from what they found in the ones with a rural upbringing."[10]

The relationship of humans with their microbiota goes back basically forever. In fact, as Professor Rook points out, microbiota symbionts are found throughout all forms of multi-cellular life. “Way back in evolution, before vertebrates had evolved, obviously there were already organisms in the guts of the creatures that were there, insecty things and wormy things and so on,” he said. “But they were in a sort of sack of chitin, the stuff which insect bodies are made of. So those organisms were separated from the gut of the animal in which they lived. Then when you get to the vertebrates, the chitin barrier started to disappear, to be replaced by a mucous barrier. And then when you get to the mammals, the chitin barrier has disappeared completely, and there's just a great big mucin barrier. Many of the organisms live in, and on, the mucin; they're actually being fed by the mucin, and they interact, send signals to the host. If you think about it, that's what happens with plants as well. Plant roots secrete polysaccharides onto the surface of the roots, and there are these bacteria that embed themselves in this polysaccharide; they eat the polysaccharide, and they send signals to the plant that alters the functioning of the plant. So this is a symbiosis: the plant feeds the bacteria the things that it wants, and the bacteria feeds things that the plant needs, helps to fix nitrogen, or whatever. So it's rather fascinating that we carry the soil around with us. We even call it ‘soil’ in some circumstances. It's extraordinarily similar: a plant root in the soil is like our gut epithelium on the poo. We just have our soil inside.”

As mammals developed their very complex microbiota and became dependent on it, a means had to be found to pass on the useful microbiota from one generation to the next, from mother to young. “The one I really love is the koala bear,” Rook commented. “Do you know about koala bears?” All I really know, and I told him, was that they eat eucalyptus leaves and look cute. “Exactly!” He agreed. “Eucalyptus is poisonous. That's why only koala bears eat eucalyptus. But how can they? The answer is: mum's poo. The baby, when he's born, he's tiny. He crawls up – this little grub-like thing – and hops into the pocket and hangs on to the teat. Once he's big enough, he pokes his head out, and mum understands and she twists her pelvis forward and poos a special kind of poo, which the baby eats. And now the baby has the microbiota that is able to detoxify the eucalyptus. As soon as you think of an animal where there's a problem getting the right microbiota on board, it gets interesting. What about dolphins? They give birth in the sea. So, how do they cope? Well, at the same time the mother gives birth, she poos, prolifically and magnificently, and that seems to do the trick. So every species has its way of passing on the microbiota."

For human infants, the first colonization happens at birth, either from the birth canal or from the mother's skin. That population is then bolstered and expanded through breast-feeding and later through the diet. "It turns out breast milk is not sterile," Rook commented. "In fact, certain organisms are picked up from the gut... You know about dendritic cells of the immune system? They poke processes through into the gut and sort of fish things out. And they circulate in the blood and get into the breast tissue, so they're passed on to the baby. Presumably, these are bugs that are difficult to get to the baby in other ways. And breast milk also contains prebiotics. That's absolutely astonishing: there are polysaccharides in breast milk that are not digestible by the human. It encourages the relevant bugs."

Clearly, microbiota must have some importance to drive us to these lengths. The gist of the Old Friends Hypothesis is that they are essential for training the immune system. "The adaptive immune system evolved in parallel with the brain," Rook said. "It's remarkably similar to the brain; they're the two learning systems. These two things that have to learn come up in the first months of life. Of course, in the early months of life, the data for the immune system to learn comes particularly from mother. Basically, at birth, the immune system is really like a computer with hardware and software, but no data." The immune system is trained from three sources of data, he explained. "It comes in the thymus gland, which expresses versions of all the proteins in the body and trains the T-cells to not make an immune response to those proteins. And then mother's bugs, you want to tolerate those as well, and then the other thing that feeds in is the bugs from the environment. The environment offers many forms of input, the most important of which is the diet, but access to green space and the natural world is also hugely important."

Though Professor Rook dealt mostly with the effect microbiota have on our immune system, he emphasized that that was only one of the aspects of our relationship with gut microbes. As he surprised his microbiota with a spicy squid dish and I amused mine with some wine, he elaborated on some of the other elements of the relationship. He used the term "farming," which I thought was a useful recalibration of how we think about microbiota. I had come into this project thinking of gut microbes as an inevitable evil, as bugs that spew toxins all over the place and we can only hope to avoid the worst effects. Instead, they create positively beneficial substances that our bodies need to maintain health, including short chain fatty acids (SCFAs).

SCFAs are products from microbiota digesting fiber that is indigestible to humans, creating something we couldn't get otherwise. "The advantage of partnering with bacteria, rather than trying to do everything ourselves," Rook commented, "is not only that the bacteria can be adjusted according to our diet, but that gut microbiota are incredibly good at what is called horizontal gene transfer. Even species that diverged millions of years ago are capable of picking up DNA from another species. The classical one is the Japanese who can digest polysaccharides in seaweed. It's not that the Japanese genome contains sequences that encode enzymes that can digest seaweed polysaccharides, it's that their gut bugs have picked up the relevant genes from marine organisms. So if you keep eating enough Japanese food and live by the seaside for a bit, eventually you'll develop that ability too."

The Old Friends Hypothesis was formed out of what had been called the "hygiene hypothesis," with a similar intent but an unfortunate name. The hygiene hypothesis came from the observation that, when examined at age 11 and again at 23, children brought up in families with many older siblings were less likely to have developed allergic disorders than those earlier in the birth order. The concept was at first a narrow one, focusing on the notion that childhood infections somehow prevented subsequent allergies.[11] As Rook and colleagues began to see how the whole immune system was affected by early exposure to microbiota and other intestinal creatures like worms, they revised the hypothesis to better cover the benefits and consequences of different types of exposures. And, importantly, to get away from any implication that household hygiene is bad for children's health. "First, we have to abandon the hygiene hypothesis," Rook said, "because hygiene has nothing to do with it. Hygiene in the home is *enormously* important. Many people come wandering through and some of them have nasty infections. Of course you want hygiene. But you want minimal amounts of antibiotics, a varied diet; you want exposure to the natural environment, exposure to the mother's microbiota."

MICROBES AND MENTAL ILLNESS

In the 1860s, German physician Hermann Senator raised the notion that systemic disease, including mental health disorders, could be rooted in intestinal "self-infective" processes.[12] The theory became known as "auto-intoxication": toxic breakdown products in the colon resulting from bacterial action on foods were absorbed by the body and contributed to various diseases. By the late 1890s, the role of harmful intestinal bacterial-derived chemicals in mental diseases was actively investigated, with articles appearing in the *Journal of the American Medical Association* and prominent European publications. In 1904, a prize-winning essay by British researcher Arthur A.D. Townsend stated,

> *The more modern and advanced opinion of the present day, not of necessity the most correct, regards toxic action as the most important factor in the pathogenesis of insanity. According to this view insanity is not regarded as primary disease of the brain, but secondary, and due to toxins derived from elsewhere acting upon the cortical nerve-cells, disordering their metabolism, and thus affecting their functional activity, damaging, or destroying them... For a long time I have strongly held the opinion, as a result of my own observations, that a very large proportion of cases suffering from melancholia are due to auto-intoxication resulting from the absorption of toxins from the alimentary tract, for in depressed states generally there are various symptoms referable to disordered metabolic processes in some part of the gastro-intestinal tract.*[13]

Cures for mental disease caused by auto-intoxication ranged from probiotics to surgery. One particularly enthusiastic physician believed focal infections to be the initiator of virtually all forms of psychoses, mood, and behavioral disorders. Under his direction, thousands of teeth were extracted and hundreds of colectomy operations were performed in his New Jersey State-run mental hospital, with a spectacular remission rate claimed to be as high as 80 percent. What they didn't advertise, though, was the 30 percent mortality rate in the 250 patients subjected to colon surgeries over just 3 years. On the much-less invasive side, other scientists and physicians experimented with probiotics, particularly *Lactobacillus bulgaricus* and *Lactobacillus acidophilus*, to "fight microbe with microbe"; good flora against the bad. Products containing one or the other culture began to hit the market, first as a trickle and then as a flood. But, as advertising hype overtook actual laboratory or clinical evidence, these preparations began to take sharp criticism from the medical community. By the 1930s, auto-intoxication as a theory had been debunked, with such biting criticism that it wasn't until 2003 that anyone dared to raise the suggestion in a scientific journal that the administration of beneficial microbes might have an adjunctive place for the treatment of depression after all.[14]

THE MICROBIOTA-GUT-BRAIN AXIS

Over the last decade, the Alimentary Pharmabiotic Centre (APC) Microbiome Ireland at University College Cork (UCC) has established itself as the center of the universe – I exaggerate only a little – for research on the microbiota-gut-brain axis. According to Dr. Siobhain O'Mahony, it began when their small lab in UCC's Biosciences Institute, looking for funding to support PhD research in neuroscience and psychiatry, approached their neighbors upstairs: the Alimentary Pharmabiotic Centre. Happily, they found interest, and the "gut people" upstairs and the "brain people" downstairs formed what would turn out to be a very productive partnership, studying how the gut and the brain interact.

Their first project – Dr. O'Mahony's PhD research, published in 2009 – examined the impact of early life stress on the behavior, immune response, stress sensitivity, visceral sensation, and fecal microbiota in rats. Male rat pups were separated from their mothers for 3 hours a day, starting when they were 2 days old and continuing until their 12th day, while an equal number of their male littermates were left undisturbed as controls. When the rats reached 7 to 8 weeks of age, they went through several tests to see what traces that early stress had left. The (now adult) rats with early life separation stress had significantly higher levels of plasma corticosterone when presented with a novel stressor, gut pain sensitivity, and TNF concentrations than their non-stressed brothers. Plus, in the novel stress test, the maternally-separated rats pooped more and analysis of the poop showed significant differences in the microbiota, even 6 weeks or more after the stresses had ceased. The implication was that the gut-brain axis is changed by early life stress, and these changes could contribute to a susceptibility to develop stress-related disorders, such as irritable bowel syndrome and psychiatric disorders, in adulthood.[15]

"We were trying to replicate what is noted in humans," Dr. O'Mahony explained. "This period in early life, called the 'stress hyper-responsive period' during postnatal days 2 to 14 in rats, also exists in humans in the first few years, when the gut microbiota is established. During this period is that you want to keep stress down, keep cortisol down in humans and keep corticosterone down in rats. And the point of that is because there's so much developing: not only your brain, but your gut microbiota, your immune system, and all the signaling pathways between them."

The APC-UCC symbiosis bloomed, with ongoing ground-breaking research into the microbiome-gut-brain axis and its direct implications on psychiatric disorders. It was fortunate timing, as just a couple years earlier a study in Japan had drawn a link from gut microbes to the stress response. The 2004 Sudo et al study was devised to see if colonization of the gut with microbes and subsequent immune reaction – both known to take place at about the same period of early life – might affect the development of HPA axis responsiveness. To test this hypothesis, they used germ-free and specific pathogen-free mice.[16]

Germ-free mice have no microorganisms living on or inside them; they are born and raised in sterile conditions. Several lines of germ-free mice were established by removing pups from their mothers by Cesarean section to be hand-fed sterilized food in isolators. The lines continue to this day; their food is irradiated to kill any microbes, and their cages tested for DNA signatures to make sure they stay germ-free until particular microorganisms are intentionally introduced. Sudo et al measured the HPA response of germ-free mice and specific-pathogen-free (control) mice to restraint stress. When trapped in a small glass tube for an hour, the ACTH and corticosterone levels in the blood of germ-free mice soared much higher than that of the control mice, but returned to normal levels after an hour or so for both groups. This showed that the germ-free mice had a much higher stress response, but the feedback loops worked as well, reducing the levels once the stressor was removed. The researchers then tried colonizing the gut of other germ-free mice at various stages of development and measured the stress response. Germ-free mice colonized with particular mouse microbiota in an early developmental period grew up to show the same stress response as normal mice, while those colonized with the same microbiota in a later developmental period kept their exaggerated stress response. This told the researchers that gut microbiota help shape the HPA stress response during early development, in a way that persists throughout life.[17]

In 2010, a Swedish team tested behavioral responses of germ-free and specific-pathogen-free mice under various scenarios, while also testing characteristics of their monoaminergic systems, gene expression, and more. These germ-free mice showed less anxiety-like and more exploratory behaviors than their specific-pathogen-free counterparts. A very notable difference between the groups of mice was in the expression of more than 100 genes in certain brain regions. In the hippocampus, 50 genes showed evidence of either more than twice as much or less than half as much mRNA transcription between germ-free and control mice, plus 20 genes in the frontal cortex, 23 genes in the striatum, 84 genes in the cerebellum, and one in the hypothalamus. The affected genes aligned to four basic pathways: synaptic long-term potentiation (used in memory formation), metabolism of glucocorticoids and similar steroid hormones (part of the stress system), second-messenger signaling (used in synaptic plasticity), and control of the citrate cycle (in which mitochondria use glucose to generate cellular energy). They also found significantly lower BDNF mRNA expression in the hippocampus, amygdala, and cingulate cortex of germ-free mice relative to specific-pathogen-free mice.[18] Further research by other teams would show differences in BDNF levels in germ-free mice that vary by sex, brain region, and the developmental period in which microbes were introduced, making a tangled picture to tease apart.[19]

At APC, Dr. Kieran Rea showed me some pictures of the neurons of germ-free mice compared to normal mice, and the differences were striking. The neurons from the germ-free mice had longer and more plentiful dendrites with additional spiny formations, and the myelin sheath around the axon was noticeably thicker. While this looked to me like these mice had more brain – and therefore better brain – Dr. Rea and his colleagues say that, rather, it shows a less-efficient brain.

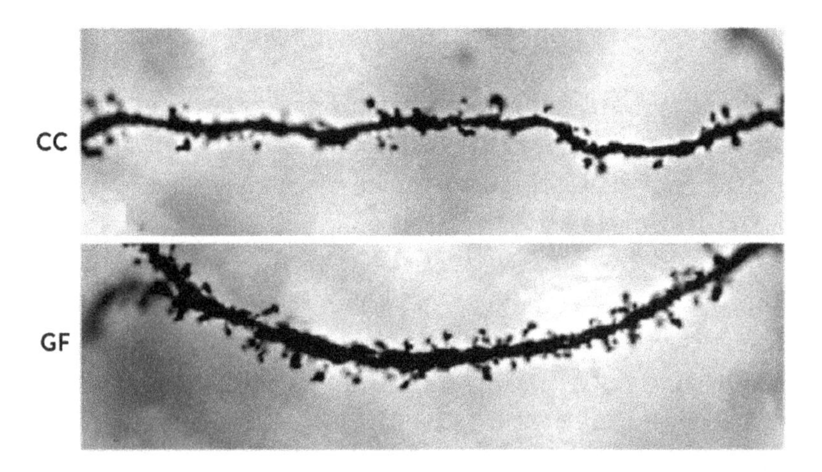

Figure 16-1: (a) Photomicrographs of basolateral amygdalar pyramidal neuron dendritic spine density in conventionally-raised (CON) and germ-free (GF) mice. Republished with permission of John Wiley & Sons - Books, from Luczynski, P., Whelan, S. O., O'Sullivan, C., Clarke, G., Shanahan, F., Dinan, T. G., & Cryan, J. F. (2016). Adult microbiota-deficient mice have distinct dendritic morphological changes: Differential effects in the amygdala and hippocampus. *European Journal of Neuroscience*, *44*(9), 2654-2666. Permission conveyed through Copyright Clearance Center, Inc.

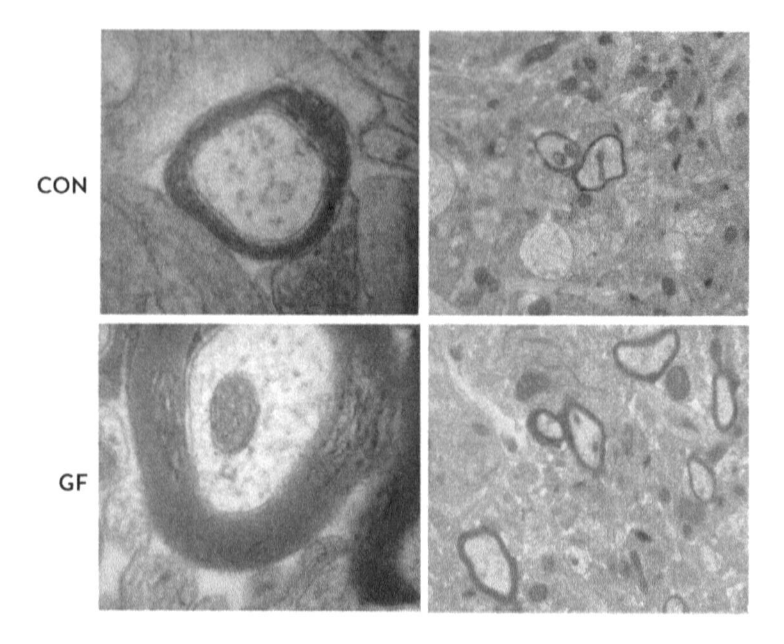

Figure 16-1: (b) Electron micrographs of axons in the PFC of male CON and GF mice, showing thickness of the myelin sheath.
From Hoban, A. E., Stilling, R. M., Ryan, F. J., Shanahan, F., Dinan, T. G., Claesson, M. J., ... & Cryan, J. F. (2016). Regulation of prefrontal cortex myelination by the microbiota. *Translational psychiatry*, 6(4), e774-e774; used under Creative Commons Attribution 4.0 International License, http://creativecommons.org/licenses/by/4.0/

"You can see the animals that are devoid of microbes, they have this kind of hyper-elongation and hypertrophy of the synaptic endings," he pointed out. "So it's like their mechanism for communication isn't as effective because there are far more synaptic endings. It's almost like there's too much information going in. It's not efficient; it's not been refined. Synaptic pruning is exactly that: the synapses are being pruned and cut off like with scissors. So [microbes] might play a role to more efficiently allow communication from one neuron to another. They get rid of superfluous endings that don't do anything, just muddy the signal." The precise mechanisms through which this is happening remain obscure, though. "We sent off some of our samples for deep sequencing, and what we saw were differences between germ-free and normal genes involved in myelin-related genes," Dr. Rea said. "So this observation is a bit unusual. When we actually went and looked at a transverse section of an axon, we have this hypermyelination in the germ-free mouse. It's effectively shielded from communication. We don't know why it's happening, but it could suggest there's some impairment either during development or afterwards to say we need to start pruning or we need to stop myelinating. That signal isn't coming."

The gut microbiota also affect development of microglia – the brain's immune system. In addition to consuming invaders, assisting the innate immune system, and turning on or off neuroinflammation, microglia are key players in brain maturation, where they tag and clear synapses for pruning, promote neuronal circuit wiring, and produce cytokines that guide neuronal differentiation throughout neurodevelopment.[20] In 2015 scientists discovered that in the absence of gut microbiota, microglia in the host's brain fail to develop and mature. To see if the effect of microbiota was restricted to neurodevelopmental stages only, the researchers treated adult normal mice with antibiotics for 4 weeks to see the effect on microglia. After antibiotic treatment, the microglia of the normal mice took on the same characteristics of immaturity they had seen in the germ-free mice. Together, their data suggest that a continuous contribution from intestinal microbes is critical for the homeostasis of microglia not only in early development, but also throughout life. When researchers added some of the metabolic products of microbiota to the drinking water of germ-free mice for 4 weeks, their microglia matured and took on the appearance and activation status of normal mice.[21]

There is further evidence that microbiota can influence brain structure and functioning throughout life. Researchers have studied the effect of giving massive loads of antibiotics to mice after weaning – targeting their adolescence and early adulthood – to see the impact on behavior and brain functioning as they mature. This intervention in normal mice significantly increased norepinephrine, serotonin, and dopamine levels in several parts of the brain compared to control mice. It also greatly reduced expression of both oxytocin and vasopressin in the hypothalamus, impacting the stress system.

Finally, the antibiotic-treated mice showed much less expression of the growth factor BDNF in the hippocampus, implying that the antibiotics impacted learning and memory. Behavioral testing of the antibiotic-treated mice confirmed that effect: they showed less anxiety-like behavior, impaired spatial learning, and reduced memory compared to the controls.[22]

Altogether, it is rather astounding the degree of effect that the presence of microbes in the gut have on development of the brain. Though physically barred from the brain by the intestinal and blood-brain barriers, microbes still affect how neurons grow, how neurons operate, and the development of glia, impacting an incredible range of systems. "One thing we're finding over and over again is that the microbes play a key role in social responses, which is a bit bizarre if you think about it: the microbes in your gut play a really strong role in social interaction," Dr. Rea commented.

MICROBIAL COMMUNICATIONS

One of the questions I asked of several people at APC was how microbiota in the gut communicate with the brain to create changes there.

Dr. Harriet Schellekens laid out some of the many possibilities. "The microbiota in our gut can directly interact with the host, but also produce metabolites based on the nutrients we provide to them. For example, they produce SCFAs, which have been investigated really extensively for their effects on the host. SCFAs are produced by the good bacteria in our gastrointestinal tract from the indigestible dietary fibers, with the principal SCFAs being acetate, propionate, and butyrate. These SCFAs are bioactive molecules and have been shown to positively influence host health. SCFAs play a role in numerous physiological functions such as gastrointestinal functionality, host metabolism, blood-pressure regulation, circadian rhythm, neuroimmune function, and stress-related behavior. Some of the mechanism by which SCFAs exert these functions include secretion of satiety hormones from the gut as well as neuroactive compounds (including the "feel good" serotonin); they can act as epigenetic regulators – histone acetylation and methylation of DNA – and hence impact gene expression; they can manipulate signaling of the vagus nerve (the longest nerve connecting the gut with the brain), and can modify receptor activity in the gut-brain axis.

"[Gut microbes] do so many things, and that's what makes it very complicated. What is it that they do that is causative in a disease process, such as depression and anxiety? So what they can do is make these metabolites, such as SCFAs, but also GABA, serotonin, and neurotransmitters, and they all have receptors within the gut. But they can get out of the gut and maybe some of their metabolites are actually neuroactive. So when they reach the circulation, maybe they can reach the brain. We have the blood-brain barrier, but there are places where it is less strict."

As Dr. Schellekens said, one of the most intriguing aspects of SCFAs are the epigenetic mechanisms they employ. All types of SCFAs inhibit the enzyme that peels the acetyl group from a histone complex, thereby leaving those gene transcription facilitators in place. Butyrate is the most potent of those inhibitors, but acetate can also be converted to a different substance that increases histone acetylation, thus facilitating gene expression. But, while this is taking place a lot in the periphery, the degree to which these gut microbiota-derived substances actually reach the brain is still being investigated. Acetate does get there, and apparently propionate helps maintain the integrity of both the blood-brain barrier and the intestinal barrier.[23]

Stretching from the brain to the gut, the vagus nerve may actually be the most important neural pathway for communication between gut microbes and the brain. Part of the autonomic nervous system, it regulates bronchial constriction, heart rate, and gut motility, among other functions. It is a two-way tract, bringing signals from the brain to those organs and from those organs to the brain. In the brain, it signals the hypothalamus, cingulate cortex, and amygdala, as well as other regions, either directly or by relay. It is the fastest and most direct route that connects the gut and the brain, with a vast variety of receptors. In animal models, microbial metabolites directly stimulate this nerve to signal the brain. Part of the vagus nerve seems to be specialized for detecting pro-inflammatory cytokines, including those released by microbiota, and stimulating the immune system as a result.[24]

Animal studies that involve cutting the vagus nerve show its relevance immediately. When normal mice were fed the probiotic *Lactobacillus rhamnosus*, they showed reduced stress-induced corticosterone and anxiety- and depression-like behavior in comparison to controls. When the experiment was repeated in a new set of mice from which a section of the vagus nerve had been removed, the neurochemical and behavioral effects did not occur, implicating the vagus nerve as a key pathway for the effects of the probiotic.[25] In addition, cutting the vagus nerve in mice reduced proliferation and survival of newborn cells, decreased the number of immature neurons and the activation of microglia in the hippocampus – all conditions that can also be found in psychiatric disorders. The vagus nerve has also been implicated in psychiatric disorders in humans, particularly depression, and vagus nerve stimulation has been used to treat depression that won't respond to less-invasive measures. In rodent models, at least, vagus nerve stimulation has been shown to increase adult hippocampal neurogenesis, increase expression of the growth factor BDNF in the hippocampus, and modulate the release of serotonin, norepinephrine, and dopamine in brain regions related to anxiety and depression.[26]

There are other possibilities as well. Since only 5 percent of our serotonin is synthesized in the brain and the rest in the body, it is important that the small fraction of the tryptophan necessary for the brain's serotonin synthesis actually makes it there. Our body's serotonin is synthesized by some of our own gut cells, so the fact that some gut microbes are also producing serotonin implies that they make additional demands on available tryptophan. Serotonin availability in the brain can also be affected indirectly, too. Some products of gut microbiota increase the activities of particular enzymes that break down tryptophan along the kynurenic pathway instead of the serotonin pathway. Kynurenine can cross the blood-brain barrier, where it presumably can be further metabolized into the neuroprotective kynurenic acid or the neurotoxic quinolinic acid. Plus, some gut bacteria can synthesize tryptophan; something our bodies cannot do.[27] Studies with germ-free mice have shown higher serotonin concentrations in the hippocampus of male mice, plus higher levels of tryptophan and a lower kynurenine-to-tryptophan ratio in the blood for both sexes,[28] implying that gut microbiota do affect neurotransmitter production in the host's brain, with consequences for mood, neurodevelopment, and neurotoxicity.

FROM MOUSE TO MAN/WOMAN

It is reasonable to wonder how directly experiences with mouse gut microbiota translate to the human condition, but the links are there. In one interesting experiment, gut microbiota were taken from fecal samples of five non-medicated major depression patients and used to colonize the guts of germ-free mice. Separately, gut microbiota recovered from the fecal matter of five healthy controls were used to colonize the guts of other germ-free mice. After the second week, the mice colonized with microbiota from depressed donors were showing the classic mouse signs of depression and anxiety-like behavior in laboratory tests when compared to the mice colonized with healthy-donor microbiota.[29]

With such experiments able to induce depression-like behaviors in mice using human gut microbiota, it seems like we ought to be able to isolate which gut microbiota are related to the development of depression. Unfortunately, the many studies looking for the answer tend to come up with very different, often opposing, results. Part of the issue seems to be that "normal" human gut microbiota varies greatly among healthy individuals, with each of us hosting about 160 of the over 1,000 species known to inhabit the gut. Plus, human microbiota samples are taken from fecal matter, and what comes out the end is not necessarily the same as what is active in the middle. Several studies point to an overall reduction in *Bacteroidetes* and association of *Lachnospiraceae* (in the *Firmicutes* phylum) with depression, but even that association is tenuous and goes the other way in some studies.[30]

Some probiotic and prebiotic supplements attempt to influence the composition of the gut microbiome. A "probiotic" contains living microbial populations with the intent of establishing them in the intestines. A "prebiotic" contains indigestible (to humans) fiber; it is intended to feed the microbes so that they produce desired metabolites. As they say, we are what we eat, and our gut microbiota share that fate as well. After our gut microbiota are established at birth, they reach a characteristic profile by 3 years of age that remains relatively stable through adulthood, until aging wreaks its havoc on them as well as on us. In adulthood, diet is the strongest driver of change in our gut microbiota profile. More than 50 percent of the variation of gut microbiota has been related to dietary changes, and major changes in diet during adulthood can modify the microbiota in a matter of days.[31]

Some human studies on the effects of probiotics show encouraging results. A 2011 study that administered *Lactobacillus helveticus* and *Bifidobacterium longum* to 25 healthy human volunteers for 30 days found it decreased their depression and anxiety scores compared to controls.[32] This effect was also seen in functional imaging of 12 healthy women given a fermented milk product containing *Bifidobacterium animalis*, *Streptococcus thermophiles*, *Lactobacillus bulgaricus*, and *Lactococcus lactis* twice daily for 4 weeks. The women, along with 11 controls who took an unfermented milk product matched for taste but without probiotics, and 13 others with no intervention at all, underwent resting state fMRI and further imaging as they completed a task that probed attention to negative context (faces displaying fear and anger), before and after the 4-week intervention. The study concluded that the probiotics reduced the women's responses to negative images, and affected activity of brain regions that control central processing of emotion and sensation.[33]

There have been several studies investigating the effects of dietary fiber (prebiotics) on health and mood. Numerous studies support the idea that diets rich in plant fiber may promote the diversification of gut microbiota, with overall benefit to the human host. One study showed that three diets with different fiber-rich whole grains (barley, brown rice, or combination of both) increased microbial diversity and the *Firmicutes/Bacteroidetes* ratio. Further, the administration of whole grain barley induced an increase in *Bifidobacteria* which is considered a positive indicator of prebiotic activity.[34] In another study, healthy volunteers taking Bimuno-galactooligosaccharides (B-GOS) for 3 weeks had a significantly lower salivary cortisol awakening response compared with volunteers on placebo, and also showed decreased attentional vigilance to negative versus positive information in a task battery compared to the placebo group.[35]

There is still a very long way to go before anyone understands what the microbiome even is, much less manipulates it to intervene in a complex psychiatric disorder. Dr. Schellekens commented, "It becomes more clear now that it is the whole combination that you have, and not necessarily who's there but what they do and what do they make. And it could be the ones you can't detect yet. What we're doing now in microbial analysis is going deeper and deeper. We used to do 16S sequencing, which would tell us something about the composition, but now we're looking more at whole genome sequencing to be more in detail of if you have a particular species, what types of strains do you have, and variations thereof. And bacteria can change, really quickly, you know, with horizontal gene transfers. We're talking now about the microbiome, which includes the microbiota, the bacteria, but also the phageome – the phages and the parasites – and everything. We have to see the microbiota and human cell, the human physiology, as a whole culture, a whole organism, a holobiont, beyond all the genes and mechanisms combined. So it is very exciting, but we're still at the early stages of this new field."

It is exciting to think of beneficial applications coming from this research. Dr. O'Mahony, who works with perinatal stress and the microbiome, gets to see the eventual results of her studies come into use in the form of infant formulas and prebiotics. Dr. Schellekens, though, commented on the downside of that excitement: "We have to be careful, because at the moment it's such an exciting area that people say, 'Oh! It's my microbes!' and step away from it. And also, you have videos you can find on-line of people doing their own home-made fecal microbiota transfer, which, of course, carries *huge* risk of infection. This is really, really bad. But, I guess, the good and the bad go hand-in-hand together when there's an exciting scientific frontier happening."

Chapter Seventeen

UNSEEN

"I'm here," read Josef's text. "I'm in a yellow shirt."

It was a beautiful, sunny, holiday weekend morning, and the coffee shop was packed. There were several men wearing yellow shirts inside, and I kicked myself for not having set up a more definitive meeting location. Josef would be my fifteenth interview with someone suffering from depression, and it was driven home to me, yet again, that I could never pick them out of a crowd. Every time, I came to the appointed location and looked around for someone of the right sex who was depressed, and every time I fail to identify them; they find me.

No difference here. The first yellow-shirt man I approached said, somewhat apologetically, no, he was not Josef. Another yellow-shirt man on the other side of the room saw my approach and rejection, and made his way over to me.

It is hard to say why I was still surprised at failing to recognize number 15, or would be again with others. I suppose that even when you have depression you still do what you've always done. If you're going to work (or meeting with a writer for an interview), you get dressed and you get out there and do what you have to. The CDC's 2013 to 2016 National Health and Nutrition Examination Survey reported that 8.1 percent of U.S. adults had depression in any 2-week period.[1] That implies that one out of every 12 people you pass on the street or see in a grocery store, a classroom, or an office, is currently depressed. I don't think I am the only one who can't see it; I think this is one of the great diseases that hides in plain sight.

Josef's schoolmates, co-workers, and parents didn't see it. "It's been my whole life that I've been dealing with it," he said. "I first noticed when I was in high school that something just triggered something in my brain, and I've had it and been struggling ever since. It wasn't so much a mood, but a sense that I was off, not really socializing; isolated. I felt like it was my fault for having these symptoms and illness. I hid it from everybody, pretending that things were fine, but I had problems in school; I couldn't finish certain things. I couldn't finish college."

One summer, still in college, Josef told his mother what he was experiencing and she took him to see a psychiatrist. The psychiatrist diagnosed depression and put him on Zoloft, and he has since been diagnosed with anxiety disorders as well. "I didn't find much relief," he commented, "but it kind of took the edge off."

Josef is still looking for the right medications, or treatment of some sort; his depression and anxiety have been chronic and unremitting for 2 decades now. And, for that time, he's hidden his illness. "I went from job to job," he said. "I can't hold down a job for very long; I feel overwhelmed with even simple tasks. I can't concentrate, and 8 hours seems like forever. I really don't want to tell my co-workers how I'm feeling, though, so I kind of just have to bear with whatever is going on. But I have to leave early for doctors' appointments, and finally they'd figure out something was wrong."

Even his family remains largely in the dark about what he is going through. "They don't get it," he said. "They think I should be doing more, or going to therapy more, doing this and that. I try not to talk to them. I just say 'I'm feeling bad.' They're still like, 'Oh just get over it, go for a walk. Think positive.' Stuff like that. It's just not the right thing to say sometimes."

It's been a long search for relief, for a normal life. "I'm still looking for the right meds; I've been on like seven meds now," he said. "I'm on something for a couple weeks then they change it; increase it or add something. I do feel hopeless a lot, and I just take a bunch of pills every morning and hope for the best. I try not to rely on them; I just take them and try to do the best I can." He even tried a few of the procedures available. "I saw a new doctor, and she said, 'Let's try TMS and see how it goes.' Afterwards I think I was doing better than I realized, but she was like, 'Do you want to do ECT?' And I said, 'OK, sure.'"

In 2012, Josef had 20 TMS sessions followed by 16 sessions of ECT. It definitely impacted his memory, but he knows that, for a while, he did feel better. "Honestly, 2012 was a good year for me. I'm a furniture builder, a woodworker. I take pictures of my projects before I sell them, so I keep a journal of every project. That's how I know I'm doing well: when I'm busy with my projects or when I work all day and I'm not fatigued. I'm motivated, I guess. I was doing good that year. When I look back now I'm like 'Wow, I made all that, and sold all that.' I'm kind of trying to get back to that point where the illness is down here" – he gestured toward the floor – "and my productivity is like here" – a wave in the air.

Josef feels his illness is inherited. "It runs in the family on both sides. It's more, I'd say, genetic, because I don't have a source. It's just there. It just all affects you physically where you feel very sluggish," he said. "Sometimes I don't like to travel too far outside my home. It's hard to communicate with other people due to my social anxiety. I'm still wary of meeting new people; I'm a very isolated person."

He's tried psychotherapy but didn't find it helpful. "Mine's more like a physical sensation," he said, "Like my illness, it's not more therapy that's going to bring me out of it, I feel; it's more on the chemical, somatic side."

Instinctively, he may have stumbled on the best way to manage his symptoms. Rather recently, Josef noticed that if he goes to bed early and gets up early, he feels better and is productive at least until noon. That makes me wonder if he is one of the depression-sufferers with low cortisol, and he's harnessing that morning surge that peaks around 8:00 am. Or maybe he's managing his circadian rhythms through partial sleep deprivation.

The ironic thing about our discussion is that at the same time I was thinking, "I just can't see who has depression and who doesn't;

it's like they're hiding in plain sight," it turns out he was thinking the opposite. "It affects your whole life, like you can't work sometimes, and you can't communicate very well," he said. "You feel like it's obvious to other people. You think that everybody knows."

The summary conclusion of the scientific community is that vulnerability to depression comes from the interaction of genetic and environmental factors, and that early life adversity is a very powerful environmental factor. Recent advances highlight the influence of other forces operating in the current day, however, like diet. Are the latter-day influences strong enough to overcome an unfortunate genetic or early life legacy? The heavy prevalence of depression throughout the world provides an opportunity to probe for an answer. Though diagnostic criteria and record-keeping are far from consistent across every country and timeframe, data collected over decades shows meaningful patterns reflecting the strength of some genetic and environmental factors. Comparing the factors that don't change (like sex or the population genetics of a community in some part of the world) to those that do change (like when a chunk of that small community immigrates to a different country with different local microbes, diet, and social practices, it tells us something about depression.

SEX

Across all countries, ethnicities, and cultures studied, women are much more likely to have major depression than are men, with 1.5 to 3 times the prevalence. This predominance holds for the lowest measured countries like Taiwan (female prevalence 1.8 percent, male prevalence 1.1 percent) to the highest, like France (females 21.9 percent, males 10.5 percent).[2] In comparison, bipolar disorder has a much lower prevalence overall and is usually split about evenly between men and women. One of the broad epidemiological studies documenting these data in the U.S. was the National Epidemiological Survey of Alcohol and Related Conditions (NESARC). This 2001 to 2002 survey involved face-to-face interviews with 43,093 people aged 18 years and older.[3]

The NESARC found the lifetime prevalence of major depression in the U.S. to be 13.2 percent overall: 17.1 percent for women and 9.0 percent for men. The 12-month prevalence* was 5.3 percent overall, including 6.9 percent for women and 3.6 percent for men. Both sexes reported the same average age of onset (30 years), the average number of lifetime episodes (4.7 for women, 4.6 for men), and median duration of longest episode (about 22 weeks). The only significant difference was that the women sampled were 15 percent more likely to be treated for the disorder than were men.[4]

A female preponderance of the disorder could reflect functional or structural differences in the brains of men and women, actions of sex hormones like estrogen and testosterone, and/or differences in neurochemical systems such as serotonin, vasopressin, and oxytocin.

The sex hormones affect how the brain develops: a man's brain ends up structured differently than a woman's. Some sex differences affect multiple areas of the brain, such as ratios of white matter to gray matter and thickness of the cortex. Sex differences are seen in every brain lobe, including several areas strongly implicated in depression: hippocampus, prefrontal cortex, and amygdala.† The structure of the hippocampus differs between males and females, and when adjusted for total brain size, a woman's hippocampus is larger than a man's. The amygdala goes the other way – larger in men than women relative to total brain size. In the prefrontal cortex, imaging and lesion studies have produced evidence that that in men, but not women, lesions in the right hemisphere of the prefrontal cortex impair performance on decision-making tasks; in women, but not men, left hemisphere lesions impair decision-making performance.[5]

---

* The 12-month prevalence is the proportion of people who had started a major depressive episode within 12 months of the survey

† More recent meta-analyses of MRI studies dispute the findings of relative volume differences between men and women in the amygdala and hippocampus, and it will probably take more studies with better machine resolution to really decide.

There are also sex-based differences in a variety of neurochemical systems, including GABA, vasopressin, opioids, serotonin, and other monoamines. Women have significantly higher monoamine oxidase levels in several brain regions compared to men. Sex differences are reported in rates of serotonin synthesis in healthy human brains and in levels of serotonin metabolites in postmortem tissue as well.[6]

The neurochemicals oxytocin and vasopressin have received more attention recently in association with depression. Oxytocin is released during positive social interactions; it calms anxiety and tones down HPA axis over-activity, affecting the stress response. As a "pro-social" hormone, it increases support within a group; however, that increased cooperation and trust exhibited for group members is accompanied by increased distrust and defensiveness towards out-group members. Vasopressin is implicated in animal models in male-typical social behaviors, including aggression, pair-bond formation, scent marking, and courtship.[7]

Though there are intriguing links between both substances and factors underlying depression, the picture is incomplete. Studies of oxytocin levels in people with depression have produced confusing and contradictory results – sometimes higher than normal, sometimes lower, sometimes showing greater variability either direction compared to healthy controls. Oxytocin is related to hormone interactions involved in the stress response, monoamine activities, growth factors, and inflammatory processes. Oxytocin and CRH (which initiates the stress response) may have reciprocal effects – if one increases the other decreases – and administration of oxytocin reduced stress reactions in humans. Evidence for a role of vasopressin in depression is also sparse and inconsistent. Both vasopressin and oxytocin modulate social behaviors, including social bonding, and so it is possible that they have only indirect roles in depression – that they help form supportive social connections that may be protective against depression.[8]

So, being born with two "X" chromosomes confers higher odds of having major depression as an adult, and since science points toward impacts on the developing brain, that's not going to change in anyone's lifetime. One's race also doesn't change; however, one's nationality and location certainly can.

NATIONALITY AND RACE

Epidemiological surveys around the world show differing incidence of major depression by region and country, though considering the different languages, cultural norms, methodologies, and instruments at work, these study results are more meaningful for relative standings than for absolute values. In the early 2000s, the WHO coordinated community epidemiological surveys involving 60,463 adults in 18 countries, using a common instrument to consistently measure lifetime and 12-month prevalences of psychiatric disorders.[9] The high income countries surveyed (Belgium, France, Germany, Israel, Italy, Japan, The Netherlands, New Zealand, Spain, and the United States) had an average lifetime prevalence of major depression of 14.6 percent of their populations, and an average 12-month prevalence of 5.5 percent.* Within this group, the highest lifetime prevalence was seen in France, at 21 percent, followed by the U.S. at 19.2 percent. The lowest lifetime prevalence was measured in Japan, at 6.6 percent, with Germany and Italy tied for next-lowest at 9.9 percent. The low-to-middle income countries surveyed (Brazil/Sao Paulo only, Colombia, India/Pondicherry region, Lebanon, Mexico, China/Shenzhen only, South Africa, and Ukraine) had an average lifetime prevalence of major depression of 11.1 percent of the population and 12-month prevalence of 5.9 percent. Within this group, the highest lifetime prevalence was measured in Sao Paulo, Brazil (18.4 percent) followed by Ukraine (14.6 percent), and the lowest was measured in Shenzhen, China (6.5 percent), followed by Mexico (8 percent).[10]

The most current and extensive data on rates of major depression around the world can be reached through the Global Health Data Exchange, provided by the Institute for Health Metrics and Evaluation, an independent global health research center at the University of

---

* These studies typically measure the occurrence of major depressive episodes, which occur in bipolar disorder as well as major depressive disorder, so some amount of the former is mixed in with estimates of the latter.

Washington. Analyzing the most current age-standardized* incidence rates of major depression around the world from that data catalog, researchers confirmed and extended the observation that countries in Asia (most particularly Southeast Asia) and Latin America have the lowest rates of major depression in the world.[11] So what happens when a population moves from one of those countries to somewhere like the U.S., where rates of depression are approximately twice as high as in their country of origin?

In the U.S., the NESARC epidemiological survey reported prevalence of major depression by ethnicities as well as the overall rate. In order of highest prevalence to lowest, depression hit Native Americans hardest (lifetime prevalence 19.2 percent; 12-month prevalence 8.9 percent), then White/Caucasian (lifetime prevalence 14.6 percent, 12-month 5.5 percent), then Hispanic (lifetime prevalence 9.6 percent, 12-month 4.3 percent), Black (lifetime prevalence 8.9 percent, 12-month 4.5 percent), and Asian or Pacific Islander (lifetime prevalence 8.8 percent, 12-month 4.1 percent).[12] So, current U.S. populations of Hispanics and Asian-Americans maintained significantly lower rates of depression than Native Americans or populations originating in Europe, but didn't keep their entire advantage.

THE IMMIGRANT PARADOX

It's an observation that has been consistently reported since the early 1980s, termed the "immigrant paradox": immigrants to the U.S. have lower risk for mood and anxiety disorders than the U.S.-born population of the same national origin. Researchers in Canada took it further: foreign-born populations have superior health profiles than native-born Canadian populations, including lower mortality rates;

---

* The age distribution of a population (the number of people in particular age categories) can change over time and can be different in different geographic areas. Age-adjusting the rates ensures that differences in incidence or deaths from one year to another, or between one geographic area and another, are not due to differences in the age distribution of the populations being compared.

fewer chronic conditions, disabilities, and overnight hospitalizations; and less mental illness.[13]

Apparently, being a child in the U.S. is unhealthy. People who immigrated to the U.S. from certain regions before age 13 – spending a good chunk of their childhood here – tended to lose a lot of their immigrant health advantage, much as if they were born here. Those regions include Mexico, South or Central America (significant for anxiety disorders only), Eastern Europe, Africa or the Caribbean, and Asia. In contrast, the people who arrived at age 13 or older from these locations were much less likely to get any mood or anxiety disorder compared to someone born in the U.S. For any of those regions, people who immigrated to the U.S. at age 13 or older showed a lifetime risk for mood and anxiety disorders of about 8 percent, while their compatriots who were born in the U.S. or arrived at age 12 or younger showed a lifetime risk of about 20 percent for a mood or anxiety disorder. In contrast, immigrants from Western Europe, Cuba, and Puerto Rico showed no significant differences in risk for these disorders no matter where they were born or age at immigration.[14]

To some extent, data from the 2002 to 2003 National Latino and Asian American Study shows the effect on mental health of an immigrant population as it acculturates to the U.S. This survey found that 7 percent of the first generation Asian-Americans (born outside the U.S.) surveyed had met DSM-IV criteria for "any psychiatric disorder" during the preceding 12 months, while 15 percent of the second or later generation Asian-Americans had met those criteria.[15] In comparison, a 2005 estimate of prevalence for "any psychiatric disorder" over a 12-month period for the U.S. population as a whole was 26 percent.[16] This doubled diagnosis rate between immigrants and their U.S.-born children was not due to a language barrier or higher propensity to seek out psychiatric care, as second generation Asian-Americans (born in U.S. and at least one parent is an immigrant) matched their parents' low rate of seeking out any health care while third generation were more than twice as likely to seek out such care.[17] So, Asian-Americans showed lower prevalences of depression and any other psychiatric disorder when compared to the U.S. population as a whole, but being born in the U.S., whether from immigrants or generations later, ate away a lot of that advantage.

The immigrant paradox was disconcerting for many sociologists, who assumed that the process of relocating to a new country with a new culture and unfamiliar language, implied lower socioeconomic status, loss of native support groups, and higher incidence of discrimination would tend to increase the exposure of adult immigrants to many psychological disorders. The studies show otherwise, though. Being born and spending one's early years in a lower income country is more protective against depression than being born and growing up in the U.S. or many other high income countries.

SETTING THE TABLE

One's genetic inheritance and early life exposures go a long way toward shaping how the brain develops and thus how it will respond to experiences throughout the lifetime. By themselves, though, those factors aren't enough to actually cause major depression. That takes interaction with influences later in life – a long period of stress, chronic inflammation… something to push that delicate balance just over the line. Major depressive disorder typically manifests in adulthood, while many other mental disorders have a much earlier age of onset – they seem to be more strongly driven by early life and genetic factors. The vast majority of anxiety disorders and disruptive behavior disorders start in childhood or adolescence; the vast majority of substance use disorders start in late adolescence or early adulthood. In the WHO surveys, though, the median age of onset for major depression in high income countries was 25.7 years of age; for the low-to-middle income countries surveyed it was 24 years.[18]

Though genetics and early life experience may set the table, there is a lot that can be done to affect the course of depression. The broad epidemiological surveys are valuable in that they show what happens across wide segments of a population, but by the same token they don't highlight what happens to the subset of people who take protective measures, like eating a healthy diet, maintaining beneficial gut microbes, staying physically active, and keeping supportive social connections. In those areas, the smaller studies are starting to speak, to show that one can influence the probability and course of depression.

At the same time, other experiences affecting millions of people worldwide show that resisting the illness is worth a lot of effort. Depression is bad enough – the WHO recently ranked it as the single largest contributor to global disability,[19] and depression's cumulative impact of co-morbid medical illnesses, lack of adherence to treatment regimes, and influence on lifestyle factors like drug or alcohol dependence cost its sufferers an average of 8 years of life.[20] But what is even worse is depression's ugliest companion: suicide.

Part Three

IN THE SHADOW OF A MONSTER

Chapter Eighteen

SUICIDE

Appropriately enough, I was working on this section when Carolyn decided to take her own life. She didn't succeed this time either; a different friend picked up on it and called the police to request a wellness check. They got to her in time and took her to the hospital, where she was admitted for a week. I found out about this when I was finally able to reach her sister-in-law the evening after her attempt. I had called and texted Carolyn with no answer for over 24 hours, worried about her state of mind.

Learning what she had done left me with long-lasting feelings of fear and guilt. I had not picked up on the fact, when we talked the day before her attempt, that she was about to do this. And I should have; looking back, the clues are plain to see. After a week or two, the anger kicked in. She had tried to put her family and friends through hell. She and I had very different perspectives on what she had done. She had become convinced that suicide was the only way to end her current troubles, and felt that the rest of us would quickly come to appreciate that fact. I saw it as the attempted murder of her mother's daughter.

Of course all those emotions have subsided now, and I revisit them only to acknowledge them and put them aside. I need to talk about the science of suicide, and while it is an emotionally-charged subject, there's a lot to learn by looking it coldly in the eye.

Before I began researching this book, I thought of suicide as a sort of "stage 4" of depression. Suicidality has been a prominent feature of Carolyn's depression from the beginning. I see now that I was wrong in conflating the two. Suicide has different demographics and inheritance, and its own neurobiology with unique risk factors. There is significant overlap, though. About 90 percent of the people who commit suicide have been diagnosed with some psychiatric disorder. More than half – about 50 to 60 percent – are by people in a major depressive episode with either unipolar or bipolar depression. That leaves a significant number diagnosed with something else, and about 10 percent with no known psychiatric illness, even after the fact.[1] Major depression is a very potent risk factor for suicide, but suicide is not a necessary progression of that illness.

SUICIDAL BEHAVIORS

I and Carolyn's other friends and family are lucky that back in 2008 when she had her first suicidal ideation, her physicians knew that such events were not random thoughts; they were a point on a path to suicide. There are several suicidal behaviors short of suicide; where they are recognized, they leave an opening for intervention. Suicidal ideation is on the lesser end of severity; it can be passive, involving thoughts about death or wanting to be dead with any plan or intent, up through active suicidal ideation, involving thoughts about taking action to end one's life with a chosen method, a specific plan, and having the intent to act. It can also be an impulsive flash just before the attempt. When the ideation is more specific, such as making a plan, there is a greater risk of suicide attempt within 12 months. An intermediate stage of suicidality found more often in the young is a self-injurious behavior with no intent to die: non-suicidal self-injury. It can be repetitively cutting, burning, or other injurious behavior, motivated by a desire to relieve distress, feel something, punish oneself, get attention, or escape from a difficult situation. Means must be procured – some way to carry out the act – before a suicide attempt, which is a potentially self-injurious behavior associated with at least some intent to die. And then the final stop on the trail is a completed suicide.[2]

Though all of these behaviors can progress rapidly from one stage to the next, more often they don't. A study of suicidal behaviors using data from the WHO World Mental Health Survey initiative reported a 12-month prevalence for suicidal ideation, plans, and attempts in the adult general population (of survivors, of course). Across the 21 countries surveyed, the 12-month prevalence was 2.1 percent for suicidal ideation, 0.7 percent for planning suicide, and 0.4 percent for attempting suicide. In 2005, the mid-point of data collection, rates of completed suicide in the countries surveyed had ranged from 3.1 per 100,000 population in Lebanon to 34.5 per 100,000 in Ukraine, with a median suicide rate of 11.8 per 100,000, or 0.0118 percent. So the vast majority of people do not progress from one stage to the next, or at least leave a long trail of less-severe behaviors.[3]

People with depression are much more likely to die of something other than suicide rather than suicide itself. In 1970, a very influential review of studies of suicide concluded that 15 percent of all patients with depression end their lives by suicide. In 2000, however, a new team re-examined the evidence and noted that the earlier study had included only hospitalized inpatients and that changes in the definition of major depression had opened the aperture to include many people with a less severe form of the disorder. In 1970, depression was defined in DSM-II to include "only involutional melancholia, the unipolar form of manic depression, psychotic depression, and 'severe depressive neuroses'," and the lifetime prevalence of major depression in the U.S. was assessed to be 2 to 3 percent of the population.[4] So, the population studied had been limited to the most severe cases and was not representative of major depression as diagnosed today.

By 2000, with the much broader DSM-IV criteria in place and a lifetime prevalence of major depressive disorder between 10 and 20 percent of the U.S. population, it was time to re-assess. Adding newer studies to the ones used previously, a research team identified tiers of suicide risk. The highest prevalence of suicide was seen in patients who had ever been hospitalized for suicidal behavior: an estimated 8.6 percent of those patients ended their lives by suicide. For affective disorder patients who had been hospitalized but without the specification of suicidality, the prevalence of suicide was 4 percent. Affective disorder outpatients* were estimated to end their lives by suicide in 2.2 percent of the cases.[5]

By starting with hospitalization, though, both the 1970 and 2000 studies neglected those who died on a first attempt. To address that gap, the team added a new statistic based on county coroner and medical records of a community sample in Minnesota, covering data from 1986 to 2010. There were 1,490 first suicide attempts during that 24-year period in that population, and 81 individuals, or 5.4 percent of the attempters, died by suicide during the study period. Of the completed suicides, about 60 percent were successful on the first attempt. Another third of the attempters completed the act within 12 months, leaving less than 10 percent who completed their suicide more than a year after their first attempt.[6] There is a definite, and frightening, suicide risk in someone who is depressed, but attempts are not inevitable and suicide is not inevitable in major depression.

And yet, we can never be satisfied that if someone attempted suicide and failed to complete the act within a year that they are somehow out of danger. With Carolyn, her attempts progressed as she felt out her method and tried to control the world she would leave behind. Her first introduction to major depression in 2008 had been when she impulsively decided she wanted to die, and started walking into her garage to commit suicide by vehicular exhaust. That time she veered off course, though, called her mother, and agreed to call emergency services for help. At the hospital, she was diagnosed with major depressive disorder, prescribed antidepressants and required to participate in first an inpatient and then an outpatient course of psychotherapy. She felt it had all really helped, that she got much better during the course of treatment and that the danger had passed.

---

* Because of the data sets involved, this category includes mostly outpatients, but also some people who were hospitalized for their disorder along the way.

About 3 months later, she made her first suicide attempt. Her psychotherapy course had ended, and her medications were not working. She didn't realize that antidepressants are often not effective in the first dosage given, and that individual responses to the different types vary widely. She was taking her medicine and considered it exactly that: the right medicine for what was wrong with her. But one day in March she woke up and felt she just couldn't take it anymore. She went to her garage and initiated her suicide attempt, but cut it short when she felt nauseous – she did not want her last memories to be of vomiting and did not want to be found that way. She went back into her house, packed a bag, drove herself to the hospital, and asked to be admitted. Again a week of hospitalization, months of psychotherapy, and a new prescription.

Over the years that followed there would be several more attempts, each one interrupted by chance. Finally, during an attempt in 2013, a neighbor noticed something suspicious and called 911; the sheriff's deputy arrived in time to haul Carolyn, unconscious, out of her car and to the hospital. This time she agreed to a course of ECT, which, along with different antidepressant and anti-anxiety medications, relieved her depression and suicidality for years.

The depression crept back, though, along with the suicidality. Moving across the country to be near family and help with her father's care, she began another depressive episode, and about the time she started working with me on this book, she was sinking deeper. She became certain that her life would end by suicide. She updated her will, established beneficiaries for her retirement accounts, and made funeral plans. She also started ketamine treatment and researched other options like transcranial magnetic stimulation. Though ECT had worked for her before, she refused to consider it again. It was during this time that she began to see suicide as inevitable and herself as a burden on everyone. To her, suicide was the clear solution.

On the day of her attempt in the summer of 2019 she had been under ketamine treatment for 6 weeks. She was unable to sleep and her mood had spiraled down into hopelessness. She planned this attempt carefully, taking her dog to the kennel and queuing up a suicide note to be sent by email after a few hours delay. After fortifying herself with anti-nausea meds, she went to her garage, started her car and waited for death. That was when a friend called, disturbed by emails she had received. Carolyn, groggy from incipient carbon monoxide poisoning, answered the phone and admitted what she was doing. The friend called the police. Hospitalization for a week. Counseling. A court appearance, which was new for her, and very unpleasant. She resumed the ketamine treatments.

So what is next for Carolyn? Figure out the flaw in the last plan and correct it? I don't know. Statistics give small odds of her completing suicide at this point, but people defy the odds every day. Statistics are just summaries of what has happened to other people in the past; they aren't a prediction of what will happen to someone else in the future.

DEMOGRAPHICS OF SUICIDALITY AND SUICIDE

Surprisingly, depression, suicidality (suicidal ideation, planning, or attempt), and completed suicide show differences at a population-level view, implying the involvement of unique factors across the three. Globally, the WHO survey data showed that suicidal ideation, planning, and attempts were more prevalent in women than in men, echoing the differing prevalences of depression between the sexes. The gender ratio reverses itself when going from suicide attempt to completed suicide, however. Throughout the world, about twice as many men as women die by suicide. The WHO survey data also showed that the odds of ideation and attempt decreased with increasing age, such that the youngest group in the survey (age 18 to 34 years) showed the highest propensity for ideation, planning, and attempts, and the oldest group the lowest. In contrast, the prevalence of depression increases with increasing age, reaching its peak between ages 60 to 64 years.[7] And some Asian countries with a lifetime prevalence of major depression typically measured at about half that of the U.S. – Japan and the Republic of Korea (South Korea) – have significantly higher suicide rates, at 15.1 per 100,000 population for Japan and 21.4 per 100,000 for South Korea, compared to 13.3 in the U.S. (2015 statistics).[8]

The WHO survey data and independent studies show that suicidal behaviors are strongly associated with the prior presence of a mental disorder. Their data ranked mood disorders (major depressive, dysthymic, and bipolar disorders) first, then impulse-control disorders (intermittent explosive, attention-deficit hyperactivity, conduct, and oppositional defiant disorders), then anxiety disorders (panic, agoraphobia, generalized anxiety, specific phobia, social phobia, post-traumatic stress, and adult separation anxiety disorders), then substance abuse disorders (alcohol/illicit drug misuse or dependence) in order of strength of association.[9]

In the U.S., CDC data also shows suicide's signature by gender, race, and age. Totaling data from 1999 through 2018 in the CDC WONDER database – over 740,000 suicide deaths – the male suicide rate was 19.6 per 100,000 population, and the female suicide rate was 5.0 per 100,000. By race, the group with the highest suicide rate was Native American/Alaskan Native at 16.4 per 100,000 population, then non-Hispanic White at 14.6, then Hispanic at 6.4, then Asian/Pacific Islander at 6.0, then Black/African American at 5.7. Males and females show different patterns of suicide risk by age groups. In women suicide reached its highest rate between ages 45 and 54, with a steep rise and decline. For men, it climbed sharply to plateau by ages 20 to 24, and leapt up again from about age 75 on.[10]

There can be nothing more distressing than the suicide of a child; sadly, suicide is the second leading cause of death among children aged 5 to 19 years in the U.S., accounting for 35 percent of the deaths in that population in the year 2018 (the most recent year of data in the CDC database).[11] Across all the years of CDC data, young females in the U.S. saw a rapidly rising rate of completed suicides until about age 16. From that point, the rate continued to increase with age but at a slower pace, reaching 5.6 per 100,000 population by age 30. For males in the same age groups, the rates of completed suicide grew explosively starting in the early teen years until age 21, where it flattened out and showed no more significant growth through age 30. That top rate, however, stayed between 22 and 23 per 100,000 population, about four times the rate in women of the same age.[12] Just like in adults, approximately 90 percent of all youth suicides are diagnosed with a psychiatric disorder, and the highest risk factor is a previous suicide attempt.[13]

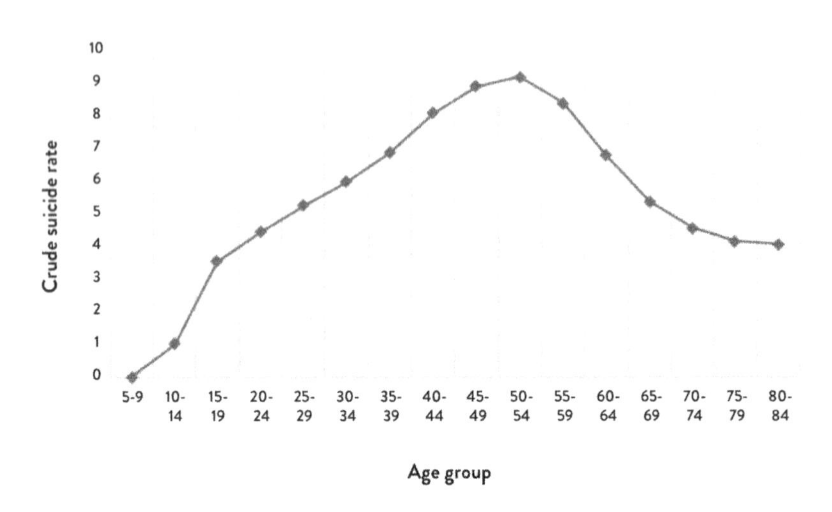

Figure 18-1: **Crude rate of death by suicide in U.S. population by 5-year age groups 1999-2018, (top) females; (bottom) males.** From Centers for Disease Control and Prevention, National Center for Health Statistics. Underlying Cause of Death 1999-2018 on CDC WONDER Online Database, released in 2020. Data are from the Multiple Cause of Death Files, 1999-2018, as compiled from data provided by the 57 vital statistics jurisdictions through the Vital Statistics Cooperative Program. Accessed at http://wonder.cdc.gov/ucd-icd10.html on May 21, 2020.

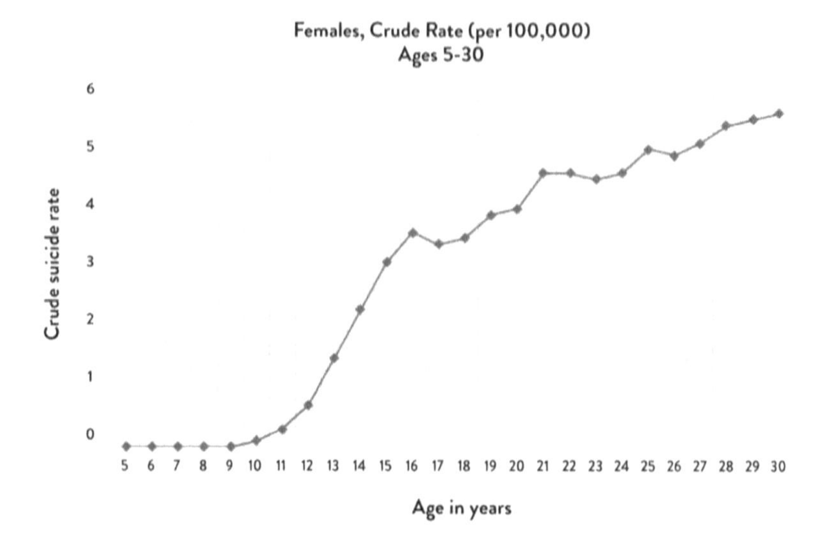

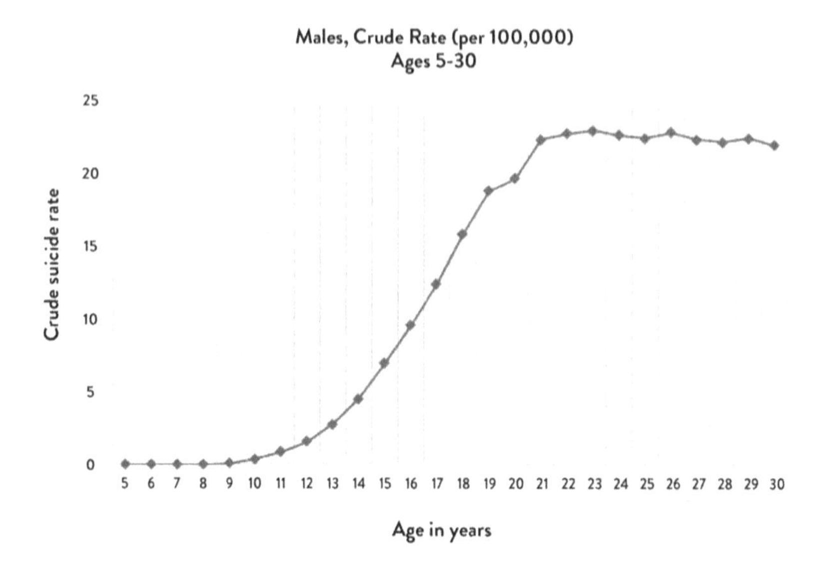

Figure 18-2: **Crude rate of death by suicide in U.S. population by 1-year age groups, 1999-2018, ages 5 to 30 years for (top) females; (bottom) males.** From Centers for Disease Control and Prevention, National Center for Health Statistics. Underlying Cause of Death 1999-2018 on CDC WONDER Online Database, released in 2020. Accessed at http://wonder.cdc.gov/ucd-icd10.html *on May 21, 2020.*

Looking across from depression to suicidality to suicide we can see there are very significant differences from one end to the other by gender and age, and differences of some degree by nationality and race. It may be that the three simply aren't statistically comparable because one can have many major depressive episodes and many instances of suicidal ideation and attempt, but only one completed suicide per lifetime. Research is starting to identify other significant differences between depression and suicide in genetics and neurobiology, though, affirming that they are not the same. It is not clear to which extreme the intermediate suicidal behaviors – ideation and attempt – belong. And since these behaviors are not the same, are there additional points to intervene, to prevent suicide even if a battle with depression is running a slow course?

HERITABILITY

Suicidal behaviors do run in families – they have a genetic component, and the genes involved do not appear to be the same as the ones involved in depression or bipolar disorder. As with depression, heritability factors have been deduced from family, twin, and adoption studies.

Because psychiatric illness is diagnosed in such a high percentage of suicide victims, it is reasonable to wonder if researchers are seeing the heritability of a psychiatric illness rather than heritability of suicidal behavior. Family studies help to disentangle the two factors. Records kept by the Old Order Amish of southeastern Pennsylvania show psychiatric illnesses clustering in some families over many generations. In some of those families there are also clusters of suicides across generations, while in other families with the same psychiatric diagnoses there are no suicides. Other studies that compared suicide rates in psychiatric patients with particular diagnoses found that patients with a family history of suicide were about four times as likely to commit suicide as patients with the same diagnosis and same clinical history but no family history of suicide.[14]

Twin and adoption studies help to peel out the degree of heritability for suicide and suicide attempts. A 2001 review of five twin studies showed that 15 percent of identical twins (who share 100 percent of their DNA with their co-twin) committed suicide following the suicide of their co-twin, while only 0.7 percent of the fraternal co-twins (who, on average share 50 percent of their DNA with their twin) did so. The much higher rate in identical twins did not appear to come from greater bereavement, because the risk of a suicide attempt following the non-suicide death of a co-twin was similar among identical and fraternal twins.[15] In 1979, researchers using a Danish registry of adoptions identified 57 cases of adoptees who had committed suicide, and matched them to a control sample of 57 other adoptees. When biological families were traced, there were significantly more suicides among the biological relatives of the suicide adoptees compared with the biological relatives of the control adoptees.[16]

All in all, estimates of the heritable component of suicidal behaviors range from 30 to 50 percent; a wide range indicating the uncertainty posed by the presence of heritable psychiatric illnesses. Those heritable psychiatric illnesses also add to the story, though. Researchers are finding that risk of suicidal ideation tends to be inherited along with a mood disorder, while higher risk of completed suicide is inherited along with impulsive aggression or the sort of personality traits seen in antisocial, borderline, histrionic, or narcissistic personality disorders – a tendency toward dramatic, emotional, or erratic responses to life events.[17]

Attempts to identify the particular genes involved in suicide have proceeded much like the genetic studies of major depression. Researchers began by focusing on candidate genes but are moving now toward the more objective GWAS approach. By 2017, 755 genes – almost 4 percent of the human genome – had been associated with suicide risk with some level of statistical support. An analysis of DNA samples from 4,585 suicide victims in Utah identified 43 families at significantly high risk for suicide, and revealed 207 gene variants with genome-wide significance.[18] (None of them matched any of the 153 genes identified in the recent major depression GWAS.) A GWAS involving 3,413 completed suicides and 14,810 matched controls identified 32 genes of significance, with functions including bioenergetics (mitochondrial activity), neuronal development, inflammatory responses, and methylation (an epigenetic process). Again, no matches to the depression GWAS genes. Just as in depression, though, researchers see a polygenic process at work, where multiple genes add together to heighten vulnerability to suicide.[19]

NEUROBIOLOGY OF SUICIDE

As noted before, more than half of all suicides are by committed by people in a major depressive episode, which makes it challenging to assess what neurobiological factors apply to suicide instead of depression. Extensive research supports the idea that there are factors that are common to both major depression and suicide, though there is often a particular direction seen in one that is not observed in the other. There are also factors that appear to be unique to suicide, not observed in major depression alone.

Low Serotonin

Serotonin in our brains affects mood, cognition, anxiety, and much more, and there is a lot that is unknown about the full repertoire of the serotonergic system. Depending on the type of receptor activated and its location, release of serotonin can prompt good mood or bad, or a variety of other effects. In suicidal behaviors, serotonin metabolism takes a particular slant: a low serotonin level is strongly associated with suicide.

Because scientists cannot directly measure levels of serotonin in the brain, they measure 5-hydroxyindoleacetic acid (5-HIAA), a break-down product of serotonin that makes its way into the cerebral spinal fluid. Both prospective and postmortem studies have shown that completed suicides tend to have lower 5-HIAA levels in their cerebral spinal fluid years before they died. A meta-analysis of such studies estimated that the odds of suicide completion in someone with major depression are four times higher in the low cerebral spinal fluid 5-HIAA group than in the group with high 5-HIAA levels.[20]

Studies of suicides of patients with other psychiatric illnesses such as schizophrenia or a personality disorder without concurrent major depression provide more evidence of low cerebral spinal fluid 5-HIAA, indicating that the biological trait belongs to suicidal behaviors rather than to depression. Most postmortem brain studies have found that suicides have lower levels of serotonin and/or 5-HIAA in the brainstem serotonin neurons compared with matched groups of patients with the same psychiatric diagnosis but no history of suicidal behavior. Low cerebral spinal fluid 5-HIAA has also been shown to correlate with lifelong aggressive behavior and can predict acts of impulsive aggression like arson or homicide.[21]

Dr. John Mann of Columbia University is a world-renowned expert on suicide and serotonin. "The cerebral spinal fluid 5-HIAA actually shows a stronger relationship with suicide and highly-lethal suicide attempts than it does with the presence or absence of major depression," he commented. "Now, when we look at the 5-HT1A autoreceptor in depressed people who make suicide attempts, we find that the more lethal the suicidal behavior, the higher the 5-HT1A autoreceptor binding. So, just as you find a relationship between suicide attempt lethality and the level of cerebral spinal fluid 5-HIAA, you also find a relationship or correlation between higher autoreceptor binding and the degree of lethality of suicide attempts." Because 5-HT1A autoreceptors reduce the firing rate in serotonin neurons, higher 5-HT1A binding would lead to lower serotonin levels in the brain.

Inflammation

Dr. Raison of the University of Wisconsin-Madison had commented on the tight relationship observed between inflammation and suicide, and there is a lot of evidence now that neuroinflammation is involved in suicide risk, even without major depression. Postmortem studies of suicide victims saw elevated messenger RNA for pro-inflammatory cytokines and activated microglia in some brain areas, both indicators of neuroinflammation. A recent meta-analysis of studies into pro-inflammatory markers in patients showing suicidal behaviors found significantly increased levels of pro-inflammatory cytokines IL-1b and IL-6 in blood and postmortem brain samples of the patients with suicidal behaviors in comparison to psychiatric patients without suicidal behaviors and healthy control subjects.[22]

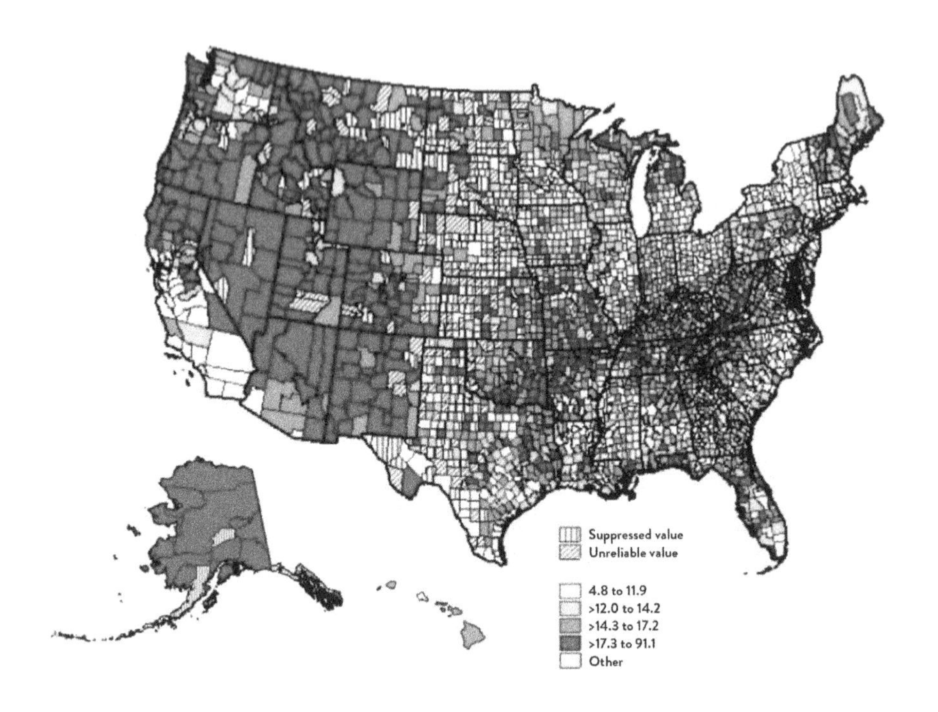

Figure 18-3: **Age-standardized suicide rate by U.S. county, 1999-2017. From Centers for Disease Control and Prevention, National Center for Health Statistics. Underlying Cause of Death 1999-2017 on CDC WONDER Online Database, released December, 2018.** Data are from the Multiple Cause of Death Files, 1999-2017, as compiled from data provided by the 57 vital statistics jurisdictions through the Vital Statistics Cooperative Program. Accessed at http://wonder.cdc.gov/ucd-icd10.html on Feb 2, 2019.

Hypoxia

Suicide also has its unique factors. Recently, researchers mapping suicide rates by U.S. county noticed something startling: people living in the mountainous regions of the U.S. showed noticeably higher rates of suicide death than those living in coastal areas or other lowlands.[23] Suicide data from the CDC's WONDER database paints a picture that somewhat resembles a topological map of the U.S., with the mountainous west painted in dark tones and the same coloring along the Appalachian mountains in the eastern part of the country.

Certainly, this could be a result of small population numbers having an outsized effect, since many more people live near the coast than on mountaintops. And that factor would be especially potent if the mountain-dwellers have a strong representation from the groups with the highest suicide rates in the U.S. population: Native Americans and White/Caucasians. The pattern is not replicated the world over, though there could be many additional, country-specific factors weighing in. Switzerland, a high-altitude country, has an above-average suicide rate; but that of Belgium, a lowland country, is much higher.[24]

In the U.S. at least, studies support the hypothesis that suicide is related to high altitude. An analysis of all U.S. counties matched to altitude using 20 years of county-specific mortality data found a strong positive correlation between altitude and suicide rate. This was in spite of a negative correlation between county altitude and all-cause mortality, and controlling for percent of population over 50 years of age, percent male, percent white, median household income, and population density of each county. A 2018 review of altitude and suicide studies identified six similar studies, four from the U.S., one from South Korea, and one from Turkey. All but the one from Turkey found that individuals living at high altitude have greater risk of suicide; in Turkey, no such relationship was observed.[25]

The mechanism proposed for translating high altitude into higher suicide risk is hypoxia – lack of oxygen. In 1982 Dr. Ira Katz published a paper postulating a "hypoxic affective syndrome" in various illnesses that are accompanied by mild to moderate hypoxia. In particular, he noted that some of the symptoms of depression are also seen as effects of exposure to high altitude. As a possible mechanism for that association, he pointed to tryptophan and tyrosine metabolism, noting that the enzymes that transform those substances into serotonin and dopamine are usually not fully saturated with oxygen. He suggested that the relative poverty of oxygen in hypoxia may decrease synthesis of serotonin, dopamine, and norepinephrine in the brain.[26] Though his paper seems to have been largely ignored in the interim, recently researchers noting the exaggerated rate of suicide in high-altitude counties in the U.S. resurrected his work and started to build out the hypothesis. A 2005 study showed that serotonin synthesis was 50 percent higher on average when study participants were breathing an air mixture with 60 percent oxygen than when they were breathing an air mixture with 15 percent oxygen.[27] With a low serotonin level consistently associated with suicide, reduced serotonin synthesis over an extended time period would be a plausible link from high altitude to suicide.

Seasonality

Another way in which depression differs from suicide is its seasonality. Seasonal Affective Disorder (SAD), a form of depression, is common in winter, particularly in northern latitudes, when low sunlight conditions are thought to affect vitamin D production and circadian rhythms. Depression and depressive symptoms peak in the winter months, and not only for people with SAD. A longitudinal study of people in central Massachusetts, all of whom screened negative for SAD or any other psychiatric illness, found that participants experienced the most depressive symptoms in the winter and least in the summer.[28]

Suicide, however, is more common in the spring and summer. In the U.S., the suicide rate is consistently at its lowest from December through February, rises from March through July and then declines slowly down to its winter low.[29] In Australia, the low season for suicides is May through August – their winter months – after which it rises to its highest point in February. The difference from seasonal high to low is about 10 to 15 percent, but it is still a notable contrast to what is seen in depression. A study using 10 years of mortality data from the state of Victoria in southeastern Australia found the amount of daily sunshine correlated positively with the number of suicides: more sunlight, more suicide death.[30]

Cholesterol

Another possible factor for suicide that does not appear in studies of depression is cholesterol metabolism. When two large, randomized, placebo-controlled trials of cholesterol-lowering statin drugs both showed that lives saved by reduced mortality from coronary heart disease were offset by a higher rate of death from accidents, homicide, and suicide, concerns were raised about possible side effects of those drugs.

Cholesterol had already been linked to a tendency toward violence. In 1979, Finnish scientist Dr. Matti Virkkunen demonstrated that male subjects with antisocial personality disorder – a high risk group prone to violence and suicide – had lower levels of serum cholesterol than a control group of male patients with other personality disorders. A 1990 meta-analysis combining data from cholesterol-lowering drug trials concluded that there was, in fact, a significant increase in mortality from accidents, suicide, or homicide in the treatment groups versus the placebo groups.[31]

The relationship between cholesterol-lowering drugs and violence has been challenged and is more controversial today. Data from more recent statin trials don't show significant increases in violence-related deaths, but pharmaceutical companies have also modified their exclusion criteria to eliminate subjects at risk of such behavior.[32] Making the situation more complex, statins can be either fat-soluble (e.g., atorvastatin and simvastatin) or water-soluble (e.g., rosuvastatin and pravastatin). Fat-soluble statins are known to cross the blood brain barrier, while water-soluble statins are thought to be blocked, though this may depend on dosage.[33] Brain cholesterol is synthesized in the brain and not transported from the periphery,[34] so a statin that cannot cross the blood-brain barrier is unlikely to affect the state of cholesterol in the brain. Thus, the type of statin involved in these trials could also be a factor in the results observed.

There are several studies that associate suicidal behavior with total serum cholesterol. A recent meta-analysis involving over 500,000 patients found that those exhibiting suicidal behaviors had a significantly lower serum total cholesterol than did non-suicidal patients. When compared to the group with the highest total serum cholesterol, the group with the lowest total serum cholesterol had a 123 percent higher risk of suicide attempt and an 85 percent higher risk of suicide completion.[35] Cholesterol is the building block of cortisol, and therefore it is intimately related to the body's stress system. Beyond that link, scientists are still searching for why cholesterol levels would be related to violence or suicide. Speculation covers a role by way of PUFAs, in which lowering cholesterol decreases absorption of omega-3 but not omega-6 PUFAs, thus raising the omega-6/omega-3 ratio and supporting an inflammatory state. Low cholesterol has been linked to low serotonergic activity in animal models, but again, the mechanism is unknown.[36]

CAN WE REDUCE SUICIDE RISK?

Genetics and other factors set the stage for a person's baseline suicide risk, but even families with high suicide rates over generations have many more members who die from something other than suicide than from suicide. Where does the next part of the risk come from? Researchers say that the strongest predictors of suicide attempts are previous attempts and the existence of a psychiatric disorder. More specifically, the highest risk for suicidal behaviors across the lifespan exists when a mood disorder that is associated with suicidal ideation co-occurs with other conditions that either increase distress (such as panic disorder or PTSD) or decrease restraint (such as conduct and anti-social disorders or substance abuse).[37] The key, the nexus for suicide, seems come from the convergence of three elements: a psychiatric disorder, a state of impulsive aggressiveness, and available means – a firearm, a vehicle, controlled medications, etc. Perhaps it is because major depression can provide all three elements at once that it is such a potent risk factor.

So depression, suicidality, and suicide do not all come from the same biological state, and the differences among them do provide a few more points for intervention. A 2016 review of suicide prevention efforts around the world found that restricting access to deadly means, school-based awareness programs, and some medications (lithium and clozapine in particular) were proven effective in lowering suicide risk. Other interventions, such as various forms of psychotherapy and education programs for people likely to come into contact with someone at risk of suicide, were promising as well.[38]

There was one recent development that the suicide prevention review group objected to, though, suggesting that it made suicide prevention harder: the warnings against use of antidepressant drugs for children and young adults issued by authorities in the U.S. and Europe. Their report commented, "Since psychiatric disorders are a major risk factor for suicidal behaviour, their pharmacological treatment contributes substantially to the prevention of suicide... In children and adolescents with depression, evidence does not support avoidance of use of antidepressant medication because of increased risk of suicidal behaviour, although there is evidence to suggest an increased risk of suicidal ideation in this population."[39] With rising suicide rates in the U.S. and around the world, it is easy to see why those warnings remain very contentious, even 14 years after their issuance. Do antidepressants, or does the warning itself, increase suicide risk?

Chapter Nineteen

THE WARNING

There's an argument that plays out in the anti-vaccination movement in America: Which is worse – an error of omission or an error of commission? Would you rather be the parent who refused to vaccinate a child only to see the child contract a deadly disease? Or would you rather be the parent who did vaccinate a child only to see the child suffer a life-threatening adverse reaction to that vaccine? It is a vicious dilemma, for the parents but also for the people charged with providing the factual information and recommendations intended to help them navigate these choices. That same dilemma is faced by the parents of a child with depression.

In May 2003, GlaxoSmithKline submitted an analysis of suicide-related adverse events in pediatric trials of paroxetine (an SSRI) to the U.S. Food and Drug Administration (FDA). Recognizing a statistically-significant higher level of suicide-related behaviors in children taking the active drug over those on placebo, the FDA issued a public warning about the risk of increased suicidality in children and adolescents treated with that medication. They then requested that the sponsors of eight other psychotropic drugs tested in children and adolescents search their databases for similar findings in their drug trials.

A total of 23 pediatric drug trials conducted over a period of almost 20 years fed the resulting analysis, covering five SSRIs, three atypical antidepressants, and an SNRI. The drug trials involved over 4,400 children and adolescents and had durations of 4 to 16 weeks. The results gleaned from those trials, down to the individual data sets containing adverse events, were analyzed and each adverse event categorized by an independent team of suicidology experts. There were no completed suicides recorded in any of the trials. There were, however, 78 reports of definitive suicidal behaviors (27 suicide attempts, 6 preparatory actions for a suicide attempt, and 45 suicidal ideation events). The data was eye-opening: adverse events involving suicidality were reported for 2.1 percent of the children taking the active drug compared to 1.2 percent of the children taking the placebo and 2.1 percent of the few children on an active control arm* of a drug trial. The resulting report, called the Hammad Report after its principal author, was completed in August 2004.[1] Based on these results and with advice from the Psychopharmacologic Drugs Advisory Committee (PDAC), the FDA concluded that there was a statistically-significant increase in risk of suicidal behaviors in the treatment groups compared to the placebo groups, and expanded the warning to encompass all antidepressant medications for children and adolescents 18 years and younger.[2]

In December 2006, the FDA's Center for Drug Evaluation and Research called another meeting of the PDAC to consider whether any sort of antidepressant warning was appropriate for the adult population as well. Using information from 372 placebo-controlled antidepressant trials involving almost 100,000 adult patients, the FDA analysis showed a graded risk of suicidal behavior or ideation in these trials along age groups, again from differences in drug and placebo arms.[3] The analysis indicated that patients under the age of 25 had a higher risk of suicidal ideation or behavior on the drug arm

---

* An "active control" is a potentially effective treatment to be compared to the efficacy of the drug under trial. In this case it could be psychotherapy or a different drug.

versus placebo, while above that point the risk decreased by age, such that the drugs showed statistically-significant protective effects in the oldest populations.[4] The antidepressant safety warning was amended to include young adults through age 24.

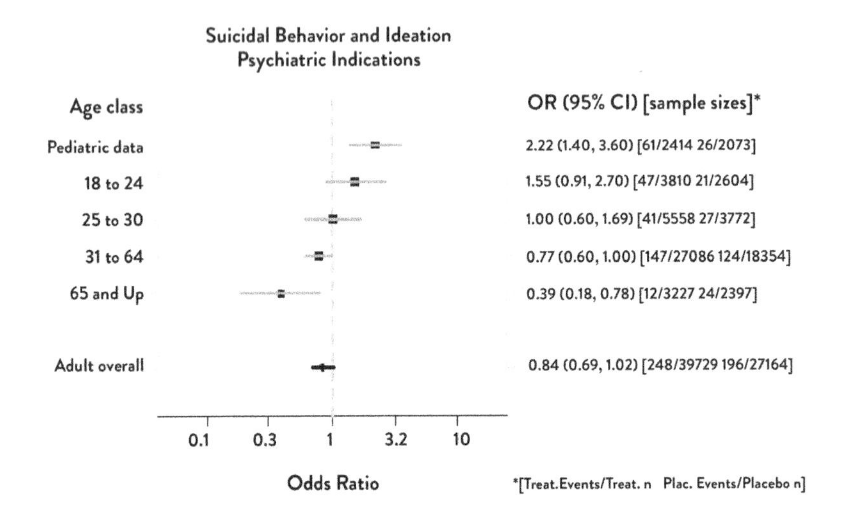

Figure 19-1: **Odds ratios by age group for suicidal behavior and ideation (psychiatric indications).** From Laughren, T. P. (2006, November 16). Memorandum to Members of PDAC. Retrieved from https://wayback.archive-it.org/7993/20170405070114/https://www.fda.gov/ohrms/dockets/ac/06/briefing/2006-4272b1-01-FDA.pdf, pg. 112 of 140.

A "black box" warning is the most serious type of caution the FDA issues, the last stop before banning a drug outright. The FDA was not alone in their concerns; the warning was repeated and amplified by the UK Committee on Safety of Medicines and authorities in Europe.[5]

CONTROVERSY

In the clinical community, at least, the black box warning has been very controversial, and many feel it is a mistake that has cost lives. It had far-reaching and unanticipated impacts. As expected, antidepressant prescription rates for children and young adults went down after the warnings were issued. What was not expected were decreased antidepressant prescriptions for adults over 24 years of age, decreased rates of depression diagnoses in all age groups, and a lack of diversion to alternative treatment modes for young people. In 2009, a paper in the *Archives of General Psychiatry* reported that new diagnoses of depression from primary care providers had decreased 44 percent for children and 29 percent for adults. SSRI prescriptions decreased by 10 percent for children and 15 percent for adults. There was no compensatory increases in talk therapy or pharmaceutical alternatives for children, and only a small compensatory increase for adults. Large and significant declines in depression treatment had spilled over to the adult population, even though the FDA analysis indicated a significantly protective effect against suicidality in older patients. The authors of the article concluded that the unintended effects of the FDA black box warning were "non-transitory, substantial, and diffuse in a large national population."[6]

Other analyses of the impacts of the warnings disagreed, though. One assessment challenged the finding of reduced antidepressant prescription rates for children, noting that there had been a steady increase in prescription rates from 1999 to 2004. The authors reported that the effect seen amounted to reduced *growth* in prescription rates, not a reduced rate overall. They found that after falling off for a short time following the FDA warning, antidepressant prescriptions for children started to rise again, along with the child suicide rate. And in a cost-benefit analysis, they felt antidepressant usage in children just didn't show compelling efficacy. A study of 13 pediatric antidepressant trials showed the pooled absolute response rate for youths treated with antidepressants was 60 percent, compared with 50 percent response rate for placebo – a relatively slender difference.[7] And indeed, a 10 percent response rate over placebo is not impressive... but what does it really mean?

The Placebo Effect

Dating from the earliest antidepressant clinical trials in the 1950s, physicians and researchers noted a significant placebo effect: patients unwittingly taking a non-active substance showed a substantial response rate, usually cited as 30 percent. An antidepressant's efficacy is measured in terms of its effectiveness above the floor value set by the placebo response. So, if 50 percent of the patients taking the active drug show significant improvement and 30 percent of the patients on placebo show significant improvement, then the efficacy of the drug is 20 percent. Though, in actuality, the effectiveness of the intervention was 50 percent in the medication group and 30 percent in the placebo group. It was actually the mechanism of the drug, versus the mechanism of the placebo effect, that had 20 percent more efficacy.[8]

There is a mechanism to the placebo effect – it has a physical basis. There is a naturally-existing opioid system involved in the placebo effect both for pain and depression. Functional imaging studies of pain show that a placebo activates the same regions as an opiate pain reliever, and mu-opioid neurotransmission has been observed in several areas of the brain implicated in depression based on the expectation of relief. These results suggest that there may be a general system for self-regulation that applies to both emotions and pain that can be triggered by either internal expectation – placebo effect – or by external factors, such as a pain reliever or antidepressant.[9]

Being in a placebo arm of a drug trial involves more than taking a pill, too. In his book *The Placebo Effect in Practice*, W.A. Brown notes, "The capsule [patients] receive is pharmacologically inert, but hardly inert with respect to its symbolic value and its power as a conditioned stimulus. In addition, placebo-treated patients receive all the components of the treatment situation common to any treatment, i.e., a thorough evaluation; an explanation for distress; an expert healer; a plausible treatment; a healer's commitment, enthusiasm, and positive regard; an opportunity to verbalize their distress."[10]

At 30 percent the placebo effect was already strong back during the early antidepressant trials, and it has been growing since. A study of placebo responders in 75 antidepressant trials between 1981 and 2000 found that the proportion of patients who responded to placebo increased by about 7 percent each decade. The effectiveness of the medication arms of those antidepressant trials also increased by an equal proportion, indicating that increasing placebo response was lifting response rates for the medication.[11] A placebo is not nothing; it is something. Being in a placebo arm in a drug trial is not the same as being offered no treatment at all, which apparently was happening to many children and young adults with depression, following the warning.

METHODOLOGY

Critics of the FDA black box warning also call out several issues in how the analyses were conducted. The first methodological element they question is giving suicidal ideation the same weight as a suicide attempt or completion. The risks for those behaviors are not equivalent. The WHO worldwide surveys show the adult gender demographics of suicidal ideation and attempts to be opposite that of completed suicide; U.S. surveys show the same result in adolescents as well. The CDC conducts a Youth Risk Behaviors Survey every few years, surveying students aged 13 to 18 years across schools in the U.S. on a wide variety of risky behaviors, including suicidal behaviors. From their first survey in 1991 and continuing to present, their results consistently show high school-aged girls out-representing boys in their responses about seriously considering, planning, and attempting suicide. In the most recent data set, from 2015, 23.4 percent of girls and 12.2 percent of boys reported seriously considering attempting suicide. That ratio of twice as many girls as boys exhibiting suicidal thoughts and behaviors persisted through intermediate behaviors, and on the most severe end of the behaviors surveyed, 3.7 percent of the girls and 1.9 percent of the boys had made a suicide attempt that resulted in an injury, poisoning, or overdose that had to be treated by a doctor or nurse.[12] That same year, according to the CDC WONDER database, 515 girls and 1,252 boys aged 13 to 18 years committed suicide in the U.S. These tragic statistics, including a greater than 2 to 1 ratio of boys to girls dying by suicide, are consistent across the years. So the ratio of twice as many girls as boys showing every degree of suicidal behavior short of completed suicide matched the higher prevalence of depression in females and more than reversed itself when it came to completed acts of suicide.

The FDA's 2004 Hammad Report shows that same female preponderance in the adverse events reported in the pediatric drug trials. Patient data used in the FDA analysis covered the 78 events considered definitive suicidal behaviors, plus 42 "possible suicidal behaviors" (self-injurious behaviors with unknown intent or with no intent, and events without enough information to classify them). The 120 children who had experienced these events included 57 girls and 30 boys on the drug arms and 21 girls and 12 boys on placebo arms of the trials.[13] With no suicides in the studied population, the FDA study may have been measuring frightening and distressing symptoms that go with the underlying psychological disorder rather than an actual higher risk of suicide.[14]

Another criticism of the methodology used is that both treatment arms and active control arms are more likely to generate side effects than are placebo arms of any trial, resulting in more contact with providers and hence more opportunity to report an adverse effect.[15] That implies that had the results been measured as reports of suicidal behavior per contact hour rather than per patient, it may have shown a very different picture. Even the FDA report's lead author, Dr. Hammad, caveated the analysis in a paper summarizing the information used in assessing the need for a warning for children and adolescents:

> *There exist alternative explanations for this finding. First, the apparent increased risk of drug-induced suicidality may actually represent a greater likelihood of reporting of suicidality events by patients rather than an increased rate of the events themselves. Suicidal ideation and attempts are often characterized as secretive in pediatric patients. Several antidepressant agents have been found to be effective in treating social anxiety, resulting in increased verbalization and communication with others. Thus, it is possible that antidepressant drug therapy leads to differentially greater reporting of suicidal thoughts and behaviors in pediatric patients compared with those receiving placebo.*[16]

Dr. John Mann of Columbia University is among the experts who considers the FDA methodology used to determine whether these medications place patients at greater risk for suicide to be flawed. When I asked his opinion on the subject of the black box warning, it was clearly a sore issue. "Makes no sense," he said shortly. "It doesn't make sense. They never found an increase in suicide rates; in fact, the data that there was actually an increase in attempts was not very good. They had to combine attempts with ideation just to get a signal. But – and this is a very big 'but' – they ran those analyses using a quick and dirty method, both in the adult studies and in the kiddie study. They did a meta-analysis by looking at the overall results of a series of clinical trials. They did not do what we advised them to do. We advised them to take the data from all the studies and do a patient-level analysis. In other words, do a little extra work and combine the data from all the studies more properly. That gives you a much more precise meta-analysis." The "we" had been the American College of Neuropsychopharmacology (ACNP). Dr. Mann had chaired ACNP committees that met about the child data and later about the adult data.

I asked, "I don't understand why they are seeing these increases in suicidal behaviors when someone starts antidepressant treatment. Is it a matter of timing?" In my mind I was thinking that maybe when antidepressants first hit the brain, 5-HT1A autoreceptors would be among the earliest to detect the increased extracellular serotonin, and respond by stepping up their binding to turn off serotonin release. After a few weeks, the autoreceptors desensitize and allow more serotonin release to resume, but that intervening period of extra-low serotonin would set up a state of impulsiveness seen in suicide. So I braced myself for a complicated technical discussion of receptor and transporter action.

But his answer was not complicated at all; in fact, he had something staggeringly simple in mind. "The way to understand this best is to look at a study by Simon et al in the *American Journal of Psychiatry*, 2006," he said. "Have a look at that study and you'll see that the highest suicide attempt rate is in the month *before* starting antidepressants. They went back four months before they started the antidepressants, and you'll see the suicide rate jumps up precipitously one month before they start treatment. And then, if you look after treatment, the suicide rate starts coming down. But if you look at Figure 4 in this study, you'll see that the first month, the suicide attempt rate is clearly higher, especially in people under the age of 18, more striking than in older patients. You'll see that the suicide attempt rate in the first month is much higher than in the second month, so that it looks like it's dropping. If you didn't look at the month *before* they started treatment, you'd think, 'Wow! It's really bad in the first month, and then they start to get a bit of benefit.' But really what's happened is the suicide attempt rate in that first month is already one third or less of what it was in the month before. It's a colossal benefit!"

I wanted to slap my forehead. My entire perspective on the impact of antidepressants on suicide risk had started from the time that someone entered a drug trial; I hadn't even thought about what brought them to that point: what motivated them to go to the doctor – or bring their child to a doctor and allow him or her to be used in a drug trial – in the first place. A suicide attempt, or frightening thoughts and behaviors just short of that, would do it.

I found the studies. In 2006, Simon et al analyzed computerized medical and pharmacy data starting 6 months before and extending 6 months after an initial antidepressant prescription in a population of approximately 500,000 members of a health care plan in Washington and Idaho. In that population, between 1992 and 2003, there were 82,285 new episodes of antidepressant treatment. More than 5,000 episodes were among patients aged 17 years or younger. Through medical records and death certificates, the researchers identified 31 suicide deaths and 76 suicide attempts leading to hospitalization during the 6-month follow-up period. But there were also 73 suicide attempts leading to hospitalization during the 3 months before the first prescription, with the highest number for all ages *in the month before* the first prescription. More specifically, the highest number of suicide attempts were in the week before the first prescription. Duh. I guess that's *why* they sought medical treatment. In the 6 months following the prescription, suicide attempts declined each month, while suicide deaths stayed pretty stable following the first prescription.[17]

Like the FDA, I had only considered the data from start of treatment onward, and so I had not realized I was looking at a downward slope; I thought I was looking at a peak. Another study showed that this effect was not limited to medications. Analyzing data from 54,123 adolescent and adult patients being treated with psychotherapy instead of antidepressant medications, researchers again found the highest number of suicide attempts in the month before beginning treatment, then a sharp decrease in the first month of treatment and continuing decreases in the months following.[18] In all the cases, there were suicide attempts in the first month of treatment. However, neither antidepressants nor psychotherapy could have been said to have *caused* those suicide attempts; by all appearances, both treatments had a significantly positive effect in reducing suicidality and became more effective as time went on.

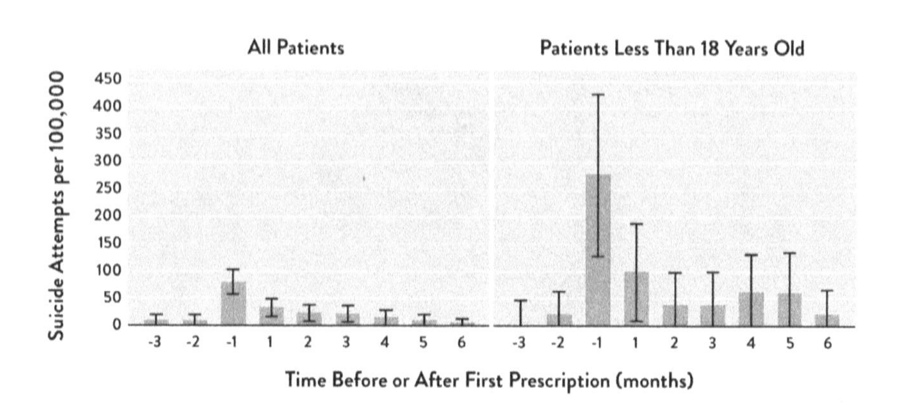

Figure 19-2: **Rates of suicide attempts during 3 months before and 6 months after initial antidepressant prescription. Bars indicate 95% confidence intervals.** From Simon, G. E., Savarino, J., Operskalski, B., & Wang, P. S. (2006). Suicide risk during antidepressant treatment. *American Journal of Psychiatry, 163*(1), 41-47. Reprinted with permission from the American Journal of Psychiatry, (Copyright ©2006). American Psychiatric Association. All Rights Reserved.

Ketamine was not among the drugs tested in 2004 or 2006, but Spravato (esketamine) also carries the black box warning. Carolyn had been treated with ketamine for 6 weeks before she attempted suicide this most recent time, with a treatment just 2 days before her attempt. If I were considering her case using the FDA method, I would put a tick mark in the box that says "She started ketamine treatment and then attempted suicide." But, in preparation for this attempt, she had updated her will, established beneficiaries on her retirement accounts, and arranged for the care of her dog... all completed long before she had even heard of ketamine treatment, much less started it. Using the Simon et al method, I would note that her suicidality existed long before her ketamine treatment started. Ketamine may, in the end, save her life, even if it didn't act quickly enough. Ironically, ketamine has shown promise as a treatment to rapidly reduce suicidality,[19] so we may see indications saying "Use this drug to reduce suicidality in depressed patients" sitting right by the label saying "Caution: this drug may increase suicidality."

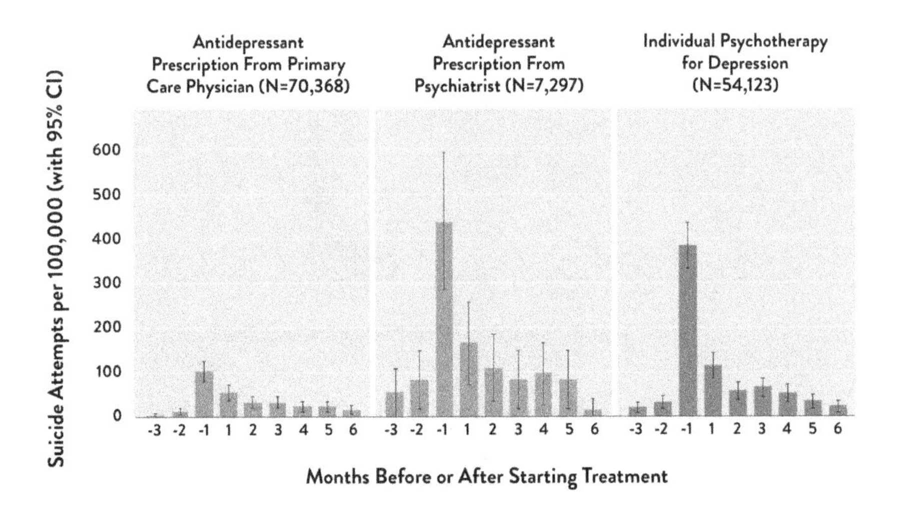

Figure 19-3: **Risk of suicide attempt or possible suicide attempt before and after starting treatment among adolescents and adults (a) receiving new antidepressant prescriptions from primary care physicians, (b) receiving new antidepressant prescriptions from psychiatrists, or (c) starting individual psychotherapy for depression.**

From Simon, G. E., & Savarino, J. (2007). Suicide attempts among patients starting depression treatment with medications or psychotherapy. *American Journal of Psychiatry, 164*(7), 1029-1034. Reprinted with permission from the American Journal of Psychiatry,

I also saw what Dr. Mann meant by his criticism that they should have "combined the data more properly" in a patient-level analysis. In 2012, Gibbons et al published a meta-analysis of 20 randomized controlled trials of fluoxetine across geriatric, adult, and youth populations and 21 adult trials of venlafaxine that did combine more data. In this analysis, each patient got a suicidality score every week; in children, it ranged from 1 (understands the word 'suicide' but does not apply term to self) to 7 (has made suicide attempt within the last month or is actively suicidal). Adverse events like self-injury were included in the data when they occurred as well. Notably, the analysis concluded that there was no evidence of increased suicide risk with treatment in any of the groups: adult, geriatric, or youths. Quite the opposite – for all groups there was a major reduction in suicide risk over time.[20]

The FDA saw similar data in 2004, though. They also analyzed time-based progressive suicidality data provided by the pediatric trials, but did so separately from their analysis of adverse events. "Objective 2" of the 2004 analysis was to investigate whether the drugs led to the emergence or worsening of pediatric suicidality based on scores on the suicidality items in depression questionnaires. For most trials they had the same sort of data as the 2012 Gibbons et al meta-analysis: week by week level of suicidality. The results the FDA got for Objective 2 was a risk ratio of 0.92 to 1, on the face of it indicating that the active treatment groups saw just a little *less* emergence or worsening of suicidality than the placebo groups. This result suggests that antidepressants may actually help protect children from developing suicidal thoughts or behaviors. With a 95 percent confidence interval of 0.76 to 1.11, though, these results were not statistically significant; there just wasn't enough data to come to a firm conclusion.[21]

When I threw away the frame of reference the FDA used, it called all sorts of conclusions into question. For example, looking at the age-based results from the FDA meta-analysis in 2006, I started thinking, what if we're not seeing antidepressant-driven suicide risk here; what if we're seeing access to effective medical treatment? Maybe the older adults have less suicidality because they get treatment promptly when symptoms first show up. They are already seeing a doctor for a dozen other things, so they toss these symptoms in during their next appointment. Going to the next younger group, they have health insurance, so they will see a doctor but they are going to have to get time off from work to do so. The next younger group has a few more hurdles to overcome; they are slower to get to it and more likely to have more severe cases before they do so. And a child needs to convince a parent to take them to a doctor: they have the least ready access to medical treatment. So maybe these results are not saying antidepressants increase suicide risk among young people. Maybe, instead, they are saying the longer the delay in getting effective treatment for depression, the higher the risk of suicidality. Or maybe they are saying nothing at all. The results of the 2004 and 2006 FDA meta-analyses echo the age-based propensity for suicidal ideation and attempts that WHO and CDC surveys revealed. So maybe we are seeing people on an active arm of a trial having more opportunity to report what everyone is experiencing.

TO TREAT OR NOT TO TREAT?

This whole controversy takes place against the backdrop of an alarming increase in U.S. suicide rates over the last 20 years, especially in children. From 2000 through 2003, the suicide rate for U.S. children aged 5 to 19 years declined, from 3.1 per 100,000 population to 2.8, but in 2004 jumped back up to 3.2. From 2004, when the warning was first issued, to 2007, U.S. child suicide rates declined back down to 2.7 per 100,000. During these same time periods, the suicide rate for the people not included in the warning, ages 20 on up, climbed from 13.6 per 100,000 in 2000 to 14.4 in 2002, then pretty much held steady until 2007 when it jumped to 15.0 per 100,000.[22] The FDA must have felt secure in the appropriateness of its warning for children when in December 2006 they extended the warning to young adults through age 24.

From that point, however, the outcomes have not been good. From 2007 to 2016* the suicide rates for all age groups increased year after

* In March 2017, Netflix released the series "13 Reasons Why", featuring a teen suicide.

year; in young people most of all. According to CDC data, the age-adjusted rate of suicide for ages 5 to 24 years grew by 43 percent, while the rate for adults aged 25 and older grew by 16 percent in the same 10 years. A higher rate of suicides in a growing population is a tragedy of epidemic proportions. In the U.S. in 2016, 6,156 children and young people and 38,717 adults aged 25 and above were lost to suicide; that's 10,350 more suicide deaths that year than the annual total 10 years before.[23]

But – and this is another big "but" – neither the 2004 nor the 2006 FDA analyses had addressed overall levels of suicidal behaviors in the study population or the U.S. population as a whole. Rather, they analyzed the differences between placebo arms and active arms of the same drug trials. Plus, a lot happened between 2000 and 2016; there are many influences on suicide rates. An important factor that really got underway around 2007 is the growth of social media, with a presumed stronger impact on young people than on older adults. Social media brings connectedness across populations, but those connections can open a person to greater distress as well as greater support. For young people especially, it opens the door to cyber-bullying, information about suicide means, and glamorization of suicide.

An example of the latter two factors can be seen in the response to the Netflix streaming series 13 *Reasons Why*, built around a teen suicide revealed in the first episode. Following the release of the first episode of the series in March 2017, researchers found a substantial increase in suicide-related internet queries. Total queries about suicide were cumulatively 19 percent higher for the 19 days† following the release of the series, reflecting 900,000 to 1.5 million more searches about suicide than would be expected based on the search patterns of the preceding 3 months. There were significant increases in searches supporting suicidal ideation (e.g., "how to commit suicide" – 26 percent

---

Following that release, the youth suicide rate jumped to a new high. To avoid a possible influence of that event on suicides, I'm just taking the discussion to 2016.

† The authors of the study ended the search as of April 18, 2017, due to the well-publicized suicide death of American football player Aaron Hernandez on April 19, 2017.

increased), as well as awareness and prevention (e.g., "suicide hotline number" – 21 percent increased).[24] The year 2017 also shows a sharp rise in suicide rates; in particular, the younger age group saw a 10 percent rate increase that year.[25]

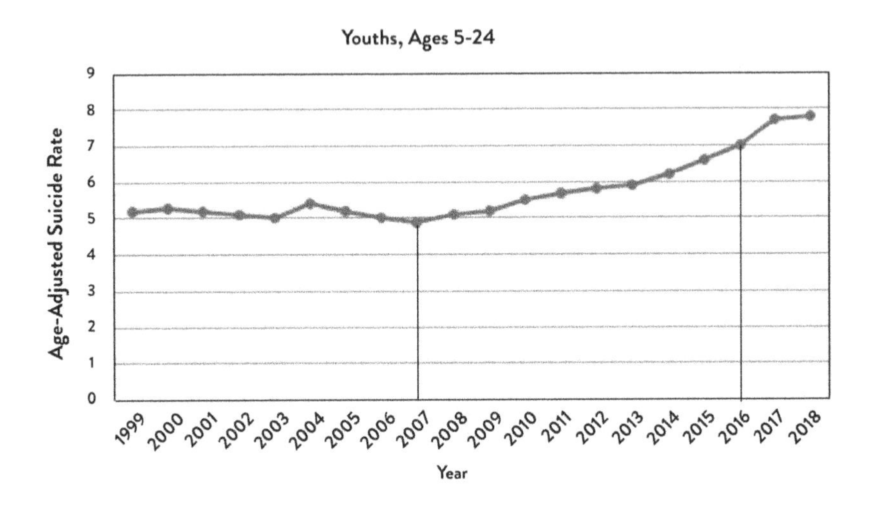

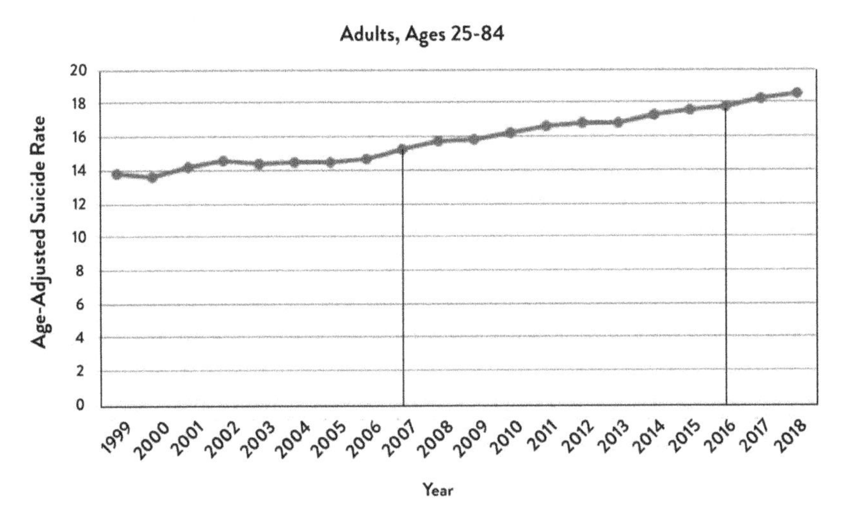

Figure 19-4: **Age-adjusted suicide rates in the U.S., by year 1999-2018, (top) Youths aged 5-24 years, (bottom) Adults aged 25 years and older.** From Centers for Disease Control and Prevention, National Center for Health Statistics. Underlying Cause of Death 1999-2018 on CDC WONDER Online Database, released in 2020. Accessed at http://wonder.cdc.gov/ucd-icd10.html on Aug 31, 2020.

Does the existence of the black box warning affect any of this? A warning is just a warning after all. Presented with enough information about what was considered, and aimed at an intelligent and interested audience, it does not prevent someone from providing appropriate care for children, adolescents, or adults. Many assert that it does matter, though, and I tend to agree with them. Especially for a parent bringing in a child who has attempted suicide or expressed thoughts about doing so. No matter what the doctor says to reduce the impact of the warning, the parent will think if the FDA issued it then they had good reason, and would not want to pile additional risk on their child. The Simon study, the Gibbon study, and the FDA's Objective 2 results all make it look like antidepressant medications may actually be protective against suicidality in children. It has been 14 years since the warning was last visited, and there is a lot more information now from drug trials, clinical experience, and advances in understanding the neurobiology underlying suicide risk. It is time for a whole new analysis.

Part Four

A MATRIX OF INTERVENTIONS

Chapter Twenty

THE REAL WORLD

"It was my proudest moment, in my bizarro world of pride," Simone remarked with wry amusement. Our conversation had made its way around to her plans and preparation for suicide. "I forged a lethal prescription for a barbiturate," she continued. "This was, like, 30 years ago. I stole part of a prescription pad from my doctor's office. And it was successful: I got the pills. My boyfriend later flushed them down the toilet. I was pissed; I was keeping them for when I was really ready."

Simone has struggled with depression since at least age 8, when she remembers first meeting with a psychologist about it. "My mom claims that her depression started – she's outdoing me – by age 2." Growing up an only child with her parents in Austria, Simone felt both privileged and abused. "They used to leave me alone without a babysitter at the age of 5, which nowadays is child abuse. It's more of a European thing; it was acceptable there, at that time at least. But it was horrible, being alone, because I would never know when they were coming back."

When Simone was 12 years old, her family resettled in the U.S. "My next therapeutic intervention was when I was 13. My mom and

I had strong fights, and my dad was very subservient; to this day it upsets me. We went to family therapy once, and right afterward we had a huge fight. So instead of family therapy we decided to try individual therapy. The therapist, he was lovely. He really was, and I had him for years, up until I went to college. He helped me learn to appreciate my qualities."

Starting in her first year of college, though, Simone would live with the constant knowledge that something was wrong. It finally 'clicked' for her after her freshman year that she had clinical depression, and she tried to get appropriate treatment. "My therapist, he recommended someone," she said, "and I started with her right before sophomore year. She put me on amitriptyline [a TCA], 10 milligrams. It was minuscule; it didn't do anything. So I was recommended someone else, and she ended up giving me Xanax because I presented with anxiety as well. But that did nothing for depression." Over time, Simone would be diagnosed with borderline personality disorder, PTSD, and several other psychiatric conditions, in addition to depression and anxiety. "In the mid-90s, somehow ADD and ADHD became a thing, so I asked one of my psychiatrists if it was possible that I had one of those learning disabilities. I didn't have that label, but I sought it out. I got tested at a university, by a PhD candidate because it was cheaper, and I didn't fit criteria. However, I know that I have it, and I've had other people say that."

Simone's life has been marked my medical illnesses as well as psychiatric conditions. "I have PCOS – polycystic ovarian syndrome – which makes my hair fall out. At one point I was so frustrated I just shaved my head," she said. And later in our conversation, "I have inflammatory bowel disease. I developed these weird lesions, they're called erythema nodosum [a type of skin inflammation]. And also stomach issues. I finally saw a gastroenterologist 15 years ago. Then it was colitis; it wasn't Crohn's, she thought it was ulcerative colitis, which was better because that can be cured. But in 2007 I took a blood test and also with more information found out it was Crohn's most likely, which has extra intestinal symptoms... I also have arthritis."

Hearing Simone's determined collection of psychiatric and medical diagnoses, I could almost let myself think that she was trying to out-do her mother, in number of illnesses if not in age of onset. Almost, but not quite; I believe the real reasons lie elsewhere. For one thing, when a fundamental biological pathway is disrupted, the consequences could show up in a variety of ways. I think there's a stronger factor at work here, though. Simone feels bad. She has sought relief from pain and discomfort for decades, and any new diagnosis may open the door to a different suite of treatment options.

"I went through a portfolio of medications in the early days," she said. "Tricyclics, tetracyclics, mirtazapine, and venlafaxine. Those took over 9 months to work. I tried an MAOI, but I had at least three hypertensive crises in the decade I was on it. But I was afraid to shift because every time I went off it, it was horrendous, the onslaught of depression." She pursued psychotherapy and volunteered for ECT. "Unfortunately that was a bad idea," she said. "I had eight treatments, bilateral, and that's more impairing. I did it at a local hospital, almost 30 years ago, and I forgot chunks of months before the treatment." Ketamine was better. "This year, I tried – and to be honest, it was more drug-seeking than anything – ketamine infusion. I had a lot, I did it over 3 months, and at first I really enjoyed the process, it was very dreamy. I also have addiction in my history, and I was very cagey how recent my addiction was when I talked to the psychiatrist. But I would sometimes have rebound depression right after the treatment, and then the next time I'd be OK. But after the insurance switched and I was no longer 50 percent covered, I stopped."

The real world tends to intrude on our best intentions. In the real world – unlike in clinical trials – people often have co-morbid illnesses; they have demands on their time and energy from work and family, and they may lack insurance or access to specialized care. In the early 2000s, the Sequenced Treatment Alternatives to Relieve Depression (STAR*D) study went looking for what happens in the real world to someone diagnosed with major depressive disorder.

The STAR*D study was a nation-wide public health clinical trial sponsored by NIMH to determine the effectiveness of different treatments for people with major depressive disorder who did not respond to their initial antidepressant. The trial actually involved very few modes of treatment – cognitive behavior therapy was the only non-pharmaceutical option offered – but the experience served to clarify some important points.

First, reality intervened: comorbid conditions complicated recovery, and many patients failed to show up for scheduled visits. Over the 7 years of the trial, a third of the 4,041 patients who entered the first level left for non-medical reasons. The remaining 2,876 adult participants completed up to four levels of treatment and were then monitored for 12 months. The medications offered included two different SSRIs, a TCA, a MAOI, an SNRI, two atypical antidepressants, the anti-anxiety drug buspirone, the thyroid hormone triiodothyronine (T3), and lithium. All the patients started Level 1 with the SSRI citalopram. If they did not become symptom-free or if they experienced intolerable side effects, they were encouraged to move to Level 2, then eventually to Level 3, then Level 4. Level 2 offered nine different options, running the gamut from switching to or augmenting citalopram with cognitive therapy or one of several different classes of traditional antidepressant. Level 3 brought in options for lithium or T3 and a couple traditional antidepressants, and Level 4 brought back some of the same traditional antidepressants seen in Levels 2 and 3 and an option for an MAOI. Patients were allowed to accept or decline the "switch" or "augment" strategies recommended to them at each level. Encouragingly, fully half of the patients achieved remission (a score of less than 7 on the Hamilton depression rating scale) in 14 weeks with Levels 1 or 2. Few of those who progressed to Levels 3 or 4 achieved remission, however, and those who did were more likely to relapse within the year of follow-up. The patients who did not remit within two steps were considered treatment-resistant.[1]

To their surprise, the researchers did not find meaningful or statistical clinical differences in remission rates, response rates*, or time to remission among any of the treatments used. Switching to a second

---

* "Response" means they lowered their depression scale score by half.

SSRI when the first one didn't work was just as effective as switching to a different class of medication, and the results were the same when the patient switched to or added cognitive therapy instead of switching to or adding a new medication.[2] It all served to emphasize several points. First, traditional antidepressants have good standing as the first-line treatment for depression – they are easy to administer and lead to remission for about half of depression's sufferers. Second, no one really understands how these medications work, or why they often don't work. That applies to other modes of therapy as well – even though so many of depression's treatments were discovered first, opening a window into some new aspect of the brain and depression, scientists have never been able to fully close that loop. Seeing patients recover after taking an anti-tuberculosis medication, or gain relief from their symptoms after skipping a night's sleep sparked intense research, marvelous advances, and more effective treatments. It's never led through to fully understanding *why* that treatment works, though – what it is actually doing to remediate depression – or why it fails to work for someone else. Finally, about half of depression's sufferers don't achieve remission through these medications, and those treatment-resistant patients need other options.

If depression comes from abrasive processes that overwhelm the brain's ability for repair, then you would think that effective therapies would counteract one or both sides of that imbalance: reduce abrasion and/or enhance repair. The established treatments don't align cleanly into that framework, though. Perhaps that is just our lack of understanding at this point; perhaps the framework is wrong. Looking at the modes of operation of successful therapeutics, though, I see them acting along two lines: those that enhance regrowth, and those that help rebalance hypo- and hyperactive parts of the brain. That's how I've organized this section. My "rebalance" therapies are psychotherapy and non-invasive interventions from neurofeedback to transcranial magnetic stimulation. Whether targeting left-side versus right-side or "sad" subgenual cingulate versus "happy" pregenual cingulate, they seem to make activity levels approximate normal for a while, perhaps retraining the brain or simply giving it space to heal. My set of "grow" therapies includes medications, seizure therapy, exercise, and chronotherapies. That's an easier choice, as most of those interventions have been shown to result in growth of new brain cells and connections in animal models. With chronotherapies, though I personally speculate that wake therapy may act by causing a temporary rebalance, the other therapies in the set, like light therapy, more likely work through growth. Most, possibly all, of the therapies employ both mechanisms – that could be why they are often individually successful – but I put them where I felt they have the stronger influence.

I am not a medical professional and I make no recommendations about the treatments that follow. So much of what science has learned about depression comes from observing what helps people recover from it, though, and these treatments still have more to tell us.

Simone has tried a lot of different therapies to restore her health, and has found ways to get by even through difficult times. "I find that medications coupled with certain supplements, coupled with exercise, coupled with pets, coupled with behavior work helps... It's a matrix of intercessions, basically."

Every new approach offers hope, but comes at a price in terms of the effort and optimism necessary to give it a try... qualities that are in short supply for someone with depression. "Living with depression is really hard work," Simone commented. "Even with lived experience, we don't appreciate that; we don't realize it until we talk to other advocates. My husband, he was maybe the first one who made me think about this a little differently, 20 years or so ago when we met. I was upfront; I had just gotten dumped a few months before that when I showed signs of depression, so I told him right away what was going on with me. And he said, 'So you've really been fighting depression.' And I guess the term 'fighting depression' at the time was novel to me, because usually it was 'You're afflicted,' like something passive. Sometimes I'm really not that functional, and that can be 'Can I get out of bed by a certain time? Can I do the laundry, or feed myself, if I'm so mired, and I'm just getting by minute by minute?' You'll find people who, on the outside, are extremely functional, and they might be putting in 5,000 percent of effort. But you don't know. They may also have good skills and they learned early. I have friends with depression, and sometimes I forget, like I can be like, 'What? You couldn't check my email or just contact this person?' I've been there, and I even *am* there, and I can quickly forget."

Chapter Twenty-One

PSYCHOTHERAPY

Audrey laughed when I accused her of trying to be her own case study. She was studying psychology in college when she first sought help for depression and anxiety. "It was really interesting going through the therapy process, and also learning about it," she agreed. "There were times that I was like 'I know what trick you're trying to use, but I'm going to let it work because it is working.' I learned a little bit more but I was careful; I didn't want to self-diagnose."

Audrey's depression and anxiety started in high school with predominantly physical symptoms. "I had these really bad stomach issues that would keep me from eating for a full week," she said, "and I was losing 10 pounds in 5 days. I was already pretty underweight at the time, so it was really unhealthy and I didn't really know what was going on." Late in her high school years she became aware of the depression. "It really started with – when my friends were dealing with things – being able to support them and feeling like I wasn't a good enough friend, and very much blaming myself for things that were going on," she said. "Those were the first things that I noticed and it was not until college that I was really aware of it. It was a self-blame that led to a negative self-talk. It started as 'You're not a good enough friend; you're hurting your friends,' and it expanded to other areas. I was competing in gymnastics, and when I didn't do well there was a lot of self-blame for that. It definitely initiated with not being able to support others, not feeling like a good friend, and expanded to more and more of these negative thoughts, blaming myself for everything, feeling really down, starting to feel very lonely. I've always had an easy time making friends, and I've always had a very good support system, but I wasn't always able to talk about the way I was feeling because experiencing what I did in high school with some of my friends, I thought talking about it would make other people feel bad. And that's part of the anxiety, those worries that go ahead and spiral out of control.

"When I got to college, I knew that I was struggling but I didn't think I needed help. I didn't think it was a condition because I had seen how serious it was in my friends, how intense it was, and it was like, it's not that bad. Then, at the beginning of sophomore year I reached such very low point that I contemplated suicide for the first time." As a college student with limited resources, Audrey considered taking the deadliest drug she could find: ibuprofen. "It wasn't an attempt, and it wasn't even an actual plan," she laughed. "For me, it was less about dying. In fact, I didn't want to die because I was terrified of that. But I just wanted to stop feeling the way that I was feeling. I just wanted a break from everything. At one point I think I tried to calculate how many pills could get me to pass out and go to the hospital but not put me too far. I just felt so down, I felt hopeless, like nothing could make me feel better. And I did often have the thoughts that people would be better off if I wasn't around. Once I started talking to people I felt like a burden, especially to my family. They were dealing with another family member who had mental health issues. I thought mine aren't that bad, and I'm just making things worse, stressing them out more. I felt like a burden a lot of the time. My answer to that was typically to just stop reaching out to people, which is when I got to a worse place."

To find her thoughts first touching, and then dwelling on suicide, or at least, self-harm, was the wake-up call for Audrey. "That was the moment that I realized that it *is* that bad. So I went through my school's counseling system, and they referred me to somebody outside. That's when I started seeing a therapist weekly."

PSYCHOTHERAPIES

Placing psychotherapy as the lead of my "rebalance" set is a matter of organization, not dogma. Scientists are still trying to figure out how psychotherapies work – the mechanisms through which they act remain a mystery.[1] I see a strong vein of exercising control over affect as a core of this practice, perhaps deliberately moderating under- and over-active parts of the brain. Certainly, some elements of psychotherapy may draw effect from reducing chronic personal stress, which presumably reduces glucocorticoid signaling and consequent damage. Environmental enrichment has been shown to increase growth of new brain cells and connections in laboratory animals; by engaging in new experiences, contacts, and learning, a depressed person could be increasing growth factors from psychotherapy too. Though its mechanisms are obscure, many types of psychotherapy have proved their effectiveness in clinical testing, and a few of them are discussed below.

COGNITIVE BEHAVIOR THERAPY

In a 40-year retrospective, Dr. Aaron Beck recounted how he came to describe a cognitive model of depression, from which he would invent cognitive behavior therapy (CBT). A practicing psychiatrist, Beck saw many depressed patients and noticed that they each experienced specific types of thoughts that he could perceive by their effect on them, but of which the patients themselves were only somewhat aware. He called them "automatic thoughts."

> *These thoughts (cognitions) tended to arise quickly and automatically, as though by reflex; they were not subject to volition or conscious control and seemed perfectly plausible to the individual. They were frequently followed by an unpleasant affect (in the case of the depressed patients) that the patients were very much aware*

> *of, even though they were unaware of, or barely aware of, the preceding automatic thoughts. When I directed the patients to focus their attention on these "automatic thoughts," they began to report a string of them, particularly in response to a cognitive probe, "What are you thinking right now?" Connecting these thoughts brought out certain negative themes such as deprivation, disease, or defeat. Grouped together they fell into the category of a negative view of the present, past, and future experiences. Later, in working with more severely depressed patients, I noted that these types of thoughts were no longer peripheral but occupied a dominant position in consciousness and were repetitive.*[2]

Through the 1960s and 1970s, Beck developed and refined a cognitive model of depression, proposing that people with depression have a negative bias in information acquisition, retrieval, and processing, and that this bias has a primary role in the development and maintenance of depression. People with depression pay more attention to negative stimuli than to neutral or positive ones; they have enhanced recall for negative events and an over-generalized memory for positive events, and tend to assign a negative aspect to benign events. Since Beck's formulation of the cognitive model, he and others have extended and refined these frameworks and applied them in psychotherapy practice.[3]

Beck developed CBT specifically to treat depression, and with deliberate intent to produce a practice that could be tested and practiced consistently in laboratory and clinical settings. It has evolved into a practice today consisting of three major elements: behavioral activation, modifying automatic negative thoughts, and modifying dysfunctional assumptions and core beliefs. CBT is problem-oriented – it addresses the patients' immediate difficulties and helps them develop the tools they need to break and re-form biased perceptions and schemas. In the behavioral activation component, the patient learns to monitor daily activities and assess the pleasure and mastery – the degree to which he or she is able to assert control over their situation – involved in those activities. The therapist and patient assign new daily activities to increase pleasure and mastery and the patient learns to mentally rehearse activities before undertaking them, engaging problem-solving to address any obstacles. With that process underway, the therapist and patient work on assessing and modifying the negative automatic thoughts that characterize depression. They identify automatic thoughts arising during the therapy session, use daily thought records and homework assignments to assess the validity of their negative interpretations, examine the evidence for and against automatic thoughts, and look for attributional biases in assessing successes and failures. The final element targets core beliefs and assumptions, getting to the structure of the internal schema supporting this negative bias. Identifying these core beliefs and assumptions, the patient and therapist cast them against alternative beliefs and assumptions that aren't inherently negative. It encourages an introspective process, in which the patient learns to assess the validity of their negative interpretations and recast their ongoing experiences into a more positive context.[4]

CBT quickly asserted itself as equally effective as antidepressant medications. It maintains that distinction today, and has even been shown to be superior to antidepressants in preventing relapse.[5] After proving successful against depression, CBT was adapted to address suicide and then other psychiatric disorders, including anxiety disorders, phobias, panic disorder, personality disorders, and substance abuse. With each additional disorder addressed, therapeutic protocols and packages based on the principles of CBT were adapted to the character of the dysfunctional behaviors and schemas of that disorder.[6]

INTERPERSONAL PSYCHOTHERAPY

Interpersonal Psychotherapy (IPT) began as a university research experiment and proved to be so effective it was adapted for practice after the study ended. In 1969, Dr. Gerald Klerman needed to add a psychotherapy branch to an 8-month randomized, controlled trial of an antidepressant medication. The study wanted to assess the effectiveness of the antidepressant by itself and also with psychotherapy. To do this in controlled, experimental conditions, he needed a time-limited but effective psychotherapy practice that could be applied consistently to all of the subjects and be faithfully replicated in future trials. Klerman engaged Myrna Weissman, a newly-graduated social worker, to design the therapy. Guided by Beck's cognitive therapy manual, case studies, and observations from clinical practice, Weissman, Klerman, and colleagues developed a manual defining the structure and content of a new mode of treatment.[7]

The underlying theory for IPT came from observations of the importance of interpersonal context and environment in psychiatric illness; particularly that social and interpersonal stress tended to provoke depressive episodes. The originators saw in practice that social supports protect against psychopathology, and that whatever the immediate cause of a depressive episode, it occurs in an interpersonal context, usually involving disruption of significant attachments and social roles. The death of a loved one (complicated bereavement), antagonistic relationships (role disputes), life disruptions or losses (role transitions), and lack of social support (interpersonal deficits) placed vulnerable individuals at risk for a depressive episode. To support the needs of the university experiment, the clinical research group constructed a time-limited, supportive, patient-friendly intervention. Patients learned to name their feelings, understand them as social cues, and express them effectively in order to improve their social situations. The hope was that this intervention would improve patients' social functioning and thereby relieve their depressive symptoms.[8]

The first treatment trial compared amitriptyline, pill placebo, IPT (at that time called the "high contact" condition), a "low contact" alternative to IPT, and combined IPT and amitriptyline. Finding that random assignment to the full IPT treatment actually improved symptoms of depression, the team conducted another, 16-week trial comparing amitriptyline, IPT, amitriptyline plus IPT, and the low contact condition among acutely depressed patients. At the conclusion of that second trial, the patients who had taken amitriptyline and also gone through IPT had the most relief from their symptoms, while each of those interventions was more effective than the low contact alternative. With publication of the IPT manual and study results, the therapy was put into wide practice. It was eventually adapted for use in other types of disorders and patients with specific characteristics.[9]

IPT is intentionally time-limited, providing pressure on the patient to take action. Therapy starts with up to three sessions during which the therapist identifies the target diagnosis (for example, major depressive disorder) and the interpersonal context in which it presents (complicated bereavement, role dispute, role transition, or interpersonal deficit). Structured diagnostic criteria like the Beck Depression Inventory are employed to assess the patient's starting point, emphasizing that they have an actual illness to treat and keeping the process objective and evidence-based. For the next 10 weeks, the therapist uses specific strategies to deal with whichever of the four potential problem areas is the focus: appropriate mourning for bereavement, resolving an interpersonal struggle, helping a patient to mourn the loss of an old role and assume a new one in a role transition, or decreasing social isolation for interpersonal deficits. In the final three sessions, the therapist and patient review his or her accomplishments during the treatment, encouraging the patient to feel more capable and independent, and helping them adjust to the end of therapy as a role transition in itself.[10]

Just as Beck had designed cognitive therapy under the self-imposed mandate that the framework and practice needed to be subject to independent testing and validation, IPT was conceived to be subject to objective measurement. A meta-analysis of nine studies comparing IPT to medications found IPT was consistently superior to placebo and equally effective as the antidepressants against which it was compared (TCAs, SSRIs, and an MAOI).[11]

MINDFULNESS-BASED COGNITIVE THERAPY

The application of mindfulness – the awareness and nonjudgmental acceptance of one's moment-to-moment experience – has been part of the Buddhist and other spiritual traditions for centuries, finally reaching the mental health and medical realms of Western medicine in the 1970s. Zen Buddhism had been making its way into American culture, and interest in using meditation techniques in psychotherapy began to grow. In the late 1970s, pioneering clinicians like Dr. Jon Kabat-Zinn began applying mindfulness meditation to clinical use as Mindfulness-Based

Stress Reduction (MBSR). Mindfulness-Based Cognitive Therapy (MBCT) followed, helping participants to view thoughts as mental events rather than facts, recognize the role of negative automatic thoughts in maintaining depressive symptoms, and disengage the occurrence of negative thoughts from their negative psychological effects. Unlike the CBT approach that emphasizes evaluating the content of thoughts and developing alternatives, MBCT seeks to change the patient's awareness of and relationship to thoughts and emotions.[12]

Since their introduction, mindfulness-based therapies have been used extensively for depression, anxiety, pain, substance abuse, and many more disorders. A meta-analysis published in 2013 reviewed the results of 209 studies in which MBSR or MBCT had been used in the treatment of any one or more of those disorders. MBCT and MBSR showed large and clinically-significant effects in treating anxiety and depressive disorders, and maintained those results at follow-up.[13]

DIALECTICAL BEHAVIOR THERAPY

Dr. Marsha Linehan developed Dialectical Behavior Therapy (DBT) for chronically suicidal individuals in 1993, and later adapted it into a treatment for multi-disordered individuals with borderline personality disorder. DBT is a form of cognitive behavior therapy that combines the basic strategies of behavior therapy with mindfulness practices. The "dialectic" part refers to resolving the contradiction of simultaneously embracing acceptance and a need to change behaviors.[14]

In DBT, the therapist works with the patient to find ways to hold two seemingly opposite perspectives at once, promoting balance and avoiding all-or-nothing styles of thinking. It promotes a "both-and" rather than an "either-or" outlook.[15] Randomized trials comparing DBT to treatment as usual or another active treatment for patients with bipolar disorder have shown greater effectiveness than the alternative treatments in reducing suicidal behaviors. There have been few studies of the effectiveness of DBT for major depression *per se* (not related to suicidal behaviors). One study in 2003 of medications alone or medications with DBT in elderly adults found a significant benefit of the DBT-adjunctive treatment.[16]

There are many more types of psychotherapy, but not as well supported with laboratory or clinical testing. Very often, the two treatment modes – psychotherapy and medication – are prescribed in conjunction with one another. Combined treatment has been shown to be more effective than antidepressant medication alone, with moderately large effect. Plus, patients tend to maintain their progress longer on combined treatment than on antidepressant medication alone.[17]

Early on, Audrey didn't want to take any medications for her depression and started with psychotherapy only. "We did a lot of CBT," she said. "That was the focus, especially because the anxiety led to the depression a lot of the time. They're very connected. So, addressing those irrational thoughts was what I was trying to do. It took me a really long time to get there. Part of the problem was that in college, I was in a relationship that was not the healthiest. Not a bad person, just he was not equipped at the time to be dating somebody who was working on their mental health. So I had a lot of anxiety around that relationship, feeling insecure, feeling like I was putting too much on him."

Even working at it and meeting with her therapist weekly, it took a couple of years for Audrey to make real progress. "I started sophomore year of college," she said, "and I honestly don't think I made much progress until after senior year, after that relationship ended. I feel like to some extent I relied too much on the support system rather than addressing those thoughts that were causing those problems. I was looking for the external validation rather than, in myself, trying to address those thoughts. And then when the relationship ended, there was the problem of 'Oh, that was a thought this whole time, that the relationship is going to end because my problems are too much for this person.' So I think after that is when I really started to work with my therapist. Like, let's break down why this happened so that it would stop being about me and more about that fact that this was not a good fit between two people. Really, taking some of that blame off of myself and realizing that there are two people in every relationship. And some people are not in a place where they can support someone with mental health issues."

With her progress underway, Audrey also found over time that she needs medications as well. "Last year, I fell off the parallel bars and had a concussion for about 6 months. One of the tough things about concussion is that post-concussion syndrome can worsen depression or create depression in people who haven't experienced it before. So I had a very hard time with that at first. They put me on a medication that is an older antidepressants but it's been used to treat pain – amitriptyline. And I happened to notice it was improving my mood a little bit. After the concussion was through I was still feeling a lot better than I had been before. I talked to my doctor and I said 'This seems to be making me feel better' and she said 'Why don't we try putting you on the actual antidepressant dose?' At that time I had already seen the effects of it, and the side effects had for the most part gone away or weren't that serious, and I thought if this is already making me feel better, it doesn't hurt to continue on it. It's not a medication that causes a lot of changes. There's always side effects, but they're not major, and it didn't negatively impact me in many ways. And that was my turning point with the medication. So we increased the dosage a little bit, and I've been on that for over a year now."

COGNITIVE STRATEGIES

The people who devised the models and guidelines for each of these types of psychotherapy wanted to create something that could be practiced consistently with results measured. Years of observing these therapies in action have also helped psychologists to identify some of the basic underlying patterns of healthy and unhealthy mental processing.

Thoughts are individual, but the way we think often falls into one of many patterns. Rumination – a train of thought narrowly focused on one's distress and its causes and outcomes, dwelling on self-focused and self-critical experiences – is one of those patterns. It is a negative, maladaptive strategy that can make a person feel hopeless and immobilized, perpetuating psychological distress. Two other negative strategies are avoidance and suppression. Counter-intuitively, avoiding thoughts, sensations, emotions, and memories, as well as anything perceived as a trigger for a bad emotional state, actually increases those negative thoughts. One way in which this might work is through operant conditioning. If an experience or encounter resulted in a negative outcome, then we learn that this type of experience equals a bad outcome. As the individual continues to avoid the "bad" stimulus, he or she never gets to see a not-so-bad, neutral, or even positive outcome from the same experience – the one experience stands unchallenged. Attempts to suppress unwanted thoughts, emotions, or expressions also, conversely, lead to an increased accessibility of and hypersensitivity to the thought or emotion.[18]

The positive strategies encapsulated in most methods of psychotherapy are reappraisal, acceptance, and problem-solving. Psychological tests have shown that people with depression tend to perceive and remember events through a negative lens; they are biased to see most events and emotion-eliciting stimuli in a negative light. Reappraisal asks them to step out of that mode and look again at the cause of their distress in an objective way. It recasts the meaning of such stimuli to a neutral or positive form, eating away at the core of their psychological distress. Acceptance is a non-judgmental, present-centered awareness in which thoughts, feelings, and sensations are accepted as they are. It helps separate the individual from thoughts of guilt or blame for their condition or other factors that are unhelpful to their recovery. Problem-solving, the third positive strategy often employed in psychotherapy, does not by itself regulate an emotional response. Rather, it gives the patient a sense of control or mastery and helps to eliminate smaller issues that could otherwise pile up to seem insurmountable.[19]

Different types of therapies have a stronger emphasis on each of the three positive strategies: CBT emphasizes problem-solving and reappraisal, IPT emphasizes problem-solving, MBCT emphasizes acceptance, and DBT emphasizes acceptance and problem-solving. When it comes to negative strategies, people with depression show the strongest tendency toward rumination, then avoidance and then suppression. In fact, high levels of self-reported rumination over the period of a year can predict the onset of major depression and the worsening of depressive symptoms where they already exist. In terms of the positive strategies, people with depression have been shown to be weakest in problem-solving and reappraisal.[20]

IMAGING PSYCHOTHERAPY

Using functional imaging, the effects of structured psychological intervention strategies – CBT and IPT in particular – have been investigated for insights into the process of recovery from major depression. A 2013 study analyzed the results of 6 resting state functional imaging studies involving 70 subjects, and 5 emotional-cognitive task studies involving 65 subjects. These were longitudinal studies with patients imaged before and after their course of psychotherapy, so each patient was compared to him- or herself. Across the resting state studies, therapy generally increased activation in an area in the left temporal lobe and decreased activation in the left dorsomedial prefrontal cortex and the right parietal lobe. In the task studies, therapy generally resulted in increased task-based activation of several areas in the left dorsomedial prefrontal cortex and one in the right precuneus, and decreased task-based activation in both sides of the occipital lobe and the cerebellum. Looking at areas with the greatest degree of difference between a resting state and an emotional-cognitive task state, several clusters in the left dorsomedial prefrontal cortex and cingulate cortex and the right precuneus became less activated in the resting state, yet more activated in an emotional-cognitive task state following psychotherapy for depression. These were small studies so there are a lot of limitations to what can be taken from their results, but they did show a functional effect of psychotherapy, with enhanced differences between the resting state and task state.[21]

It will take many more, larger scale and consistently conducted and assessed functional imaging studies of psychotherapy to identify its operating mechanisms with any degree of assurance. Even these limited studies provide interesting information, though. They showed that the changes in brain activity before and after psychotherapy differed from the changes wrought by antidepressants, and also differed from the changes in brain activity seen when patients responded to placebo. Psychotherapy is not just an interactive placebo, and though it may have comparable effectiveness, it does not have the same mode of action as do antidepressants. [22]

Audrey followed her undergraduate degree in psychology with a 2-year master's degree in the same subject. Throughout this time, she studied psychology in school and went through psychotherapy on her own behalf, so we can consider her well-informed on the topic and process. None of that made it easy for her to deal with her depression and anxiety, though. It was still a long, difficult, and very personal journey.

Even after objectively looking at the relationship that had ended and reappraising the negative thoughts it had sparked, she found herself in another unhealthy relationship that also ended badly. Though it initially threw new obstacles in her path, in the end, that experience helped to crystallize her understanding of herself. "When that relationship ended I had the same depression thoughts coming back, of 'This is my fault because I have a mental health condition,' and 'Nobody's going to want to date me because of this.' But that's when I had a turning point with the anxiety and depression. This past year, that's when I feel like I have really come into my recovery. After that relationship ended, that's when I decided to stay on the amitriptyline as an antidepressant. I feel like being on the medication put me on a better baseline so that I could more easily work on things. When you're at a low, it's very hard to say 'Let's try positive self-talk.' And I spent a lot of time talking with my therapist about those relationships, about what was happening in them that made me feel this way, and really being able to distinguish what was me and what was that person. That was a really, really tough process because this had been a person who seemed to be very good with mental health. He had his own experiences with mental health conditions, but looking back I realized that in many ways he actually used my mental health conditions to manipulate me. And that was a really tough realization. It was that whole process, that whole year that I was working on that and having these realizations, and something like that, 3 or 4 years ago, would have destroyed me. But because I had been working so hard on this, because I was in a better place with the medication, I was able to take those and process those and very much separate 'This is not my fault.' I kept literally telling myself ten times a day, 'This is not your fault.' I think the toughest part when I had the realization that I had been sexually assaulted in that relationship. Part of the manipulation made that hard to see, and at first I became extremely anxious and depressed. For a week, I couldn't go to the places that I had seen him, but I was able to come out of that quicker than I would have a few years ago because I have those tools ready and I had been already been reminding myself of that separation: 'This is not your fault. This is the fault of another person.'"

Getting back on track also demanded that she reconnect with her family and friends. "It was the support system that I had for so long, but felt nervous about reaching out to, because I didn't want to put too much on them," she said. "I spent a lot of time talking to my therapist about finding ways to do that, not relying on one person, and also talking to them about what it is that I really needed. Because I know at first a lot of them had not really had any experience with a friend who had gone through this. So, I would say, 'This is how I'm feeling,' and they would say 'Just look on the bright side!' I feel like everyone's instinct is 'Let's be positive!' and at first that's so hard to hear. But I reached a point that I could talk to them and say 'Hey, this is actually what I need.' And my parents spent a lot of time educating themselves in order to help me and my other family member. That was really great; I felt a lot more comfortable knowing that. I talk to my mom all the time now."

All of these experiences have given Audrey a deeper insight into her new profession. "Often people have this view of mental health conditions as this terrible thing that breaks people, like 'These people are fragile; they're broken.' And that's something I've felt from people around me a lot, was 'I have to be careful what I say around you because I'm going to hurt you. I have to tiptoe around you; I can't tell you things.' And that almost felt worse, and there were times when I was like 'I can't hear everything right now,' but being able to decide when I was ready to hear something was so important to me; that autonomy is so important and I think people either have that or they have a fear of it. A lot of times I think that people think of stigma as thinking that people are violent or that fear component. Most of the stigma I've experienced is that other end, 'You're broken; I can't break you anymore.' And everyone struggles with mental health. The people with mental health conditions need extra support, but I have plenty of friends who go to therapy now, not for a diagnosed mental health condition but because it's good to talk to someone. And if I could go back and tell myself, 'If you're feeling that way, even if it wasn't a diagnosable mental health condition, get help because you deserve to feel better. You're not going to be happy all the time; no one is happy all the time, it's not possible. But you can feel good about yourself and cope with difficult things.'"

Chapter Twenty-Two

EXTERNAL REBALANCING

"I have an origin story," Dr. DuBose said. Dr. Don DuBose owns a psychiatric practice named Future Psych Solutions, in Columbia, South Carolina. We had been introduced by a mutual friend, and I was intrigued by the practice.

"The practice was birthed by an awful experience I had with one of my clients. I did my adult psychiatry training at Morehouse, and then I came back to Columbia to do child psychiatry training. Usually, when you're in training for another specialty, you can also work outside the residency to make actual money. One of my first ten patients was a gentleman that struggled with depression. That was not uncommon, a lot of people have depression, but in this particular case, he talked me into doing his talk therapy as well. I only had maybe three or four talk therapy cases; psychiatrists typically don't do that anymore.

"I started seeing him once a week, in my last year of training. I worked with him for several years, and I watched him graduate from college, get married, have his first baby, and I was doing weekly therapy with him. I watched him go from young adult to man with responsibilities, and depression was always in the background for him. We basically did what the book tells us to do, prescribe meds and do therapy, and he maintained, but he didn't thrive. Unfortunately, the depression was having a toll on his marriage, and at some point, they separated. At this point, I was a couple years out of fellowship, thinking I was experienced. After the separation, his depression got worse; he went downward and downward. So we did what the book told me to do. We increased the meds, changed the meds, add another med, switch out this med, more intensive therapy, seeing him twice a week. Then we were doing couples therapy, and all this other stuff. I did everything I thought I was trained to do, and, unfortunately, his depression just worsened. Until one day, we made one more medication change, and he came in for his visit, and he said, 'Doc, I'm starting to feel better, bouncing back.' I was seeing him twice a week at the time, and he said, 'Let's go back to once a week therapy, and I'll see you next Friday.' And I was like, 'Yes! Finally, a breakthrough!' That was the last time I ever saw him. I think he told me that because he had already made up his mind that he was going to take his own life, and he didn't want me to hospitalize him or worry about him.

"His wife, since I had been doing couples therapy, called me for an emergency visit. She brought all of his emails, all of his texts, leading up to what we think is maybe 15 minutes before his death. And I could see how his hope was just fading away. That really devastated me, to the point that I thought I didn't want to do psychiatry anymore. It was the first time I had that situation happen, and I really didn't know how to handle it. Fortunately, one of my therapists talked me into going to the young man's memorial service. I didn't want to go, because I felt I had let the family down. So, after much coaxing, I decided to go. The gentlemen was a white gentleman, and I'm, like, 300 pounds, a big black guy; I'm the only black guy in the room. So my heart was pounding in my chest, I was really nervous because I felt like the family would blame me that I didn't catch it. I had been there maybe 5 minutes, his mom sees me from across the room, beelines towards me, and I'm like 'Oh, no; she's so pissed.' Instead, she grabbed my hand with both her hands, and she says, 'Don't worry, Dr. DuBose.' I guess she could see it in my face. And she said, 'My son always told me the things that you said to inspire him. And I know that you did your best to try to help him. We don't blame you.' Part of me was absolved, and I felt better. But another part of me was just tortured by the statement, 'I know you did your best.' Something in me was saying, 'Did you really do your best? Did you turn over every rock; did you try everything possible to help this person?' And when I really evaluated that question, the answer was yes and no. I did everything I was taught to do. I was fresh out of training; I was up to date on everything we were taught to do. And no, because I didn't explore outside the box. We didn't look at the things that were not traditional, and most clients don't want to do shock therapy. When you explain it, they usually don't. So between that, I felt like there were other things I could try to do."

He started researching other options, and came across transcranial magnetic stimulation (TMS). "I had heard of it before, and there was some evidence that it worked but they didn't have that much data, so it was kind of like fringe science at the time. So once I found it, I thought, 'this is interesting; this might be something.' And this is how I know God is real. The very next day, a young lady came to the checkout window at my office upstairs, super-pregnant; she looked like she was 12 months pregnant. She was struggling, sweating in the heat, with a bunch of boxes of pizza. She said she was looking for the office of another psychiatrist next door. So I'm walking her downstairs and asked her what she does. She says, 'I'm a rep for a company called NeuroStar, we sell TMS.' And I was like, 'What are the chances? I just researched this last night!' She said she'd come back and do a lunch for me. So the doors just started opening up for bringing TMS to our practice. I tried to talk my partners into it, and they were just not interested. They were traditionalists, and the machine is expensive, and so I was like 'I'll do it on my own.' At this point I was driven.

"A law office used to be in these spaces here, and they had been here probably 20 years. And once again, a sign that God is real: I came down here that next week, just randomly, and they'd put in a notice to leave the building. I was like, 'I'll take it! I'll take the space!' I didn't know how I was going to afford it; I didn't do any market research; I just knew that we needed to do this. It was myself and one worker I had poached from upstairs. And I purchased the machine and I got started, thinking, 'Yes! I'm going to save the world, and everyone's going to love to do this, and I'm just going to do TMS... ' and I almost went bankrupt. People didn't know it existed. And, we ran into a barrier: the insurance just would not cover it. Even though it's FDA-approved, even though the insurance had policies for it and you would follow that policy to a T, they would find a way not to pay for it. They wrote their criteria to be ambiguous, so that if the patient wasn't sick enough, 'We're not paying for it.' Then you write the criteria just like they said the last time and now 'They're too sick; they need to do ECT.' So we had so many barriers to getting people approved that it nearly ran me out of business. For every 20 patients that would meet the criteria, I'd probably get two through. So, I really struggled in the very beginning. Fortunately, the manufacturer was very motivated, so they would bring in experts to talk to Blue Cross. But it took about 2 years before they would release a little bit, and I would say they were marginally fair, and probably 3 and a half years before they followed their own criteria."

REPETITIVE TRANSCRANIAL MAGNETIC STIMULATION

Transcranial magnetic stimulation (TMS) was introduced to the scientific community by Dr. A.T. Barker in a 1985 letter to the editors of *The Lancet*. The technology was an off-shoot of experiments with transcranial electrical stimulation (TES), in which a pulse of several hundred volts of electricity is passed through the skull between electrodes, most of the current traveling along the scalp rather than through the brain. Scientists noted that TES was uncomfortable for its subjects, resulting in what must be a very much understated "brisk" sensation. Attempting TMS instead, they got much less "brisk" results. TMS was originally applied to the motor cortex in brain-mapping studies, but when researchers observed that it had psychological effects too, they began to investigate it for use in psychiatric disorders.[1] In 1993, researchers at the University of Bonn first reported the use of TMS to treat two cases of drug-resistant depression. Though their attempts were unsuccessful – possibly because the patients had psychotic depression – their work stirred interest and further research into the new technique.[2]

In TMS, an insulated metal coil is placed on the scalp and an alternating electric current in the coil generates an alternating magnetic field perpendicular to the current flow. The alternating magnetic field passes through the skull and induces a secondary electric current in the brain. Through the configuration of the coil and parameters of the initial electrical field, scientists can sculpt the magnetic field and the precise location in the brain that it is converted back to electricity.[3] Though penetration into the brain is very limited – it can only directly affect the outer cortex – the electrical current is strong enough to make neurons fire. Its focused but temporary effect on neurons made TMS a great tool to map the brain: researchers create temporary interference in one spot and see what happens. Unlike lesion studies, it is transient and non-injurious and can thus be used on healthy volunteers.[4]

A non-invasive brain stimulation technology, repetitive TMS (rTMS) is offered in two modes: high and low frequency. Through the application of fast and consecutive stimuli through the magnetic coil, rTMS modulates excitability of the cortex, creating an effect that lasts beyond the period of stimulation.[5] As rTMS is administered day after day, the duration of the changes in brain activity last longer and longer following the treatment session, from minutes to hours to days after the end of a session.[6] The two modes for rTMS are based on the frequency at which the magnetic field oscillates during magnetic stimulation. Low frequency rTMS operates at up to one Hertz – one cycle per second – and is administered to the right prefrontal cortex of a patient with depression. It decreases excitability of the cortex where it is applied – an attempt to quiet the hyperactive right side. With high frequency rTMS, a 3 to 20 Hertz oscillating field is administered to the left prefrontal cortex, increasing its excitability with a resulting increase in activation of the hypoactive left side.[7] Unsurprisingly, for the effects to last beyond the active treatment session, neurotransmitters, neuronal growth factors, and processes of synaptic plasticity are also activated under this treatment. When rTMS is therapeutically applied, day after day, these factors create a lasting effect on the brain tissue and functioning.[8]

Of the two modes, high frequency rTMS to the left dorsolateral prefrontal cortex has the most evidence proving its efficacy on depression. More than 89 individual trials have been conducted to test rTMS in depression, and 4 large multi-site trials all found statistically and clinically significant effects of daily prefrontal rTMS for 3 to 6 weeks, compared to sham treatment, in treatment-resistant depressed patients, though it was less effective in patients with psychotic depression and in the elderly.[9] The multi-center randomized controlled trials demonstrated response rates varying between 15 to 37 percent and remission rates of 15 to 30 percent.[10] In 2014, the European Chapter of the International Federation of Clinical Neurophysiology found sufficient evidence to assess high frequency rTMS to the left dorsolateral prefrontal cortex as having "definite efficacy," while they also found enough evidence to rate low frequency rTMS to the right dorsolateral prefrontal cortex as having "probable efficacy."[11] A couple of studies have attempted to combine the two modes at the same time – bolster the left side while simultaneously quieting the right – but they just didn't get positive effects beyond what they already saw with left-side only treatment. Buoyed by these positive results, rTMS has been accepted by the APA, the Canadian Network for Mood and Anxiety Treatments, and the World Federation of Societies of Biological Psychiatry as an evidence-based treatment option for depression.[12]

There are very few adverse effects reported for rTMS. In a few cases it induced mania or hypomania,[13] but in Dr. Don DuBose's experience, it has a very benign side effect profile. "TMS is clean, in my opinion," he said. "No memory loss, no major issues with TMS. The only thing we've observed, and the only thing I really tell the clients when they come in, the most common problem people will have is discomfort at the treatment site. Because clients can feel that magnetic pulse being generated. It's like a little woodpecker; it's kind of a weird sensation. I've had it done before. It feels like a slightly electrical, intense tapping sensation. It can be uncomfortable but clients acclimate to it over the 36 treatment sessions. And sometimes people who are vulnerable to headaches or migraines will have headaches. That can be problematic sometimes. There is an exceedingly rare chance of seizures, less so that one of our common antidepressants, Wellbutrin. At the time, I think last year, [the manufacturer] put out the statement they had done over 4 million TMS treatments, and there had been 12 documented seizures. That's very low."

I asked him about the response rate he has seen in his years of using rTMS in his practice. "In the clinic, it outperforms the research data," he said. "Mostly because I think if you look at the research, especially for FDA approval, it's got to be just the machine. So you can't do meds, you can't do therapy, you can't do other stuff. That's so you don't have confounders when you're getting your evidence together. But in a clinical setting, you're putting everything together: you're nurturing the patient, encouraging the client, and yada yada yada. I've talked with other TMS providers too, and they say about the same thing. We're seeing about a 70 percent response rate for clients with treatment-resistant depression. But it's still not mainstream, which is bananas to me. So, TMS has been very successful. If I treat ten clients, I expect seven to get better, and I expect about five to have remission."

In the treatment room, the device looked like a dentist's chair, with an arm suspended above the headrest through which the magnetic field would be applied. "Basically, the patient sits here, the coil sits on their head," Dr. DuBose explained. "I'll approximate where I think the motor strip is, that's the part of the brain that controls movement. We'll find where the hand is and then we'll find the thumb. So we give them one pulse and the opposite hand will move. Then I can move it and walk down the hand until I find the thumb. It's a brain-mapping technique. Then we find the left prefrontal cortex. When we find that location, we record it so when the patient comes in each day they're placed in the exact same position. Then the patient gets their treatment. The treatments are about 17 minutes for the average client. There's also a form of rTMS called theta-burst; NeuroStar can do theta-burst but they haven't unlocked it because it hasn't been FDA-approved. Other companies have machines that do it off-label. Theta-burst is 3 to 5 minute treatment; you're getting a lot of rapid sequences of pulses at once." By experimenting with antenna shape and placement, researchers have also created a form of TMS that reaches deeper-lying tissues like the dorsal anterior cingulate cortex or the dorsomedial prefrontal cortex. This method is called dTMS for "deep" TMS, and is offered at some clinics.[14]

TRANSCRANIAL DIRECT CURRENT STIMULATION

TMS and TES are just part of the long history of attempts to use electricity to promote or restore health. The first records of electrical therapy date back to ancient times, when Greek and Roman physicians used torpedo fish (also called electric rays) to relieve headaches. In 1804, physicist Giovanni Aldini successfully used electrical current well below seizure thresholds to cure a 27-year old farmer of melancholia. This technique was picked up again in the 1950s through 1960s as potential treatment for pain, depression, epilepsy, and a host of other disorders and diseases. Controlled, systematic research using brain stimulation continued throughout that time, including some clinical trials of transcranial direct current stimulation (tDCS), for major depression. The research lapsed, however, in the 1970s and 1980s, edged out by the availability of antidepressant medications. It saw a resurgence at the turn of the millennium, as the limitations of the newer remedies for depression became apparent.[15]

In tDCS, two electrodes are positioned on the scalp at different locations depending on the nature of the current desired. In the treatment of major depression, the anode is placed over the area corresponding to the left dorsolateral prefrontal cortex and the cathode is placed over either the right dorsolateral prefrontal cortex or the right supraorbital region (above the eye socket). A weak direct electrical current flows from the cathode to the anode. Unlike rTMS, the current is too weak to make any neurons fire, but it does affect the degree of polarization in the neurons it passes through so that brain areas close to the anode move closer to their threshold for depolarization and those near the cathode move farther away. As a result, it will take less neurotransmitter binding for neurons in the left dorsolateral prefrontal cortex to fire, while the neurons in the right dorsolateral prefrontal cortex or right supraorbital region become hyperpolarized – they will require more neurotransmitter binding to fire. The current, between 0.5 and 2 milliamps (mA) is only strong enough to nudge the neurons' membrane potential. (To illustrate, a 2 mA current can change the membrane potential of a neuron by 0.8 millivolts (mV). It takes a change of at least

20 mV to trigger an action potential, so depending on whether you added or subtracted 0.8 mV, you just made it a little easier or harder to reach that 20 mV level.) These effects have been shown to endure long after the end of the application of current.[16]

The tDCS current is weak enough that it only directly modulates activity in the neocortex. However, because the neurons in that area have projections throughout many other structures of the brain, the effects propagate deep inside. Since the time that it came back into active investigation about 20 years ago, researchers have shown they get differing results based on cathode placement, amount and duration of charge, frequency of treatments, intervals between treatments, and length of treatment. Adverse effects have been mild and transient, and include itching or burning at the sites of the electrodes, tingling, headache, or a general discomfort. In several randomized trials, tDCS actually caused hypomania or mania in unipolar depressed patients.[17]

While tDCS is non-invasive, has few and mild adverse effects and poses less of a burden to patients than some other options, it also has limitations. Blinded, sham-controlled trials of tDCS have had mixed results, but enough evidence has accumulated that in 2017 the European Chapter of the International Federation of Clinical Neurophysiology assessed it as having "probable efficacy." That assessment was limited to anodal tDCS of the left dorsolateral prefrontal cortex with right orbitofrontal cathode placement in patients without drug-resistant major depression. They also had enough evidence to assess that it was probably ineffective in patients with treatment-resistant depression and did not have sufficient evidence either way to assess the efficacy of tDCS when the cathode is placed over the right dorsolateral prefrontal cortex.[18] Overall, in a meta-analysis of 6 randomized, controlled trials involving 289 patients that included both cathode placements and treatment-resistant patients as well as non-treatment-resistant patients, tDCS showed a response rate of 34 percent (compared to 19 percent who responded to the sham treatment) and remission rate of 23 percent (compared to 13 percent for sham treatment). Presumably, the rates of response and remission with the more advantageous cathode placement would be higher, and efficacy also depended on the current dosage amounts and schedule.[19]

In contrast to ECT, tDCS has been shown to actually enhance cognitive performance, including memory, and it is possible to buy a tDCS system over the internet as a "neuro-enhancing" device. Studies on healthy subjects show improvements in working memory, learning, and long-term memory through tDCS. Some of the efficacy trials evaluating it on depression also reported positive gains in cognitive tasks as well, while others found no change. These results came from the electrode set-up endorsed by the European group (with cathode over the right supraorbital region) only. One study found that the other cathode placement option, over the right dorsolateral prefrontal cortex, actually *reduced* implicit learning in depressed subjects. Other surveys have reported similar results: under some electrode placements, tDCS can *decrease* working memory and other measures of cognitive performance, while the "good" parameter settings can improve them. There is another advantage to tDCS over other stimulatory forms of treatment for depression: the devices are small enough and can be manufactured inexpensively enough to allow for home use. By saving patients a daily trip to a clinic, this could increase compliance rates for treatment. Clinicians are understandably concerned that people might misuse such a device – hook it up backwards and make their depression worse, or set the cathode wrong and make themselves stupid – but there is currently no regulation in this area.[20]

NEUROFEEDBACK

Another promising tool to help people recover from major depression is neurofeedback, a process that gives patients insight into their brain activity, training them to use imagery or sound cues to reorganize that activity. Neurofeedback currently captures brain activity from either EEG or fMRI data and presents it to the patient. When patients have a way to perceive their brain activity directly, they can be trained to use cues to alter it. The earliest trials of this therapy used EEG, and have shown that participants can be trained to influence their scalp electric activity. Because it lacks some specificity in the networks and circuits being displayed, it has proved difficult to influence specific mental states or treat psychiatric disorders with EEG-based neurofeedback.

Real-time fMRI has a delay of several seconds between when the brain activity happens and when it is presented to the patient and requires access to a traditional fMRI suite, but has great spatial resolution, fidelity, and whole-brain coverage, reaching deep subcortical structures.[21]

EEG-neurofeedback (EEG-NF) exploits a left hemisphere/right hemisphere imbalance in alpha waves. Because alpha activity of the EEG is commonly linked with lower metabolic activation, left-side hypoactivity is reflected in lower alpha wave power than seen on the right side in depression. In an "alpha asymmetry" protocol, patients with depression are trained to decrease right-hemispheric alpha activity, increase left-hemispheric alpha activity, or shift an asymmetry index in order to rebalance activation levels in favor of the left hemisphere. EEG-NF has shown success in randomized (but not blinded) trials, but researchers cannot dismiss the possibility that the game-like nature of the therapy is affecting those results as much as the feedback process itself. EEG-NF provides prompt feedback on the order of milliseconds and the suite is portable, making it more widely available than fMRI.[22]

Despite its feedback delay of several seconds, real-time fMRI neurofeedback is getting a lot of attention now because it allows patients much more specificity in the networks or circuits they focus on. In fact, targets for change can be chosen at the beginning of a session. A localizer scan with emotionally-charged pictures identifies areas involved in the processing of positive or negative stimuli, and then the patient and trainer work on controlling those activations. Some of the protocols that have been tested in studies include focusing on connections of the amygdala, while another focuses on activating the salience network.[23] A meta-analysis in 2015 found 12 studies of fMRI neurofeedback that focused on 8 different target regions, involving a total of 175 subjects and 899 neurofeedback runs. They found that no matter what the direct target of regulation, the anterior insula and the basal ganglia, in particular the striatum, were consistently active during the regulation of brain activation across the studies. The anterior insula is a key node of the salience network, and the striatum holds the nucleus accumbens, involved in motivation and reward.[24]

VAGAL NERVE STIMULATION

Vagal Nerve Stimulation (VNS) was first used in the 1880s to treat epileptic seizures; it lapsed from use and was revived in 1938. When clinicians successfully treating epileptic seizures with VNS observed a positive effect on patients' mood, they began to explore its utility in relieving depression too. VNS can affect the stress response through indirect connections to the hypothalamus (part of the HPA axis), and imaging studies have shown it to cause widespread activation in several brain structures, including the dorsal raphe (where serotonin is produced), locus coeruleus (site of norepinephrine production), amygdala, anterior cingulate, anterior insula, and nucleus accumbens.[25] It may also have its effect via anti-inflammatory pathways activated through the vagus nerve. The vagus nerve conveys information both ways between the gut and the brain. Cytokines released in the gut activate receptors on the vagus nerve, carrying a signal into the brain, which in turn produces an anti-inflammatory response. Vagus nerve activity constitutes a central element of the microbiome-gut-brain axis, and the stress response system of the HPA axis is intimately connected not only to inflammatory pathways but also to production of growth factors and glutamate signaling.[26]

In 1994, the FDA approved VNS for drug-resistant epilepsy, followed by treatment-resistant depression in 2005. As of 2017, approximately 100,000 patients in the U.S. had been treated with VNS for epilepsy and about 5,000 for depression. In the form of VNS currently FDA-approved for depression, an embedded electrode is wrapped around the left vagus nerve in the neck and connected via subcutaneous cable to a pulse generator implanted in the left chest wall. A 5-year prospective study (not randomized, not sham-controlled) published in 2017 in the *American Journal of Psychiatry* compared VNS in combination with other treatments to treatment-as-usual in 795 depression patients. All the patients had chronic or recurrent depression and had failed four or more depression treatments. The results after 5 years showed a remission rate of 43 percent in the VNS-adjunctive group versus 26 percent of the treatment-as-usual group. The surgical procedure and use of the

VNS device bring a risk of adverse effects: voice changes, infection, reaction to anesthesia during the surgery, lower facial weakness, cough, headache, and pain. Additionally, the battery in the implanted devices must be changed – surgically – after 5 to 10 years. The effects of VNS on cognition and suicidal behaviors appear to be positive: patients treated with VNS have noted improvements in alertness, memory, and thinking skills, and the 5-year study showed a much lower suicide rate in the VNS group than the treatment-as-usual group.[27]

Though the FDA-approved application of VNS for depression is through surgically-implanted electrodes and controller, a non-invasive option is coming into use. Transcutaneous auricular VNS (taVNS) has been approved for use in some European countries to treat epilepsy, and is being tested to treat depression. With taVNS, the electrode is applied to the skin at the ear or left jawbone to stimulate the vagus nerve – no part of the device is embedded. A doctor I spoke with who is involved with development of the device to treat depression said "You would wear it like a hearing aid." Patients apply the treatment themselves at home for a specified duration of time. The parameters of the most efficacious treatment (duration, frequency of signal, etc.) are still being worked out, but several sham-controlled trials have shown positive results, with a 50-percent reduction of depression scores after just a few weeks. Adverse effects observed so far include a ringing in the ears and pain at the stimulation site.[28]

Dr. DuBose's practice has been up and running for over 5 years now, and he's seen marvelous successes from TMS. Reaching out for novel solutions for patients for whom it didn't succeed, he brought ketamine into his practice as well, again battling through the barriers of insurance companies. He's experimenting with neurofeedback now, seeking the training necessary to offer ecstasy for PTSD patients, and closely monitoring progress and approvals for psilocybin. "Our vision for the practice is to get out of the box," he said. "One, I want to make this a community. I want the clients to feel a sense of community because there's so much stigma in mental health. We're trying to normalize it.

Being open about it, so people don't feel like they have to hide it. If you hang around our office enough, you'll see us breaking some of the rules. We hug the clients; we support them. Because it can be therapeutic, and it's not so sterile; you feel like you're part of something when you come here. Two, I want to make sure that if something is cutting-edge, has possible efficacy, lower risk, and is cost-effective, then I'll give it a shot. I want clients to have more options, so maybe they can find a solution to their problem."

Chapter Twenty-Three

MEDICINES

The STAR*D trial's findings – showing that even under real world conditions antidepressant medications do work effectively for over half of those who use them – were encouraging, but also raised a lot of questions. One of the most notable results was that the trial revealed no statistical differences in any measures based on the brand or even the type of medication used: TCAs, MAOIs, SSRIs, atypicals, etc., and ditto for the "switch" or "augment" strategies. It really reinforced the growing consensus that all traditional antidepressant medications, no matter what their immediate target – serotonin transporters or dopamine receptors, for example – get their antidepressant effect from a distant downstream action of increasing neuroplasticity. Newer, rapid-acting antidepressants (like ketamine) and others in testing (including hallucinogens) are also thought to have antidepressant effect from stimulating growth, but along a shorter pathway.[1] They are all considered agents of "grow."

TRADITIONAL ANTIDEPRESSANTS

Antidepressants, the miracle drugs of the latter part of the 20th century, now get a lot of bad press due to the impression that they are over-prescribed and that pharmaceutical firms are taking advantage of people. Some of their side effects are significant, and many people just don't want to take them: they feel that medications just cover up symptoms rather than treating the cause of illness, or fear a stigma associated with diagnosis and treatment for a mental illness.[2] Antidepressants are the first line treatment for major depression for good reasons, though. They are accessible: insurance will pay for them and, unlike psychotherapy, they don't require continued access to a trained professional to administer. They are effective in a large swath of sufferers (given enough time, ample dosage, and often several different tries), and, usually, their side effects are more benign than the illness they treat.

There are a lot of animal studies that back up the assertion that traditional antidepressant medications work through neuroplastic processes. These studies show that even though the medication starts affecting its targets in the brain in a short timeframe, there is a long lag time before they have an actual antidepressant effect. In humans, of course, it takes weeks for someone to see an antidepressant effect. Animal studies show that chronic (but not acute) administration of traditional antidepressants results in eventual production and release of BDNF and other growth factors. Particularly with medications that increase serotonin signaling they have also observed eventual growth of GABA neurons, maturation of astrocytes, production of the adhesion molecules necessary for synaptic remodeling, and other plasticity-related activities. Each variety of antidepressant medication has a different fingerprint in the degree to which it affects this broad array of neuroplastic processes, but they all end up there.[3]

To say that all these medications eventually operate through the same mechanisms does not mean that the individual doesn't matter. Variations in a person's genetic makeup can drive diverse experiences, as can interactions with other medications taken for co-morbid conditions. For example, the cytochrome P450 family of genes affect how rapidly and thoroughly someone's body metabolizes drugs, impacting how much of a particular drug will survive long enough to get to the brain. The various CYP450 genotypes also affect into what resulting products a drug is metabolized, sometimes leading to adverse side effects or unwanted drug interactions.[4] Though individual experiences vary based on these and other factors, the immediate targets of each class of traditional antidepressant broadly leads to their near-term effects and side effects.

MAOIs – monoamine oxidase inhibitors – prevent a monoamine like serotonin, norepinephrine, or dopamine from being broken down into a waste product, thereby leaving those neurotransmitters around longer to continue signaling. In doing so, they enhance the activity of dopaminergic "reward" pathways, desensitize or reduce in number serotonin and norepinephrine inhibitory autoreceptors, and reduce or desensitize 5-HT2C and beta-adrenergic receptors as well. Unfortunately, MAOIs are absorbed into the body too, where they inhibit the breakdown of tyramine, risking a life-threatening hypertensive reaction.[5]

Most of the TCAs, including imipramine and amitriptyline, have a broad base of action. They block reuptake of serotonin and norepinephrine in the prefrontal cortex, hippocampus, and amygdala. Blocking this reuptake function means that those neurotransmitters are left available in the synapse longer to do more signaling. TCAs also block several types of receptors for serotonin and norepinephrine, and usually, muscarinic cholinergic, and histaminic receptors as well. This blockade of so many types of receptors is believed to cause their negative side effects: drowsiness, dizziness, weight gain, and perturbed cardiovascular function. Muscarinic cholinergic receptors in the body affect the heart, smooth muscle, and exocrine glands too – once again, the medicine does not limit its effects to the brain.[6]

SSRIs (including sertraline, fluoxetine, and citalopram) are the most commonly prescribed medications today – the first of the first-line treatments for depression. They block serotonin reuptake more specifically, avoiding the shotgun-blast of TCAs. The surge in synaptic serotonin availability resulting from SSRIs leads to desensitization of 5-HT1A autoreceptors, reducing their feedback inhibition function in order to increase serotonin release. At the same time, those higher serotonin levels outside the neuron prompt adaptive changes including a reduction of 5-HT1A and 5-HT2A receptors. Of course, by elevating synaptic concentrations of serotonin, SSRIs eventually engage all classes of serotonin receptors throughout the central nervous system.[7]

In the continuing effort to hit the right targets, and not the unnecessary targets – while still not knowing precisely what falls into either group – pharmaceutical companies developed and tested substances that would inhibit reuptake of both serotonin and norepinephrine, but, unlike TCAs, would not affect muscarinic cholinergic or histamine receptors. Their results became the serotonin-norepinephrine reuptake inhibitors (SNRIs), like venlafaxine and duloxetine. By targeting both serotonin and norepinephrine reuptake sites, SNRIs were anticipated to be faster-acting and more effective. Some studies have borne this out; others have not. With substantially lower activity on other types of receptors, SNRIs still have side effects similar to those of SSRIs, but are thought to be more effective in relieving pain and other somatic symptoms.[8]

Reboxetine is a norepinephrine reuptake inhibitor (NARI). NARIs operate directly on noradrenergic pathways and indirectly recruit all types of alpha- and beta-adrenergic receptors by elevating synaptic levels of norepinephrine. Animal studies indicate that reboxetine treatment increases dopamine levels in the frontal cortex but not in other parts of the brain. Some studies show reboxetine to be equally or more effective than TCAs or SSRIs in subsets of patients with severe depression and in elderly patients. NARIs have a different side effect profile, including cardiovascular issues resulting from adrenergic receptor activation.[9]

There are even selective dopamine reuptake inhibitors (DARIs), such as bupropion. Though intended to target dopamine uptake sites selectively, bupropion is rapidly metabolized into a substance that acts as a norepinephrine reuptake inhibitor instead. Through its entire lifetime of action, though, bupropion elevates dopamine levels in the nucleus accumbens and norepinephrine levels in the frontal cortex. It also blocks several classes of nicotinic receptors, making it useful as a prescription smoking cessation aid.[10] Once bupropion's many other effects became apparent and a few more pathways of action identified, it was moved to the class of "atypical" antidepressants.

In addition to bupropion, other atypical antidepressants include mirtazapine, nefazadone, trazodone, and tianeptine; it seems to be an "all others" category. Mirtazapine is a tetracyclic antidepressant (TeCA), blocking 5-HT2A and 5-HT2C, alpha-adrenergic, and histamine receptors. It is thought that the blockade of the 5-HT2A serotonin receptors not only improves mood but also reduces release of stress hormones. Mirtazapine's side effects of drowsiness, dizziness, and weight gain can actually be useful when symptoms of depression include a broken sleep cycle and loss of appetite. Nefazadone and trazodone both blockade 5-HT2A and 5-HT2C receptors as well, and act as serotonin reuptake inhibitors. Both of these drugs rapidly metabolize into other products that have a weak antagonistic effect on dopamine and alpha-adrenergic receptors, and interact with an array of serotonin receptors. These properties lead to another slate of possible side effects that can include sleepiness and mental slowness.[11]

Tianeptine is truly an atypical, atypical antidepressant. Approved for use on depression in Europe, Asia, and Latin American but not in the U.S., clinical data indicates it is as effective as SSRIs, without SSRI side effects. Tianeptine increases serotonin re-uptake – it is the *opposite* of an SSRI – and activates opioid receptors. Its potential for abuse and withdrawal symptoms that mimic opioid toxicity have raised concerns at the FDA and CDC. Tianeptine is an unscheduled substance, but the FDA has issued warnings to companies marketing it as a dietary supplement because of the medical claims they are making.[12] One of tianeptine's known mechanisms of action activates processes of neuroplasticity through glutamate neurotransmission. Animal studies have shown a strongly protective effect of tianeptine on brain structure and neuroplasticity, particularly in the hippocampus, where it prevents cell death and enhances adult hippocampal neurogenesis.[13]

Another new antidepressant affects 5-HT2C receptors and results in increased cell proliferation, survival, maturation, and neurogenesis, but is actually focused on circadian rhythm disruptions:

agomelatine. Agomelatine activates two types of melatonin receptors while blockading 5-HT2C receptors. It was developed with the intention of regulating disturbed circadian rhythms, and preclinical and clinical testing indicate it does so. When tested in comparison with venlafaxine, sertraline, and placebo, agomelatine was just as effective as the drugs, superior to placebo, and showed effectiveness about a week earlier than venlafaxine. Unlike ketamine or sleep deprivation therapy, agomelatine doesn't show immediate efficacy against depression, but its eventual mode of action may involve a circadian rhythm reset as well as downstream impact on the growth factor BDNF and other elements of neuroplasticity.[14]

Some medications, though not developed for depression, are often prescribed alongside traditional antidepressants. Lithium, for example, has long been used to treat bipolar disorder but is also sometimes used as an adjunctive medication to TCAs, SSRIs, NARIs, and the SNRI venlafaxine. It strengthens the effects of the medicines targeting the serotonergic system, and seems to have less effect on the norepinephrine system. Animal studies indicate that lithium also has a strong influence on GABA neurotransmission, with long-term use resulting in an increase of GABA-B receptors in the hippocampus. Other common, and FDA-approved, adjunctive medications come from the family of atypical antipsychotics. The atypical antipsychotics approved by the FDA (but only when used along with an SSRI) include aripiprazole, brexpiprazole, quetiapine, and olanzapine (approved with fluoxetine only). All these medications block 5-HT2 receptors; the first two weakly activate dopamine D2 receptors, and the latter two block dopamine D2 receptors.[15]

So the path to relief via traditional antidepressants involves side effects and a lag time of weeks for antidepressant effect to possibly be seen. And in the real world, many people have co-morbid psychiatric and medical conditions, and any medications they are taking to deal with them can be a wild card in their response to a new depression medication. In the real world, too, people may forget to take their meds, or take too much; they may have limited access to medical care if there is an unwanted effect, and they may be limited in what therapeutic options are supported by their insurance.

RAPID-ACTING ANTIDEPRESSANTS

The exciting story of the last few years has been the success, and FDA approval, of ketamine as a rapid-acting antidepressant. The numbers may change a bit as clinical experience grows, but it starts with 60 percent of people who take it see their symptoms relieved, usually in a few hours but it may take up to a day. They relapse from one dose in a week to 10 days, but clinical studies using repeated doses stretched the effect over time and were able to reach 70 percent of the patients. And these are people with treatment-resistant depression.

Even great breakthrough treatments face challenges in the real world, though. Dr. Don DuBose of Future Psych Solutions has seen a lot of success using ketamine, but keeps hitting barriers in making it available to his patients.

"We got into ketamine mostly because of those three out of ten patients that didn't respond to TMS," he said. "Nothing is more distressing than seeing a patient where TMS was kind of their last hope. 'This is the last thing, I've tried every medicine, I've done every therapy; I've tried everything.' Then I talk about these great results from TMS – '70 percent of people respond; it's so great! But 30 percent don't, and unfortunately, you're in that 30 percent.' That's devastating. So, watching those other clients suffer sparked me to ask, 'What else is out there?' So we found ketamine.

"When we started it, about 4 years ago, we started really slow because I was scared of it. You hear of it as a horse tranquilizer, street drug, and surgery medicine. As a psychiatrist, we're trained mostly to talk, decipher, prescribe. That's what we're taught. So to be doing a procedure... TMS helped, because I was used to doing a TMS procedure. But moving to an IV treatment was a big step for me. But ketamine has been out since about the Vietnam War, and it's cheap. You can buy a vial of ketamine for between 12 and 20 dollars, and I can probably do two or three treatments out of one vial. When we started it, you'd see these small trials, 30 people, 15 people, 10 people, 40 people, things like that. I would read *Current Psychiatry*, and it would be so skeptical: 'It might show promise, but we need bigger studies,' and 'We advise you not to do it.' But people are sitting here suffering, and there's something that the preponderance of evidence says it works. So, once I found out it wasn't a respiratory depressant, safe to do outpatient, there were other clinics popping up, and I finally found somebody to train me, we stepped out on faith to give it a shot.

"At the time, because the medicine was so cheap, nobody would do that large, multi-center trial. There was no way for them to make money. I was talking to one of the reps for one of the companies. He said, to do one of those large trials, it might cost a company 50 million dollars or more. The drug companies think, 'What can we do to earn a lot of money on this?' That's the problem, that's why drug costs are so high in America, because the FDA necessitates these large trials to get an approval. So what they did was, ketamine is what they called a racemic mixture. That means there are two similar molecules that make up the medicine. So they took one part of the medicine out of it, esketamine. They patented it and said 'This is unique.' And they made it a nasal spray instead of an IV medicine. The FDA was watching the small studies and said 'Ketamine is a breakthrough treatment; we're going to give you an easy path to approval.' So normally the approval process takes 4 to 5 years; I think they got it through in 2 years. Even now, I think a lot of psychiatrists are skeptical of them, because they got through FDA so quickly. That's why you don't see a lot of centers. And there's a lot of logistical problems to doing the nasal spray. And the nasal spray is kind of where TMS was in the beginning, insurance will fight you about paying for it."

In Dr. DuBose's experience, insurance resists paying for IV ketamine treatments also. In his practice they tend to start a new treatment-resistant patient with an IV treatment on a cash basis to see if they respond before moving to the nasal spray and commencing battle with insurers over the approximately $2,000 per session cost. The insurers are starting to come around now, though, with most of them providing at least some coverage for the treatment.

Even ketamine is not successful in all cases, and the antidepressant effects don't last, though it is early yet to know if the protocol of repeated doses is going to create an enduring remission. I had asked Dr. Duman of Yale about this. "Ketamine can reverse the synaptic deficits – we see these changes in our rodent models that also last for about a week, but then you get a relapse," he said. "The thinking is that while you are able to reverse these changes in a short term fashion, that the underlying pressures that are producing the deficit have not been removed, that there are still problems, either in some other neurochemical imbalance or the environmental, social interactions that cause pressure to produce that deficit. Those have to be removed or corrected before you can have an effect that would be maybe very long-lasting. A week is not bad! But it's really something that is on a lot of peoples' minds."

The approval of esketamine gives patients a true pharmacological alternative to monoamine-based antidepressants, and has spurred research into additional rapid-acting antidepressants, along different parts of the pathway to release of BDNF and other growth factors. Scopolamine is a candidate undergoing preclinical testing, with great hopes. A nonselective muscarinic receptor antagonist that indirectly affects glutamatergic transmission, it produces antidepressant actions in depressed patients within a few days.[16]

NUTRACEUTICALS

With so much focus on antidepressant medications derived from rocket fuel, horse tranquilizers, and other such unlikely things, it may come as a surprise that some herbs or food can have antidepressant effect. These substances, "nutraceuticals," may involve a vitamin or mineral supplement to patch a hole in the usual diet, but some go further than that. Dr. Jerome Sarris of Western Sydney University works with many randomized, controlled trials of dietary supplements. "We obviously want to say, 'Yes, have a great diet,' but in some cases with health professionals' advice, there are some nutraceuticals which may be of benefit to the person," he commented. "There's a range of reasons for that. Sometimes, even when a person has a good diet they might be deficient in some nutrients. Or, you can have people who may have very poor digestive function – there may be inflammation going on there; there might be dysregulation with the bowel flora; there might be issues in terms of the tight junctions with the intestinal lumen in regards to how the nutrients are being absorbed. So sometimes people do need to take nutraceuticals on top of a healthy diet."

There have been a lot of studies of vitamin and mineral supplementation (commonly addressing vitamins B9, B12, D, and zinc) as adjunctive treatments for major depressive disorder; many positive and many finding no significant effect. The dietary supplements with the strongest evidence of beneficial effect against depression are vitamin D and omega-3 polyunsaturated fatty acids (PUFAs).[17] A 2018 meta-analysis of 35 randomized, controlled trials of the effects of supplementation with the PUFAs DHA and EPA initially found only a small overall effect on depression. When the studies were narrowed down to those in which participants were clinically diagnosed with major depression (leaving behind studies where participants had lesser degrees of depressed mood) the results of treatment with omega-3 PUFAs were stronger, equivalent to a clinically-relevant improvement.[18] A 2016 meta-analysis found an overall beneficial effect of omega-3 PUFAs on depressive symptoms in major depressive disorder, with the most potent effects in patients also taking antidepressants and where the supplement had a high dose of EPA rather than DHA.[19]

And just like genetic factors impact how the body processes antidepressant medications, they also impact how the body processes food. As a result, even someone with a great diet may not be getting the nutrition they need for brain health. Dr. Sarris provided an example. "You get your folate [vitamin B9] from food, especially your whole grains and leafy vegetables," he said. "However, the body needs to be able to transform that into the activated form; it needs a particular enzyme. Depending on which genetic alleles you have, you may or may not be able to process or metabolize the folate or folic acid into the activated form: methylfolate." Evidence from randomized controlled trials shows that methylfolate is effective as an adjunct to traditional antidepressants in reducing symptoms of depression,[20] but obviously, not everyone would need that sort of intervention.

S-Adenosyl-methionine (SAMe) is also manufactured in the body from folate, and also available in prescribed form in case the body is not making enough. SAMe donates methyl groups for the methylation of DNA, RNA, neurotransmitters, and proteins. While its mechanisms of antidepressant action are unknown, more than 45 randomized, controlled trials of SAMe in depressed adults across Europe and the U.S. indicate it is more effective than placebo and generally equally or more effective than TCAs.[21] NAC, or n-acetyl-cysteine, is more commonly prescribed for bipolar disorder than major depression, but has also been demonstrated to reduce depressive symptoms.[22] NAC is an activated form of cysteine, so it also skips over a metabolic step that could be operating inefficiently in a patient.

Carnitine, an essential nutrient found in meats and dairy products, carries fatty acids across the inner mitochondrial membrane, assisting with energy metabolism. It also helps in other aspects of mitochondrial function, with a net effect of protecting the cell. Animal studies point to effects on neuroplasticity and regulation of neurotransmitters as well as a role in fatty acid metabolism as possible mechanisms for its effect. Several randomized, controlled studies investigating the clinical effect of carnitine on depressed patients have shown it to be significantly more effective than placebo in reducing symptoms of depression, though some studies showed no difference.[23]

Several herbs have been demonstrated to be effective against depression in randomized, controlled trials as well, with the best evidence supporting St John's Wort and saffron, and less, but still positive, evidence for turmeric.[24]

St John's Wort grows as a common weed around the world and has been used as a therapy for insomnia and "nervous conditions" for centuries. The FDA classifies it a dietary supplement, not subject to safety or efficacy testing. Researchers, however, have run it through randomized, controlled trials to compare it to placebo and antidepressant medications. A 2016 review of such trials found St. John's Wort superior to placebo and equivalent to antidepressants in effectiveness in mild to moderate depression, but lacking evidence in severe depression. Though easily available at supermarkets and health food stores, St. John's Wort can have adverse effects: it interferes with oral contraceptives, immune suppressants, and anticoagulants, and can also interact with SSRIs in a toxic reaction. Containing at least ten substances shown to be biologically active, never think St. John's Wort is "just an herb" and therefore innocuous.[25]

Saffron is an herbal mood enhancer in Persian traditional medicine, originating in Syria. Randomized, controlled trials comparing saffron with placebo and with an SSRI for 6 weeks found that saffron was significantly more effective in reducing symptoms of depression than placebo, and found no significant difference in efficacy between saffron and the SSRI. While its mechanism of action is unclear, it has been proposed that two of its component ingredients inhibit the re-uptake of dopamine, norepinephrine, and serotonin.[26]

Turmeric, a common cooking spice, has been used in traditional Chinese medicine on a variety of health conditions for centuries. Turmeric's main active ingredient is curcumin, which possesses anti-inflammatory, antioxidant, and neuroprotective qualities, and also modulates the monoaminergic system. Turmeric or its curcumin extract have shown positive results against depression in comparison to placebo and as an augmentation to an SSRI.[27]

Experience with nutraceuticals shows that they can be very potent. Natural doesn't mean innocuous, though, and effective for someone doesn't mean effective for everyone. "It really is a case of what may work for one individual might not work for the other," Sarris commented, "and that's why it's important to get good health professional advice, and to also choose quality supplements as well."

PSYCHEDELICS

As a teenager, Eddie figured it out: he could hide his pain and punish himself thoroughly in sports.

"I learned to bury my anxiety in sports, all through high school. But then I graduated, and had to look for a career. I was directionless in life, and the only thing I felt confident in was sports. Even then I wasn't the best, but I could work harder, punish myself, push the limits of pain, because sports allows you to do that in an accepted setting. You can't just do that in life outside of sports because it's self-destructive. But in sports, you can mask a lot of pain in that. But at some point you can't do that anymore. The body can't do that anymore and you have to deal with life.

"Then 9/11 happened, and someone mailed anthrax around, and you had to leave your mail box open, and that was sort of a trigger. I remember going on a rant, this anxious, angry rant. I was so angry about that whole thing. My wife was like 'What's wrong?' and I couldn't explain it. It was almost like confirming that the world was as bad as you think it is, and I just started a decline. I was always an avid cyclist, and I remember trying to ride my bike to work, or driving to work and just crying in my car, and not knowing why. And I was having this, kind of like a mental tantrum, and I didn't know why I was so sad and angry and I just wanted it to stop. I just wanted to go home and put my car in the garage. I was missing work. I'd have to call in sick. I tried to go to counseling, I tried to go to a therapist, and I'd never been in therapy before that. I didn't like it. I didn't like the guy, he said a couple things and I just don't think he had an idea what he was saying. So I tried medication; I got a prescription, but I didn't like it. It made me shaky, had other side effects, so I cycled through that one and tried another one. Again, I didn't like the physical side effects of it, so then I stopped it altogether. I never went back to the counselor; I just kept exercising.

"I remember riding my bike on long rides, and just crying on my bike, thinking about stuff, and all twisted up in my head about fantasies of escape. I had all these fantasies in my head, like I just want to get out of here. I don't want to be married, I don't want to be in Sacramento; I just want to get in my car and keep driving. I was suicidal, but it has always been more of a fantasy than an actual plan. It's always been a really strong ideation to the point of just breaking down and sobbing without knowing what to do. I've never really done anything about it, except maybe self-medicating. Back then I didn't drink much, I just did more pain through sports. It was really self-punishment in that way."

Eddie went back to counseling, read a lot about psychology, and realized he really liked the subject. Enough, in fact, to get a masters degree in counseling and change his career. He appreciates the enlightenment that process brought, but sees limits. "Even knowing things academically and intellectually about yourself doesn't translate into making things better. Which is kind of frustrating. You can logically map things out, and you can very intellectually speak about something, but if you don't dig into the root cause of the emotions and express those emotions in some manner, it's going to stay the same. Just because you read a book on depression doesn't make it go away. I've done a lot of work in that program, and I've done a lot of counseling since. And I have periods when it's good, but I know that at least once a year I'm going to have some dark-ass period that I don't know how to keep at bay. Some years it comes and goes quickly and some years it's a train wreck."

As a sufferer himself, and possibly as someone who upended his life once already and found the experience beneficial, he has been willing to experiment. "About 4 years ago I started having the dreams, the fantasies of escape again. I went to a therapist and we worked on this stuff, but it got to the point where my wife and I were going to separate. We made the decision; we had the house about ready to go; we told my son; we looked at apartments. And then I told a friend at work what was going on and she said 'I go to this retreat center in South Florida, they do *ayahuasca* retreats. Maybe you should try that.' So I went down there and I did that one day retreat, with an *ayahuasca* ceremony, and had some pretty intense experiences, like I had some conversations with my child self. It was really a physical medicinal intervention. I'd never done anything like that before. When it was done I felt so calm, so open, like open-hearted. I didn't have any of the angst anymore. Everything just feels alive; everything just feels like love. Like nothing I've ever felt before; I'm a pretty reserved person. And I remember right after the ceremony I texted my wife 'It's gonna be okay.' We stayed together."

Eddie has been to more *ayahuasca* ceremonies since, and hopes to be able to take patients through guided psychedelic experiences someday. He is still looking for answers, for his own recurrent depression and for his practice. Medicines, psychotherapy, all the approaches we have, seem to largely miss an underlying problem. "For me, part of what drives my depression is isolation. I can go to work and I have acquaintances; I have friends I ride my bike with; I have a family here, and I still feel a little isolated. There's something about our day-to-day connections that's not complete. In my practice, I work with guys, teenagers and adult males, most of them coming in between the ages of 33 to 50, and it's the same story. Emotionally isolated, depressed, self-medicating. 'There's an emotional distance in my marriage, I'm unhappy with my work and my kids are a pain in the ass. And I don't know what to do.' It's the same story, over and over. You start to see these themes and you think we're all suffering the same thing, yet nobody talks about it. They have to come in and pay someone to talk about the same problem that their three neighbors are probably struggling with or have struggled with in their life at some point. There's more common, shared experience than we know."

Psychedelics are a class of drug with the primary action of altering the state of consciousness. These drugs include LSD, psilocybin, mescaline, and dimethyltryptamine (DMT), and are currently Schedule 1 controlled substances: they have no accepted medical treatment use, high potential for abuse, and a lack of accepted safety for use under medical supervision. That appears to be changing. As Michael Pollan points out in his book, *How to Change Your Mind*, psychedelics are being used in research and clinical trials to test their efficacy against depression in terminal cancer patients. More recently, testing has been broadened to include use – with psychological support – in treatment-resistant depression without terminal medical conditions.

Though psychedelics have been part of traditional and ritual use in some cultures for hundreds of years, modern science discovered and characterized the substances in the 1930s and 1940s. In the years following, LSD and other psychedelics passed into research, medical, and recreational use. In the 1960s and 1970s, psychedelics became the subject of political action and were put under severe regulatory restrictions, which have only been slightly lessened to allow medical research today. Before their restriction, however, some studies of their utility in mood disorders, including depression, were carried out. Most of these studies suffered from poor methodological quality – it was impossible to adequately blind and control the studies, for one thing; and the severity of depression at baseline and after treatment were rarely measured along a scale that is useful today – but they can inform current research. Eighty percent of the 423 subjects in those 19 older studies showed clinician-judged improvement after treatment with psychedelics.[28]

Research into the use of psychedelics in treating major depressive disorder is resuming under modern standards. A 2015 study tested *ayahuasca* (a DMT-laden tea brewed from the stem and leaves of two South American plants) in treating depression in six subjects. The study showed significant reductions of up to 82 percent in depression scores compared to baseline, with vomiting as the only side effect reported. Two other recent studies tested psilocybin and LSD in treating anxiety associated with life-threatening diseases, finding reductions in anxiety scores at 2 to 3-month follow-up periods.[29] An open-label feasibility trial of psilocybin along with psychological support in the treatment of 19 depressed patients was concluded in 2016, with positive results. The patients, who all had moderate to severe unipolar treatment-resistant depression, were administered two doses of psilocybin 7 days apart. The effect at 1 week was a marked reduction in depressive symptoms, with much of the benefit still remaining at the 3- and 6-month follow-ups.[30]

Psychedelics operate through the serotonergic and glutamatergic systems. The active compound in psychedelics is structurally related to serotonin, and activates 5-HT2A receptors. That receptor binding leads to glutamate release and activation of glutamate receptors. The effects are modulated by serotonin 5-HT1A and 5-HT2C receptors, combining into a hallucinogenic response in the brain. As downstream effects of increased activation of these same serotonin receptors is believed to underlie traditional antidepressant response, psychedelics may be taking a similar path to the same results – just with a hallucinogenic effect along the way. Animal studies indicate that psychedelics also increase levels of the growth factor BDNF in the hippocampus and have anti-inflammatory effects.[31]

Ketamine is the big leap forward in antidepressant medications of the current day, and it feels like the world is poised for a few more such leaps, like psychedelics and extensions to the family of rapid-acting antidepressants. Whether or not these treatments are actually accessible to people is another story. It should also be noted that in all of these new success stories, the patient is kept on traditional oral antidepressants, or returned to them after using psychedelics. It seems that the ability of those traditional antidepressants to turn the BDNF spigot back on – possibly with drips, possibly with greater force – helps to keep future major depressive episodes at bay. For people who get depression, the factors that put them in that state remain potent. A *drip...drip...* of *grow...grow...* may be necessary for the rest of their lives.

Chapter Twenty-Four

SEIZURE THERAPY

The note was brief, but shattering: "I'm sorry, Dad. I'm having trouble, and I'm not sure if I want to live anymore."

Steve thought his heart might stop. He looked up at his son, 15-year old Danny, and patted the seat next to him. Danny sat down. "Dad," he started, "I'm in this bad spot; I feel really horrible. I'm hurting myself and I'm worried about what I might do." He was cutting himself, self-mutilating. He had confided in a friend, who was so worried he gave Danny an ultimatum: "Tell your Dad, or I will." The note had been the easiest way for Danny to start the difficult conversation.

As Steve would tell me many years later, he was stunned. He hadn't known his son was struggling and he didn't know anything about depression, but he did know what he was hearing required immediate action. He took Danny to the hospital that night, where he was admitted for 72 hours observation. "All they did was follow him around," Steve said. "They put him into a treatment center, so from the emergency room we went to kind of a section 8 place in the hospital and moved him over to an actual institute or something. And so he spent three days there. They gave him some meds, calmed him down, talked to him, and released him." It was the first of four hospitalizations for Danny, who was eventually diagnosed with bipolar II disorder.

It has been a long education for Steve, and Danny, on mood disorders. "Everything else aside," Steve said, "from what I've viewed over the last 10 years, is that it seems to be a disease. A cycle, like a sine wave, it comes and goes. It gets horrible, and it gets better, and does it again and again. And the frequency, I haven't determined exactly. But I think that's what's underlying this. He has good times and bad times, and sometimes they stretch out, and other times he'll have a very depressive streak for quite a while, without much let-up. And those are tough; that's when he gets hospitalized. He has a very crisp and clear-headed suicidal ideation, and he's aware of it. It comes and goes. When it comes and it's persistent, there's some other symptoms."

Like the hypomania. "It's not equal at all; it's dominated by the depression. And in his case, the mania wasn't 'I'm going to go out, spend money and get crazy socially.' Instead, he turned inward, reflectively, and figured out the meaning of life. He literally did. He wasn't super-proud about it, and he wasn't showing off. He was saying things like, 'It's weird, Dad. It all makes sense.' And I was, like, 'That's awesome, dude.'"

They tried many treatments in the early years of Danny's illness. "So we went through the hundred drugs," Steve said. "You know, you go through and you try this... From the beginning, he was hospitalized and he's been under psychiatric care, and he's been in counseling ever since then. Notwithstanding a short period of time when he stopped taking his meds, which obviously was a disaster, he's taken lots and lots of meds. Nothing really worked for long, and, if you've had anything to do with these situations, they're absolutely throwing darts. The doctors try this, they try that, you ramp on and ramp off, you can't mix this, can't mix that. That was the first couple, 3 years, trying to get the meds right. I think sometimes, that the meds get changed arbitrarily because you think that they're not working. But I have a really strong suspicion that there's a disease coming through here, and it upsets everything else. When you see the disease, then all the things you're doing to treat it – maybe successfully, maybe not – get changed. A lot of times they'll change the meds when someone's struggling. But I think they struggle despite everything else. I think the disease comes through at some frequency, and if you don't panic, well, either you'll make it through or you won't. We tended to go through a lot of change periods whenever he would have racing thoughts or he would struggle quite a bit."

On his last hospitalization, several years after the first one, Danny started a course of ECT.

ELECTROCONVULSIVE THERAPY

It is easy to get caught up in the "electro" part of electroconvulsive therapy, conjuring images of Frankenstein's monster. But it's really the "convulsive" part that matters – electricity is just a tool.

What is now administered as ECT began in Europe as seizure or convulsive therapy. Hospitalization and clinical records of psychiatric patients collected over decades showed something odd: despite high levels of co-morbidities all around, patients with schizophrenia rarely had epilepsy and patients with epilepsy rarely had schizophrenia. Postmortem studies revealed another oddity: the brains of people who had died with schizophrenia had reduced numbers of glial cells, while the brains of people who had died with epilepsy had overgrowth of glia. Psychiatrist and neuropathologist Ladislas "Laszlo" Meduna took note of those observations and decided to test the hypothesis that a *grand mal* seizure could treat patients with schizophrenia by correcting a glial pathology. Meduna injected camphor into 11 patients to induce seizures, keeping meticulous records of the patients' symptoms, dosages, seizure characteristics, and short- and long-term response. Of those first 11 patients, two improved sufficiently to be discharged from the hospital and one recovered sufficiently to begin occupational therapy.[1]

In "Electroshock Revisited" published in the journal *American Scientist*, Professor and psychiatrist Dr. Max Fink reaches into Meduna's autobiography to paint a picture of this experiment. The treatment took place in 1934 in Lipotmezo, Hungary.

> *Zoltan, a 30-year-old Budapest laborer, lay rigidly in a bed, staring into the distance. Except for his slow and regular breathing, he appeared lifeless. He had hardly spoken or cared for himself in more*

than four years. His mental condition of catatonic schizophrenia was considered hopeless. No remedy was available and none was sought; the doctors believed the illness to be an immutable genetic fault. At 10:30 on the morning of January 24, 1934, the Hungarian neuropsychiatrist Ladislas Meduna approached Zoltan's bed to inject an oily extract of camphor into his right buttock. Zoltan's heart soon raced, sweat rose on his brow, and he became increasingly fearful. After 45 minutes, his eyes suddenly closed, his jaw clenched, his breathing stopped, and he lost consciousness. With a deep, noisy sigh, his arms and legs extended, he convulsed, and his bed thumped rhythmically; attendants caught him just before he rolled to the floor. His skin became ashen, and he wet the bed. After 60 seconds, as suddenly as the spasm started, it ended. His eyes opened, and a pink color slowly returned to his cheeks. He continued to stare and was as speechless as before. He had survived an intentionally induced grand mal epileptic fit. Without any guideline as to how often seizures should be induced, Meduna adopted the schedule used in the popular malarial-fever treatment of neurosyphilis. He injected camphor at three- to four-day intervals, and two days after the fifth seizure, Zoltan awakened, looked about, got out of bed, asked where he was and requested breakfast. He did not believe that he had been in the hospital for four years, and he knew nothing of the intervening history. Later that day, he again relapsed into stupor. After each of the next induced seizures, Zoltan remained alert and interested for longer and longer periods, until after the eighth injection he left the hospital to return to his home and to work. His mental condition of four years was fully relieved. Five years later, when Meduna left Europe for the United States, Zoltan was still well and working at his job.[2]

Chemically-induced seizures had been used as far back as the 16th century, when the Swiss alchemist Paracelsus gave camphor to patients orally to induce convulsions and "cure lunacy." Camphor proved to be unreliable in inducing seizures and was replaced with metrazol, and Meduna's treatment started to spread throughout Europe. The use of any chemical had its drawbacks, though. One of Meduna's early patients developed an abscess at the injection site, and the chemicals gave the patients extremely unpleasant sensations. The seizures were often delayed, leaving a patient in discomfort and terror for up to 90 minutes following the injection. In 1938, Italian scientists Ugo Cerletti and Lucio Bini improved the therapy by employing electrical shock to induce a seizure. Using this technology, they treated an unidentified 39-year-old man who was found delusional in a train station. The man's delusions receded after several treatments and he recovered fully after 11 treatments without adverse effects.[3]

Electroconvulsive therapy came to the U.S. in 1940, flourishing through the 1940s and 1950s. In those days, before the development of antidepressant medications, ECT was a mainstay treatment for a variety of psychiatric disorders. Other available treatments, such as insulin coma and psychosurgery, had poorer effectiveness and often dire side effects. The short, retrograde amnesia patients experienced with ECT was thought to be part of its curative properties, relieving distressing memories. It was also helpful that patients had no bad memories of previous treatments as they approached another.[4]

Early applications of ECT were very rough, though, and with the advent of antidepressant medications, ECT began to fall by the wayside. Initially applied without anesthesia or muscle relaxants, ECT sessions involved significant physical discomfort and sometimes resulted in fractures. It also produced side effects, including memory loss, sometimes to a distressing degree. With a dearth of alternatives, ECT was widely and sometimes inappropriately applied, sometimes without the patient's consent. These conditions, prominent in the 1950s, were vastly ameliorated with the use of anesthesia, muscle relaxants, oxygenation, standards for informed patient consent, and through changes in waveform and electrode placement.[5]

These improvements came too late to save the reputation of ECT. In the early 1960s, writers Ernest Hemingway and Sylvia Plath both committed suicide after ECT treatment. The 1962 book and Oscar-winning film *One Flew Over the Cuckoo's Nest* depicted a brutal psychiatric institution where ECT was used on unwilling patients as punishment. Legislative and judicial actions to regulate ECT began in the 1970s in response to strong political lobbying by people who had suffered under unmodified ECT or endured severe cognitive impairments after receiving ECT. Several states passed laws restricting or regulating its use. In response to the backlash, the APA set up task forces in 1975 and 1990 to review the practice of ECT in the U.S. and make recommendations. They set standards for obtaining consent, qualifications and credentialing of psychiatrists, and other matters. A consensus conference sponsored by the NIMH in 1985 made new recommendations for practice of ECT and education, training, and privileging of providers.[6] But nothing has effectively erased the popular bad image of ECT. From its social reputation, one would wonder why ECT still exists.

It still exists because it is the most effective treatment for major depression, with – often – the most benign side effect profile. The Consortium for Research in ECT demonstrated a 75 percent remission rate among 217 patients suffering from an acute episode of depression who completed a short course of ECT, with 65 percent of them achieving remission by the fourth week of therapy.[7] ECT has been tested extensively in comparison to antidepressant medications and other forms of treatment. Control conditions similar to placebo have been put in place as "sham" treatment, in which some patients are given anesthesia but no ECT, and changes to their depression and cognition scores assessed without them knowing they were not treated. Currently, over 100,000 patients are treated with ECT each year in the U.S., and over 1 million globally. It is prominent in treatment-resistant depression and in those for whom antidepressant medications might be dangerous, such as pregnant women and the elderly.[8]

Though considered the gold standard for difficult-to-treat depression, ECT's side effects continue to raise concern; mitigating them while keeping the efficacy of the treatment has been a decades-long effort. Up to 45 percent of patients suffer a headache after the seizure, and up to 23 percent experience nausea. Muscle soreness, likely due to the muscle twitching as muscle relaxant takes effect, is sometimes reported along with a sore jaw from direct stimulation of nearby muscles by the apparatus. A particularly upsetting possible side effect is called "anesthesia awareness," when anesthesia wears off before the muscle relaxant does. When this happens, the patient is unable to move and feels unable to breathe; they are aware but cannot communicate their situation. With oxygen provided during the treatment, it is not actually dangerous, but very disturbing.[9]

By far, the side effects that cause the most concern are cognitive, particularly memory, deficits. ECT can produce four types of cognitive impairment. The first is a transient disorientation patients experience immediately after the seizure. This disorientation can clear in minutes, or with few hours of sleep, but in some cases endures a long time and only gradually disappears. A second type of cognitive effect is anterograde amnesia – forgetting what is going on as it happens, or the inability to retain information learned during and shortly after a course of ECT. The duration of this effect also varies and can last months after the final treatment. The third type of cognitive impairment is short-term retrograde amnesia: memory gaps for events that happened weeks or even months before treatment. This also usually attenuates over time, but memory of some events may never return. The last, fortunately rare, but most upsetting type of cognitive impairment resulting from ECT is a more extensive retrograde memory loss, in which a patient suffers memory deficits for events going back months or even years. Some people report that they have persistent gaps in their memory that have not been recovered years after treatment.[10]

Several technical factors in the delivery of ECT treatment, including electrode placement, electrical stimulus, and number and dosage of treatments, impact efficacy and side effects. The original device used by Cerletti and Bini delivered an electrical charge to both sides of the head simultaneously through electrodes placed on each temple. This bilateral electrode placement is still an option, with other placements also available. In bi-frontal electrode placement, electrodes are placed above the eyes just within where the hairline (usually) begins. This electrode placement is intended to avoid passing electricity through the hippocampus, in an effort to prevent memory loss. Right unilateral placement has one electrode placed on the right temple and the other about an inch to the right of the top of the skull and was also devised to limit the path of the electrical charge to reduce side effects.[11]

Electrode placement affects the level of charge necessary for treatment effect. Individuals have different seizure thresholds – the amount of electrical charge necessary to induce a seizure. Once the provider determines the patient's seizure threshold, she initiates treatment with an electrical current, commonly at least 1.5 times the seizure threshold for bilateral and bi-frontal placement, and 6 times the seizure threshold for right unilateral placement. Treatment is usually given three times a week and patients often require 3 to 4 weeks of treatment before remission. Every time a seizure is induced, the patient's seizure threshold goes up some amount, so the charge must increase as well. The level at which it is applied (1.5 times seizure threshold, 6 times seizure threshold, or other) affects how well the treatment works and the degree of side effects, with both those factors rising together as the current rises. Researchers have tested whether sub-seizure electrical shock would work but have shown that the seizure itself is necessary for therapeutic effect.[12]

Another technical factor that has developed over time is the nature and duration of the electrical stimulus. ECT was initially delivered as a sine wave stimulus, and still is in some clinics. This pattern of electrical stimulation delivers a significant amount of electricity below the threshold needed to induce a seizure, causing more cognitive side effects while not providing any improvement to the clinical outcome. In ongoing research to reduce cognitive side effects, other patterns of electrical stimulus have been developed: a brief pulse stimulus (approximately 1.5 milliseconds) and ultra-brief pulse (0.3 milliseconds).[13]

Several studies have investigated optimal electrode placement and pulse characteristics to find the patterns with the best clinical and side effect outcomes. Three large-scale studies that took place between 2001 and 2006 reported outcomes that varied widely depending on how the individual clinics administered ECT. Bilateral electrode placement with brief pulse stimulus showed a remission rate of about 65 percent, but bilateral electrode placement with ultra-brief pulse stimulus resulted in remission for only 35 percent of the patients. Bilateral treatment resulted in the most rapid decrease in symptoms but more severe and lasting retrograde amnesia than right unilateral ECT. Bi-frontal electrode placement with brief pulse stimulus saw a remission rate of 61 percent, but showed more severe retrograde and anterograde amnesia than bilateral placement. Right unilateral electrode placement with brief pulse stimulus showed remission rates of 55 to 59 percent, while right unilateral placement with ultra-brief pulse performed better, with a 73 percent remission rate. Right unilateral electrode placement also resulted in a little less retrograde amnesia than the other electrode placements. One of the studies had included clinics that used a sine wave electrical stimulus instead of pulse stimulus; that left patients with a pronounced slowing of reaction time, both immediately and 6 months later. Ultra-brief pulse had a little less impact on memory than brief pulse, but not even the more benign treatment modes were entirely free of long-term adverse effects.[14]

In the continuing quest to improve efficacy and reduce side effects of ECT, a new approach has been devised and is being tested in laboratories. Focal Electrically Administered Seizure Therapy (FEAST), controls the direction of the electrical current (right to left or left to right for bilateral; down/up or up/down for unilateral), the polarity, and the strength and spatial distribution of the induced electric field to influence the results of the seizure. Comparing this new technique to that of bilateral ECT in rhesus monkeys, unidirectional electrical current was found more efficient than bi-directional current in inducing a seizure, and the FEAST electrode configuration was more efficient than bilateral electrode configuration. In ECT, "efficiency" in inducing a seizure translates to less electrical current passing through the brain.[15]

MAGNETIC SEIZURE THERAPY

There is another type of seizure therapy that has gone into limited use and may someday replace ECT. Transcranial magnetic stimulation (TMS) is intended to treat depression by causing certain neurons to fire, while not causing so many neurons to fire that it would cause a seizure. On several occasions, though, scientists investigating the use of TMS below seizure threshold missed their goal and generated enough electrical current that the patient had a seizure. This prompted the idea of Magnetic Seizure Therapy (MST), in which magnetic waves would pass harmlessly through the brain to reach a precisely targeted point where they create an electrical charge above seizure threshold, sparing the patient the effects of electricity passing through structures of the brain along the way. In theory, at least, it would be ECT without the adverse effects on memory. A study using a computer model of all the forms of ECT and MST found that every form of ECT passed some electrical current through the hippocampus, but MST didn't. MST has been used in limited studies in comparison with ECT, showing a lower side-effect profile. At this point, MST appears to be equally effective as ECT with fewer cognitive effects, but the blinded, controlled testing in larger samples that would confirm such a conclusion haven't taken place yet.[16]

For Danny, ECT has provided relief, but not a lasting remission. "He started in 2012; that was his first ECT," Steve recounted. "And his last one was Friday, and that was his 146th. From 2012 to now, it's been every imaginable duration. We've had them every other day for part of that, and the most it's been stretched out has been 2 years or so, and it turns out that was a mistake, but we didn't know that. In general what happened was that when he was doing the ECT, when he first got introduced to it, he did six in a row. I thought it was going to be six, and 'Have a nice life.' So he did six, and he got out, and he did well. And then they said, 'Let's do one a week.' So they did that, and the goal was always the same, stretch them out to get to a maintenance level. So, he's at 6 weeks right now. Six weeks between treatments and he's doing well up until that point. Sometimes he starts fading hard, like you get a treatment Friday and a week and a half goes by and something goes wrong. So then you got to do a series again. And then stretch them out again. So the goal is to stretch them out to maintenance, which would be one every 6 months."

Steve refers to Danny's therapies as a three-legged stool: medications, counseling, and ECT. The first series of six treatments had been right unilateral, but the rest have been bilateral. With 140 bilateral ECT treatments, Danny's short-term memory is very much affected, and his long-term memory suffers also, to a lesser degree. The years of searching for a solution provides both of them perspective on what the outcome could have been, though. "Between the drugs and treatments, and the troubles he was having, he was just consumed by this;

we all were," Steve said. "So when he got on top of it, the first time he ever got his meds right, he got a job, he got a car, so he got out in the world right at that point. But he was 18 by then. Right now, he's holding down a job, he's doing well in life in general. It's a grind. It sucks; he struggles, but he gets it done. He's found things that make him comfortable and he figured out what he can tolerate and he sticks to that."

Chapter Twenty-Five

EXERCISE

I jog to beat Alzheimer's disease. I have a regular route and schedule, of course, but every time I find myself searching for a word that I know I know, or when my father – who does have Alzheimer's disease – does something extra-crazy, I want to grab my running shoes and put in another mile. Regular physical exercise has been shown to delay the onset of Alzheimer's disease, and whether it is through the same mechanisms or different ones, it can also be effective in helping people recover from depression.

It seems rather amazing, and somehow rude, that something as simple as exercise can have the same effect as antidepressant medications, without the investment of billions of dollars in research and development. Of course, even simple things are hard for someone struggling with depression or with its co-morbid medical illnesses. Vigorous physical exercise of any sort – aerobic or strength-based – has been shown in clinical studies to have a significant effect on symptoms of depression. It has been studied in randomized and controlled studies in comparison to no treatment, psychotherapy, and antidepressants, and as an adjunct to other therapies.[1]

In 2016, a research team gathered the randomized, controlled studies of exercise in comparison to or in conjunction with no treatment or other therapies in unipolar depression, finding 23 reports involving 977 subjects. Exercise stood shoulder to shoulder with other treatments. When compared to "no treatment," exercise showed a large and statistically significant effect on reducing symptoms of depression. Studies comparing exercise with "usual care" showed a moderate but statistically-significant effect in favor of exercise. When compared to psychotherapy, the effect of exercise was not statistically different from that of cognitive therapy. Three studies with a total of 236 participants compared exercise to sertraline (an SSRI), finding no significant differences in the outcomes between those two approaches. When used as an adjunct to antidepressants, though, exercise conferred only a small and insignificant advantage over medications alone.[2]

The earliest studies of the efficacy of exercise on depression used mice and rats, and because of that they speak of exercise as "running" – running on a wheel. Aerobic exercise stresses some different systems than does strength training, so it was not obvious that strength training would also reduce depression. Several studies addressed that very question, and though aerobic studies outnumber strength studies, there was no statistically significant difference in effects between the two modes. These studies were carried out across many age groups, and almost all of them in people with low to moderate levels of depression. In all the studies, the exercise programs were at least 2, usually 3 sessions per week for several months, ranging from 30 to 90 minutes in length and generally vigorous enough to get to 70 to 85 percent of the individual's maximum heart rate. Almost all sessions were supervised; when told to exercise on their own, patients didn't see the same level of response. Though possibly due to a lack of compliance, it is also possible that exercising as part of a regular group has beneficial social aspects as well.[3]

Exercise – regular, moderate, physical exercise – promotes a sense of well-being, supports restful sleep, tends to normalize appetite, fights obesity, and strengthens the respiratory system. In addition to promoting overall health, exercise affects specific antidepressant mechanisms, too: enhancing neuroplasticity, reducing inflammation, and affecting monoamines and other neurotransmitters.

Exercise may increase serotonin availability for antidepressant effect. While the mechanisms are unknown, animal studies show that exercise results in higher levels of extracellular serotonin and its breakdown product in the hippocampus and prefrontal cortex in mice and rats.[4] Such studies have also shown that exercise increases available tryptophan in the blood and the brain. Human studies showed changes in levels of serotonin, serotonin transporters, and 5-HT2A receptors in the blood following 3 weeks of exercise in sedentary men.[5]

Physical exercise is one of the primary factors inducing adult hippocampal neurogenesis. Animal studies show that when mice are placed in a cage with a freely-available running wheel, they will tend to run 3 to 8 kilometers each night and produce 2 to 3 times as many new hippocampal neurons as mice that are not permitted access to a running wheel.[6] In humans, even a single exercise session increases peripheral levels of the growth factor BDNF to a moderate degree, and regular exercise creates a greater effect.[7] Additionally, the brains of people (and mice) who exercise show greater synaptic plasticity. Research has shown that physical exercise acts through glutamate receptors in the hippocampus to enhance learning and memory.[8] And exercise has also been shown to increase mitochondrial number and function.[9]

Along with increasing neuroplasticity, physical exercise appears to have antidepressant effect via the immune system. A single exercise session immediately provokes a rise in the pro-inflammatory cytokine IL-6, followed by a surge of anti-inflammatory cytokines and decreased production of pro-inflammatory cytokines; the net result is anti-inflammatory.[10] It is also possible that the reduction in belly fat from regular exercise contributes to its anti-inflammatory effect, as excess belly fat is a chronic source of system-wide inflammation.[11] Finally, it has also been shown, in animal models, that physical exercise activates anti-inflammatory microglia in the brain, helping to tamp down an inflammatory response that would otherwise degrade brain tissues.[12]

There is still a lot to learn about how exercise gets its antidepressant effect, and in what sort of cases it is sufficient to keep growth happening. Maybe there is a component of socialization and therapeutic attention in these studies that don't necessarily exist for someone exercising on their own. Maybe it engages a pain-based placebo effect, or beta-endorphins, or is only effective on more moderate forms of depression. Whatever the underlying mechanisms, exercise is an amazing, and unexpected, antidepressant.

Chapter Twenty-Six

CHRONOTHERAPIES

It seems a long time ago that I was talking with Theresa, who had seen such benefits from a NIH program and MAOIs. I recall, though, her mention that she had heard that crossing the International Date Line lifts depression, and in fact doing so had lifted her depression, albeit temporarily. At the time, I took it as sort of a folk remedy... one of those ancient legends regarding international air travel, perhaps. On further consideration, though, someone who flies from the U.S. across the International Date Line is pretty much guaranteed to screw up her body clock. And by "screw up," I mean fix. Maybe she had accidentally experienced wake therapy.

By 1971, multiple case studies of individuals who showed rapid and unexpected improvements in depressive symptoms following a night of sleep deprivation emerged from Europe, promising a new approach to treating depression. Though the effect was replicated over and over again, in the U.S. interest quickly cooled when it became apparent that the effects were lost after a single night of restorative sleep. However, in Europe at least, use of sleep deprivation (now called "wake therapy") and other chronotherapies is reappearing now that adjunctive therapies have been shown to extend the effect for weeks or months, providing relief as other, slower mechanisms get underway. Sleep deprivation therapy creates remarkable effects seen within hours of the new day, or even during the sleep-deprived night. Its side effects include headache and fatigue, and in bipolar depressed patients it can trigger a switch to a manic state (usually controllable with the use of lithium). It can cause seizures in people with epilepsy and so is not recommended in those cases. For others, however, up to 60 percent of patients with depression show a moderate to strong response to a single night of total sleep deprivation.[1]

Efforts to milk the fullest benefit from sleep deprivation have resulted in several forms of chronotherapy. Total sleep deprivation involves keeping the patient awake for 36 to 40 hours, from the morning of the first day to the evening of the second day. Unfortunately, restorative sleep at the end of that second day drags more than 80 percent of responders back down again. Even short naps, especially in the morning hours, can cause a relapse. Partial sleep deprivation (20 hours of sleep deprivation targeting half the sleep cycle) is one of the related therapies. Even 1 to 2 hours of wakefulness in the middle of the night to block REM sleep shows antidepressant effect after several weeks.[2] One of the combination therapies shown to preserve the benefit of total sleep deprivation is adding sleep phase advance. Following total sleep deprivation, patients are started on an early sleep schedule (such as 5:00 pm to midnight), which is then pushed back an hour each day until they reach a more normal sleep schedule. Repetition of sleep deprivation therapy at short time intervals (for example, every 2 to 3 days) leads to progressively better antidepressant effects, and going through sleep deprivation once a week has also been proposed as a preventive treatment to sustain the response and prevent relapse. Sleep deprivation therapy of different schedules and durations have been shown to work well in conjunction with antidepressant medications and bright light therapy.[3]

Light itself has strong antidepressant qualities. In addition to our classical visual system that generates images of the external world, we have another, nonvisual system that detects variations in ambient radiance and prompts a variety of responses. These responses include long-term modifications of circadian rhythms and immediate changes in hormone secretion, heart rate, sleep propensity, alertness, core body temperature, retinal neurophysiology, constriction of pupils, and gene expression. In the early 1980s, scientists experimenting with providing artificial light to people experiencing SAD found exposure to bright light for times ranging from 30 minutes to 2 hours, depending on the light intensity, repeated daily for about a week, resulted in remission lasting at least an equal number of days. More recently, bright light therapy for non-seasonal depression has also shown good results. Researchers have also tested light therapy using blue-green wavelengths only, dim lights, various intensities, applied at different times of day, for different durations, and in combination with sleep therapies and antidepressant medications.[4]

A recent clinical trial involving 122 patients with non-seasonal major depression of at least moderate severity randomly assigned them to one of four treatment groups: bright light in the early morning plus placebo pill, antidepressant therapy plus sham light therapy (an inactive negative ion generator), combination bright light and antidepressant therapy, or placebo with inactive ion generator. After 8 weeks, the best results were seen in the combination bright light and antidepressant group, with 17 of the 29 patients achieving remission and an additional 5 showing response. The next best result came from the bright light therapy plus placebo group, with 16 of the 32 patients achieving remission or response. The antidepressant plus sham group and placebo plus sham group showed roughly equal success, with 9 in remission or responding in the former versus 10 in the latter group.[5] In 2010, the APA included bright light therapy in its *Practice Guidelines* to treat both seasonal and non-seasonal depression, and to maintain the antidepressant effect of sleep deprivation therapies.[6]

Newer discoveries, especially those concerning how circadian rhythms are programmed in humans, allow researchers and clinicians to deliberately combine sleep or light therapies with particular antidepressant medications that they believe will synergize their effects. In particular, melatonin and serotonin are essential to the programming and maintenance of circadian rhythms. Melatonin carries information about the light-dark cycle throughout the brain, including the hypothalamus, where that information will be used to trigger awakening levels of cortisol. Serotonin is very responsive to changes in light, showing a strong seasonal variability. Serotonin transporters, the same ones that are the target of SSRIs, more actively bind with extracellular serotonin during the months of darkness than in the summer months, with a seasonal variation of up to 40 percent in different parts of the brain. SSRIs counter that increased binding potential. Following similar lines of thinking, researchers are testing the efficacy of different chronotherapies in combination with SSRIs, agomelatine, and other medications preferentially targeting elements of the circadian cycle.[7]

CHRONOTHERAPIES FOR BIPOLAR DISORDER

One of the most exciting aspects of chronotherapies are its effect on bipolar disorder. Bipolar patients are extremely sensitive to the environmental light-dark cycle, which can trigger symptoms of depression or mania. The recent discovery of a new class of photoreceptor in the retina sensitive to blue wavelength light provided an opportunity to understand and affect this phenomenon directly. Blue-wavelength light mediates information going to the circadian master clock and mood centers of the brain. Blue-blocking, amber-colored sunglasses have proved effective in treating mania and stopping rapid cycling between states.[8] Lithium, long used to treat bipolar disorder, is known to slow down circadian rhythms in some brain tissues and alter them in others.[9] Lithium used with chronotherapy can lead to sustained remission over months, with stable euthymia obtained in the majority of patients without the need for other psychotropic drugs. The combination of repeated sleep deprivation therapy, light therapy, and lithium can produce sustained antidepressant effects. In a recent study, about one-half of bipolar patients who did not respond to several antidepressant drug trials, and who had become hopeless and suicidal as a consequence of their long-lasting depression, experienced this sustained effect.[10]

Dr. John Gottlieb of Chicago Psychiatry Associates began practicing psychiatry about 30 years ago, finding a client base that included a lot of patients with mood disorders, particularly bipolar disorders. He became first a *de facto*, then a declared, specialist in those disorders. After several years, he found himself observing the same pattern over and over again. "I would see my practice change twice a year; regularly, periodically, predictably," he said. "The change was that I would see more anergic depression in the fall and winter, and I would see diminishment of that, along with an increase in agitated, manic states in the spring and summer." This pattern showed throughout the patient base of the 7-member practice, in long-term patients with whom they worked over a period of years and in new patients coming in for the first time. It was a very prominent seasonal shift in mood states. "I wanted to understand what was going on with that," he said. "That's what led me into chronobiology, and then into chronotherapy."

For the last 15 years, Gottlieb has been incorporating chronotherapeutic interventions into his practice where appropriate. "I would say maybe 60 or 70 percent of the people I see in my practice have bipolar disorders, and for them, and my mood disorder patients in total, I'd say 50 to 75 percent of the time I will, at some point in the course of my work with them, raise the question about whether we might want to incorporate, use, blend in, a chronotherapeutic treatment along with everything else that I'm doing, which might be psychopharmacology, might be psychotherapy, might be psycho-education, might be whatever. The most common chronotherapeutic interventions that I wind up using would be bright light therapy by far, after that, dawn simulation and blue-blocking glasses. Wake therapy would be a distant fourth."

I asked why wake therapy was so far down the list. "It sounds weird to a lot of people," he responded, "and they just have a reaction of like 'That just sounds too kooky.' Some people want medication; they want to have a more medication-based treatment. And it's hard to do; people have to be willing to commit to a certain amount of time."

The wake therapy protocol he described to me was developed about 10 years ago at the University of California-Irvine, though he noted there are other models in use as well. "This is triple chronotherapy," he began. "Let me walk you through it. You got up this morning

[Wednesday] at whatever time you got up. You would stay up all night tonight; you would get bright light therapy either Thursday morning or mid-day Thursday. You would stay up until 6:00 pm on Thursday evening; you would go to sleep at 6:00 pm, get up at 1:00 am on Friday morning. You would stay up all Friday morning, get bright light Friday either morning or mid-day; you would stay up Friday afternoon; you would go to sleep at 8:00 pm on Friday. You would wake up at 3:00 am on Saturday morning. You would have bright light morning or mid-day on Saturday; you would stay up all day Saturday; you would go to sleep at 10:00 pm on Saturday, get up at 5:00 am on Sunday morning. You're done." And when I would feel better? "If you were going to stay up all night tonight, then I would expect to see a response anywhere from 4:00 am to 10:00 am tomorrow."

Chronotherapies have been so useful to him in part because people with bipolar disorder – either I or II – are most responsive to it. He listed some other characteristics of patients that respond best to triple chronotherapy. "Late chronotype – night owls – tend to do better with this treatment. Presence of diurnal variation in mood, which is a symptom that sometimes occurs and characterizes depression: people who wake up feeling worse in the morning and progressively feel better as the day goes on. That's classical diurnal variation, and that's also a positive prognostic. It's soft things that we don't have much research on, but we suspect atypical features of depression: people with exhausted, no gas in the tank, completely anergic depressive state. Psychomotor retardation, very hard to move. Effortful, slow; everything is slow. That would be another thing. Those are the kind of factors that would lead me to think this person might do well." And with his bipolar patients, he has them on lithium, both to prevent a switch to a manic state but also because it augments the response.

But of course, wake therapy is well down the list of interventions he might recommend. He has seen great responses to bright light therapy. "If I had to give you some visceral guestimate of response rates in my practice to bright light therapy, I would say a 70 percent response rate. And I would typically see the response within 1 to 2 weeks," he said.

By his off-the-cuff estimate, Gottlieb is one of perhaps five practitioners in the U.S. who regularly incorporate chronotherapeutic interventions in their practice. "I find that chronotherapy has been an orphan kind of area in psychiatry, a poor stepchild. It's really been marginalized and that's unfortunate. I feel like I've got a fuller repertoire of things I can bring to bear in trying to work with someone who is depressed. It's not uncommon for me to see someone for consultation because they're not getting better, and then I'll go through with them 'Well, what have you tried?' 'Well, I've tried Prozac and I tried Zoloft; I tried Celexa; I tried Effexor, Cymbalta, Pristiq...' And it's just nonsense, to try and fail, one after another of the *same* classes of medication that act through the exactly *same* mechanism of action, and hope to get a different response. So, to me it feels great to have something that may act, presumably does act, through a different mechanism of action to try."

WHY SO LITTLE BUZZ?

Since chronotherapies like sleep deprivation therapy have generated such startling results for more than a generation, why aren't we using them a whole lot more? Chronotherapies are just not drawing the same attention that monoamine-based antidepressants do. Because they use natural processes, they sound like alternative medicine approaches, which can prejudice serious practitioners against them. A natural substance like melatonin cannot be patented and is easily available in the vitamin aisle of any supermarket, which means there can be little anticipation of corporate profit for its increased use. Light, even "bright light," devices are difficult to patent because they constitute a prior art – they have been in use for so long no one can easily show they are providing something unique. And without sponsors for the long and expensive processes of conducting multi-center trials, sleep therapy and bright light therapy don't generate enough studies with enough data to win support by professional societies.[11]

But think of the potential in our modern age. Bright light devices can be easily purchased over the internet. We even have "smart beds" that could assist a patient or therapist assess treatment progress. Those, along with other sleep and movement trackers, could provide really useful feedback to patient and therapist. And if sleep disturbances are among the earliest symptoms to show up in depression, could a bed or watch app warn its user that a major depressive episode is looming, allowing them to treat it pro-actively?

It is really frustrating to think about Danny, who is holding it together with the help of so much ECT, but whose father recognizes the cyclical nature of the illness. His psychiatrists and therapists have probably never talked with him about chronotherapies – that "woo woo" stuff. And it is unlikely he'll hear about them from other sources. Mental illness is wrapped in silence.

Rose grew up in a family that didn't talk about mental health issues "Back home where I come from, you don't have depression. I mean, it doesn't exist. It's all in your mind," she said. It took a traumatic encounter, an assault in her thirties, for her to look back and admit she had depression, dating from her teen years. As a pre-teen she found out she was adopted, and shortly afterwards her father abandoned the family. "The first two didn't want me, and now he didn't either. I was moody; I was not concentrating well; I was always sad; I was crying; I couldn't have good relationships," she recounted. "My mom had a girl that used to assist her in her house, with the chores; she saw me cutting myself one day. They took me to the hospital. But they were just embarrassed that I had done that. So they didn't tell anybody." At a women's center after the assault, she connected with a therapist and was finally able to understand what was happening to her.

Now Rose talks about mental health issues, professionally. She's become a peer support counselor at a community facility, helping others with depression, substance abuse, and behavioral disorders. Like most of the professionals and volunteers I met at NAMI groups and other non-profit organizations, having experienced mental illness herself gives her valuable insight and empathy. She still struggles occasionally and knows that her battle will be lifelong. It is a battle she is equipped to handle, though, and she shares that perspective with others. "Recovery is possible, but you have to have the tools to do it," she says. "You can still have depression but you don't have to live with it. You can live happy. Even though you have a condition, if you treat it, and you know when your condition is declining, and you know what to do, you have the tools to do something about it."

There are actually a lot more tools to treat major depression than is commonly realized, or utilized. Medications and psychotherapies can reach a lot of depression's sufferers, but someone has to overcome their own stigma against mental illness to ask for such remedies. Other modern treatments like rTMS and ketamine show such great promise, and results, but lack availability. There just aren't enough treatment centers and the therapies haven't reached a lot of insurance schedules yet. Hopefully, the backlash against pharmaceutical drug pricing will promote some recognition of chronotherapies and nutraceuticals, but those remedies still lack acceptance among the psychiatric community at large. As experience and understanding of what each tool can do to treat depression grows, hopefully – hopefully! – more sufferers will find prompt and sustained relief.

Scientists and physicians have made an incredible amount of progress in understanding and treating depression. But the world needs more. It needs a better understanding of what is happening to an individual, in terms of the disease process and their means to fight it. It needs to understand which therapies work with which particular types of depression, so people can direct their resources for the best effect. The last 60 years have seen incredible advances in depression treatments, but what is most startling is the increase in the pace of that progress. And there is more right around the corner.

Conclusion

BRAVE NEW WORLD

If I were writing this book 10 years from now, I think it would be substantially different. There is so much new discovery going on, it feels as if we are on the cusp of not only a greater and more meaningful understanding of the factors that create and sustain depression, but also of medical and technological advances to exploit that knowledge to help people heal.

One area in which the future is among us now lies in the explosion of smart devices, wearables, even smart beds and smart household appliances. They are a rich source of captured data about our lives, including the physical course of psychiatric illnesses. To understand how this treasure trove of information might be used to help people with depression, I went to the experts on Big Data and smart devices.

Dr. Menachem Fromer leads the Mental Health Research and Development team at Verily, the life sciences company of Alphabet, Inc. Verily became a stand alone company in 2015, from its early origins as a team within Google [X], and both data science and machine learning feature strongly in their heritage. Fromer himself has degrees in biology and computer science. He joined Verily in 2016 from The

Icahn School of Medicine at Mount Sinai in New York, where he had been a professor of psychiatry and genetics.

Fromer's team actively participates in several studies concerning technology for mental health. One of those is a large, NIMH-sponsored prospective study of post-trauma outcomes led by University of North Carolina and Harvard University. "Not PTSD," Fromer clarified, "it's post-trauma. PTSD is one of the adverse outcomes that can happen post-trauma, but we're trying to look more broadly, at depression, pain sensitivity, all those types of things can happen post-trauma. Or nothing." It is a multimodal study, officially titled Advancing Understanding of RecOvery afteR traumA (AURORA). "Recruiting is happening at about forty emergency rooms when people come in with trauma," he said. "They're collecting blood and saliva for genetics and genomics, and to look at blood-gene expression over time, and a subset gets called back for brain imaging. There's also cognitive testing."

Verily's contribution to the study is a wearable called Study Watch, a smart watch that measures blood flow and many other variables. Fromer himself has been testing the device; he took the watch off his wrist to show me some of its activities. "It has the PPG – the photoplethysmogram, the shooting green light like the fitbit – to measure heart rate. You can also take an electrocardiogram, but that's an active test; you have to initiate it." The research participants also use an application from Mindstrong, Inc., to bring real-time behavioral measures into the study. "The question is can we use things like the Study Watch – or other measurements but the watch is one of the stronger instruments – to measure important signals in the background, in a passive way? Most people after a trauma will have some sort of disruption in their life that could last a week or two. So their sleep gets disrupted, their heart races more, and it depends on the type of trauma. Obviously that impact is likely to be more severe when the trauma is more severe. People that are resilient and recover from the trauma, they'll go back to their baseline. But some people won't recover. They're socially withdrawn or they have PTSD and their heart rate can spike a lot if they're reliving the trauma. The question is can we measure the physiology and behavior of what that would look like from Study Watch?"

It is a learning effort for all involved, with the data scheduled to go public in 2022. “There's a lot of different data being collected simultaneously to understand what these trajectories look like after someone suffers a trauma. Verily's focus is how we relate the measurements that we have on physiology and behavior to these longer term clinical outcomes.”

While the AURORA study is looking to the future to understand how the body responds to trauma, most of Fromer's time is spent on a nearer-term project: the Mood Study. Instead of a wearable, the Mood Study uses capabilities of Android phones to measure changes in physiology and behavior that could indicate someone going into or recovering from depression. I say “could indicate” because there is a lot of learning – both human and machine – that needs to happen for this concept to work.

The Mood Study app uses both traditional and non-traditional methods to work with its user to assess mood. Traditional methods include presenting standard mood measurement instruments, like the Patient Health Questionnaire 9-item test (PHQ-9), a self-reported scale for determining depression. Fromer showed me the app on his phone. “It has a chat interface that says ‘Hello, let's start the three daily questions,’ and asks me about how energetic I feel today, or down, depressed, or hopeless. About stress, asking me to speak a certain passage, to read a certain passage so they can listen to voice and speech and affect and stuff like that. Once a week it asks me to type out a passage, and also asks me to speak about my week in 30 seconds.”

Those latter parts begin to get into the non-traditional methods for assessing mood. Among the symptoms of depression are psychomotor retardation or psychomotor agitation, which commonly show up as slowed speech or increasingly rapid speech patterns, respectively. They don't always occur, but when they do, it would be a good indication that something is changing. “Different days it asks you different questions,” he said, “we have a schedule of them. The point is, what we're trying to do is collect what I call traditional and non-traditional clinical data. The traditional clinical data is more these types of surveys. But then the non-traditional ones, no one right now at scale uses audio diaries or something like that to assess their patients, but the goal would be can we do that, in their daily lives, using an app, basically bringing part of the clinician, or at least the assessment part of the clinician, to the patient in the real world."

"So the idea is to analyze not only what they are saying, like 'I feel really bad now' but also their voice patterns?" I asked.

"Yes, exactly," he said, "voice and speech. Looking at pitch and tone, things like that, but also the words that they choose to use. Which is where we've started to see some signal in our pilot study."

Their pilot study was a proof of concept, just a few hundred people (small scale for Big Data), and they were about to launch a much larger study, involving thousands. "Working at scale, we could really, hopefully, learn what the patterns are, build these predictive models that relate the passive data and the active data to these traditional clinical metrics. The way I think about that is the passive data includes things like location that people opt in to share, the accelerometer-based measurements of physical activity and potentially of sleep, things like that. But also these digital diaries, combining all of that to understand day to day, or week to week, what's happening with someone's depression severity. And for comparators, we have the PHQ-9s that people are taking once a week, but also other surveys. Those are the more traditional surveys, just delivered in a non-traditional format."

That's where the power of Big Data and a learning system comes in. Though Fromer mentioned they had seen a "signal" in their earlier pilot study, there is no definitive speech pattern that identifies depression. That is one of the targets of this capability: to get enough data to figure it out. "We are trying to learn," Fromer agreed. "There are not very many large scale studies that looked at this and that have definitive patterns. I think because of the heterogeneity of the disorder there is going to be some complexity in teasing it apart. There are a lot of algorithms for measuring, there's lots of signal processing things you can do with the data, but I don't think there are definitive things we're trying to apply; we're trying to learn."

Fromer carved out his own path in bioinformatics. He started by earning a bachelor's degree in biology, but, working in a cancer lab, realized he liked the analytical parts of biology more than the juicy parts. So he added bachelor's and master's in computer science, specializing in genomics and bioinformatics, and then a PhD in computer science, "applying machine learning for a problem called computational protein design, where you're trying to choose sequences of amino acids that would fold to a certain function." Eventually, he moved to Cambridge, Massachusetts, to a position at the Broad Institute of MIT and Harvard.

"They do a lot of genomics and genetics, like the Human Genome Project," he said. "We had two large scale projects, one where we sequenced many patients with schizophrenia in Sweden, many cases, and a lot of controls." After years of sequencing genetic information in schizophrenia, he became dissatisfied with the task. "We basically had sequenced a very large portion of all the schizophrenia patients in Sweden and we started to learn a lot about genes and neuropathways that are involved. But it seemed to me always the downside of the genetic approaches was that we'd take some blood or saliva from someone and measure a million common genetic variants or thousands of rare variants... You get an idea of the scale; you're looking at thousands, hundreds of thousands, or millions of data points that you measure from one person. And some of those are important; the question is which ones? But when we say 'what's important', what we're doing is effectively reducing people down to a bit of information, saying 'Are they a case (with disease) or a control (without disease)?' That would be OK if it were a picture of a cat, and most people can agree it is a cat, and you can train a machine learning algorithm to say 'this is a cat' or 'not a cat.' But something like schizophrenia or depression where there are shades... we're not even sure what is the disease. So it seems like a stretch that we could reduce it all down to genetics and say 'I'm sure.' There's really a mismatch between the amount of data in the genetics, for example, and the amount of data in the phenotype. What became clear to me was, could we be using – and I didn't invent this concept – could we be using the digital tools of smartphones and wearables to better measure the day-to-day changes in phenotype: what it really means to be depressed at the behavioral level? Ultimately those could come together with genetics, and you could learn what the causes are, what that means day-to-day, and changes in gene expression."

So now he leads the effort to build this learning system, in which those zillions of data points build a picture of what depression is. Not to perpetually reply "is a cat" or "not a cat" with greater and greater fidelity, but to fill in parts of the mosaic that, when combined with genomic, imaging, and clinical data, may someday reveal in full depression's physical course, to measure what is now self-reported, to see where progress is really being made.

I was meeting with Fromer in advance of a kick-off celebration for their much larger study, the one to generate the data and the algorithms to make this learning system a reality. "We have some control groups, but for the main part of the study we're recruiting people that are depressed and in treatment for depression. So they are engaged in care. The ideal scenario for deployment of this as a product is allowing people to engage in care, and see how that care is going. Does the doctor need to be alerted? Should you get alerted based on what's happening? Can we have objective measures? That helps us from a lot of perspectives, even from an ethical perspective: can we do a remote study like this? By being engaged in care, someone already has a go-to person to engage with if they feel very depressed. We also remind the users that it's just a study, just observational. We're just learning about all of this. This is our first use case, where people are in care. Right now, people with depression typically get prescribed an antidepressant, mostly by their primary care physician. They're prescribed and maybe they'll come back and maybe not, or come back in 12 weeks. That's a long time to not know if an antidepressant is working. If we could know that the antidepressant is working, with some clarity, that would be helping."

The cloud of data surrounding us brings the power to learn much more about depression, but also brings potential for abuse. For the Mood Study application, the user can opt-in or -out of its ability to passively collect location and other such data, and most of the interaction is intentional anyway: answering a few questions, speaking about their week. "In terms of data," he said "in theory, it could be like 'I could record all the time and learn everything about you.' But you wouldn't want that. That would be creepy, and we don't do that because of privacy considerations. So instead, the compromise is we ask people similar to the voice diary, do a text diary. It requires some more effort, but hopefully, if it is engaging enough it could be something that is relatively low touch, low maintenance."

One of the things they've had to put a lot of thought into is what happens if a person's behavior starts to trigger some alarms. "Some of the escalation could be to pop up a notice to the user, like 'Seems like you haven't been engaged in your usual social activities recently, do you mind taking this survey?' and then you might trigger a PHQ-9 for depression and if that screens positive, then prompt to go see your doctor. Or maybe you've consented early on that your doctor gets a notification." But there is a lot of new territory here, and it is not just the patterns and progress of depression they'll be learning, it is also how what they are doing affects people.

As a learning system, it's also new territory for the regulating agencies like the FDA, as Fromer illustrated. "Let's say we're done with our study 2 years from now, we built the algorithms that we think are good, and we get some sort of approval for it as a clinical decision support tool," he said. "The FDA approves that. Great, but now there's this whole idea of learning health systems. What do you do when the algorithms that support this thing are getting better? The FDA is grappling with that now, in other areas, not this one. So the algorithm they approve now isn't the algorithm in use 6 months from now. Does it need re-approval? And maybe the algorithm learns something bad and is getting worse. How do you guarantee, or at least minimize the potential for harm, but not cut off innovation? It's a hard question. You need to constantly reassess against something, to show that you're not doing harm. But who monitors that; what are the thresholds? It's a brave new world."

I came into this project wanting to understand depression in order to help. My concern was for Carolyn first, of course, but as I got further and further in to the weird and impactful world of depression science I just kept thinking, "People ought to know this!" The journey has been very enlightening but very humbling as well. No matter how much I learned and how much I passed on to Carolyn, the cold fact is that if it were not for her other friend she would be dead now. I am humbled as well by the courage of the many victims of depression who sat down with me to tell me their stories, and through me, you. Even though they knew I couldn't fully understand, that language is insufficient to express their psychic and physical misery, they still took me inside their experiences, from the first discovery of their illness, to what it has meant to their lives, to what they have been doing to fight it. And they still fight it. They get out of bed, go to work, pursue their therapies, and keep going; their struggles unrecognized by people around them. That has also been very humbling – looking back at people I worked with and in whom I now recognize depression, though I didn't at the time.

At my core, I am reductionist – I want to find the simple solution, the fundamental biological problem that *is* depression… but depression doesn't support that. It is hugely complex; an emergent outcome of a brain under attack. And yet, there are all these intriguing and unexpected pieces of the puzzle that scientists and physicians keep discovering, expanding, and exploiting to help people resist and recover from the illness. From the early days to present, these professionals have had to fight for acceptance of each new aspect of depression, and in many cases are still fighting. Because of their work, though, we know so much more about how neurotransmitters behave in depression, and the good and bad sides of stress, and the same for the immune response. We see what is happening to brain tissue in people with depression, and we're starting to understand how the whole system of the brain works together, how it organizes itself and us for life on earth. These advances wouldn't be possible without the help of modern technology, but it is the scientists' continuing drive to learn that pushes for technological advance.

Busy as they were, the professionals I spoke with answered my calls, emails, and follow-up questions because to them, it's the human element that matters. "What is depression" is not just a brain-teaser, it's an opening to helping people resist and recover from the illness. Thanks to the dedicated efforts of researchers and the people with and without depression who were willing to work with them, we know that we all have some degree of vulnerability to depression from genetics and early life exposures. Those won't change. There are also current day factors through which we can increase personal resilience, though. I can see those factors operating in my life now, especially as I hunker down for the COVID pandemic. We are all vulnerable to depression; some circumstances – and in some especially vulnerable people every day – demands extra attention to mental health, through diet, exercise, exposure to the natural environment, and supportive social connections. And by taking indicators of ill health seriously and talking to someone when we suspect something is wrong. There are some physicians out there – not enough, but some – who are willing to push the boundaries of "usual care." Though all of the physicians I spoke with use traditional antidepressant medications in patient care, they also look beyond those remedies to other treatment options in order to reach and help more people.

I still keep thinking "People ought to know this." They ought to be aware of signs of diminishing health in themselves and friends. People ought to know that major depression is a tough, even deadly, illness that needs intervention. People ought to ask questions of their medical providers about their care. And they ought to understand that depression really is an illness, not a personality trait. That last point goes beyond depression, to any mental illness. One of the unexpected outcomes I've seen in myself as a result of this project is a change in perspective about all mental illnesses. I see them now as a product of activities in the brain, as illnesses that can be treated (though in the case of many of them, science isn't there yet).

Science has come a long way in a little over 60 years; ours truly is a brave new world. And yet the journey through depression is still painful, personal, and lonely. But people ought to know there is help out there, and there are amazing and dedicated people who spend their lives working to combat depression and its misery. The story of science is, after all, the story of people: the patients, researchers, clinicians, regulators, and supporters. And all those people pushing against this one disorder have made lasting progress, with more just around the corner.

And Carolyn? As I write this, she is thriving. She continued into a maintenance program of ketamine and eventually graduated from it. She is still taking medications, of course, and seeing her therapist. But now, she's doing that as part of a happy, normal life.

And yet... I know what Carolyn did wrong in her suicide attempts – why they failed – but I won't write it down. Depression is a vicious, tenacious beast, and I can't be sure it won't make its way back into her life someday. It is a fight for a lifetime.

ACKNOWLEDGEMENTS

It will amuse some readers to hear that when I began this project in 2018, I went to the library at a nearby university, expecting to find... well, books and magazines. In particular, I remembered from my undergraduate days (in the 1980s) our college library having huge stacks of bound abstracts, organized by subject and showing which other articles had referenced them as well as the articles they referenced. It was, literally, old school. Imagine my dismay. After stumbling out of the library and going home, I contacted my nephew, Ricky Meadows, who was then in graduate school and he introduced me to some of the miracles of the modern age: JSTOR, Web of Science, SCOPUS, Google Scholar... I was off to the races. There are amazing resources out there available for anyone. I would especially like to thank the good people at the National Institutes of Health and Google Scholar for making so much valuable information freely available and accessible. The Reedsy Freelancer Marketplace was an invaluable resource for connecting with all sorts of writing-related professionals.

If you have ever run a search on the word "depression," you've seen how vast the subject is, returning zillions of items ranging from dubious or dangerous to highly precise technical detail. My mission with this book was to put the value of scientific discovery about depression into the hands of ordinary readers; at least, ordinary readers who aren't afraid of science. Toward this end, I am very grateful to the many scholars and clinicians who answered my calls and emails and shared their knowledge with me. Whether their words are recorded directly in this book from interviews or they answered my questions or corrected my misunderstandings about their area of expertise, this book would not have been possible without them. They are Dr. Jennifer Abe of Loyola Marymount University; Dr. Kirsten Berding Harold of APC Microbiome Ireland; Dr. Carina Carbia Sinde of APC Microbiome Ireland; Dr. Richard G. Boles, M.D.; Dr. Blynn Bunney of the University of California, Irvine; Dr. John Cryan of APC Microbiome Ireland and University College Cork; Dr. Don DuBose of Future Psych Solutions; Dr. Ronald Duman of Yale University; Dr. Menachem Fromer of Verily Life Sciences; Dr. Philip Gold of the National Institute of Mental Health; Dr. John Gottlieb of Chicago Psychiatry Associates; Dr. Caitriona Long-Smith of APC Microbiome Ireland; Dr. J. John Mann of Columbia University; Dr. Jason Martin of APC Microbiome Ireland; Dr. Jose Miguel-Hidalgo of the University of Mississippi Medical Center; Dr. Lisa Monteggia of Vanderbilt University; Dr. James Murrough of Mount Sinai Medical Center; Dr. Charles Nemeroff of the University of Texas; Dr. Harry A. Oken, M.D.; Dr. Siobhain O'Mahony of APC Microbiome Ireland; Dr. Marcus Raichle of Washington University in St. Louis; Dr. Charles Raison of the University of Wisconsin-Madison; Dr. Grazyna Rajkowska of the University of Mississippi Medical Center; Dr. Kieran Rea of APC Microbiome Ireland; Professor Graham Rook of University College London; Dr. Jerome Sarris of Western Sydney University; Dr. Harriet Schellekens of APC Microbiome Ireland; Dr. Yvette Sheline of the University of Pennsylvania; Professor Anna Wirz-Justice of the University of Basel; and Dr. Naomi Wray of the University of Queensland. My apologies for the alphabetical order; not something I am fond of myself.

It is important to me that the technical details in this book be accurate, and toward that end I want to thank my Most Excellent Army of technical reviewers. All of them are PhD candidates, and all are very talented individuals who plowed through various chapters finding my errors. I apologize to them if I introduced any new errors from further edits. They are: Caitlin Aamodt of the University of California-Los Angeles, Christine Foxx of the University of Colorado, Patricia

Horvath of Vanderbilt University, Dylan Kirsch of the University of Texas, Kelsey Loupy of the University of Colorado, and Yanning Zuo of the University of California-Los Angeles.

This project came alive because of the many people suffering with depression or with a loved one who suffers from depression, who generously shared their time and stories. Whether or not their stories were used directly in this book, they were all part of it. I thank them from the bottom of my heart, and wish them all the best on their personal journeys. I also deeply appreciate the help of Mr. Joe Pettit of NAMI Northern Virginia, a tireless warrior helping families cope with mental illness.

I would also like to thank the people who helped me through this journey of publishing: Mr. Jeff Shreve, who provided editorial expertise, Mr. Harry Haysom, who designed the cover, Mr. Tom Howey for the interior design, Dr. Marian Bland, who provided her editing expertise, and Mr. Thomas A. Garrett, who took the author photo. Finally, I'd like to thank my family for their loving support and helpful advice as I transitioned careers and took on this project.

GLOSSARY

5-HT	5-hydroxytryptamine, the official name of serotonin
5-HT1A	An inhibitory serotonin receptor, it slows down serotonin release
ACC	Anterior cingulate cortex
ACE	Adverse childhood experiences
ACTH	Adrenocorticotropic hormone; released by the pituitary gland to prompt release of cortisol from the adrenal glands
action potential	An electrical charge in a neuron that conveys a signal to the next neuron
active control	A potentially effective treatment to be compared to the efficacy of the drug under trial, often psychotherapy or a different drug
ADD	Attention deficit disorder
ADHD	Attention deficit hyperactivity disorder
adjunct/adjunctive	Added as a supplement; an additional treatment
adrenaline	U.K. nomenclature for epinephrine, a hormone secreted by the adrenal glands in conditions of stress
adrenergic	Applies to the system of neurons involved in epinephrine or norepinephrine (adrenaline or noradrenaline in the U.K.) signaling
affective behaviors	Invokes a person's feelings or emotions
affective disorders	Psychiatric disorders involving mood, usually major depressive disorder, dysthymia, and bipolar disorders
agomelatine	An atypical antidepressant used to treat major depressive disorder, designed to work through changes in circadian rhythms
allele	A variant of a gene
alpha-adrenergic	A type of norepinephrine receptor
amino acid	Building blocks of proteins
amitriptyline	A tricyclic antidepressant, sold under the brand name Elavil
amygdala	Commonly known as the fear center of the brain, evaluates the emotional significance of perceptions
anhedonia	Inability to experience pleasure

antagonist	Blocks a receptor to prevent binding and activation
anterior	As a brain coordinate, refers to the direction of the nose or forehead
anterograde amnesia	Inability to form memories; forgetting events as they happen
anti-inflammatory	Tends to reduce inflammation
antioxidants	A molecule stable enough to absorb a free radical without being functionally altered in the process
antipsychotic	A drug used to manage or relieve delusions, hallucinations, paranoia, or disordered thought
anxiety disorders	Disorders that share features of excessive fear and anxiety and related behavioral disturbances
APA	American Psychiatric Association
APC	Alimentary Pharmabiotic Centre
apoptosis	Programmed cell death; cell suicide
astrocyte	The most numerous type of glial cell; provides support to neurons and neuronal signaling
ATP	Adenosine tri-phosphate, the energy-carrying molecule of the cell
atypical antidepressant	An antidepressant medication that acts in an manner that is different from that of most other antidepressants
atypical depression/ atypical features	A type of major depression characterized by mood reactivity (i.e., mood brightens in response to actual or potential positive events), significant weight gain or increase in appetite, hypersomnia, leaden paralysis, and a long-standing pattern of interpersonal rejection sensitivity
augment	Adds to an existing treatment
autonomic nervous system	The body's control system that acts largely unconsciously and regulates bodily functions, including heart rate, digestion, respiratory rate, pupillary response, urination, and sexual arousal
autoreceptor	A receptor that sits on the neuron releasing a neurotransmitter in order to catch some of that release
axon	The part of a neuron that extends to the next neuron to transmit a signal to it
ayahuasca	A hallucinogenic tea brewed from the stem and leaves of two South American plants
BDNF	Brain-derived neurotrophic factor; a protein that helps the proliferation, survival, and adaptation of brain cells
benzodiazepines	Common treatments for anxiety disorders

beta-adrenergic	A type of norepinephrine receptor
bioenergetics	The process of producing energy for a cell
bipolar disorder	A manic-depressive disorder; involves periods of depression and periods of either mania (in bipolar I disorder) or hypomania (in bipolar II disorder)
blood-brain barrier	The semipermeable border of endothelial cells that prevents substances in the blood from non-selectively crossing into the extracellular fluid of the brain
brainstem	A part of the brain made up of the midbrain, pons, and medulla oblongata; it holds structures that bring sensory information in and send information relating to arousal and awareness out
catatonic	Characterized by an immobile and unresponsive state
catecholamine	The family of neurotransmitters that includes dopamine and norepinephrine
caudal	As a brain coordinate, refers to the direction of the tail
CBT	Cognitive behavior therapy, a form of psychotherapy
CDC	U.S. Centers for Disease Control and Prevention
Celexa	Brand name for citalopram, an SSRI
cell body	The part of the cell that contains the nucleus and most organelles
central executive network	The large-scale intrinsic neural network specialized for task execution monitoring
cerebellum	A part of the brain that modulates the force and range of movements and the learning of motor skills
cerebral cortex	The thin, sheet-like structure that forms the outermost part of the brain; responsible for the integration of complex sensory and neural functions and the initiation and coordination of voluntary activity in the body
cerebral spinal fluid	The colorless body fluid found in the brain and spinal cord
chronic inflammation	A long term, continuing state of immune system response, usually denoted by redness, swelling, heat, and pain
chronic stress	A long term, continuing state of stress (denoted by circulating cortisol and activation of the "fight or flight" response)
chronotherapy	A diverse set of therapies for depression and bipolar disorder that all act on circadian rhythms
cingulate cortex	An ancient part of the brain, a component of the limbic system; part of the cerebral cortex, it is involved with emotion formation and processing, learning, and memory

circadian genes Genes that are expressed on a schedule determined by circadian rhythms

circadian rhythms A natural, internal process that regulates the sleep-wake cycle; any biological process that displays an endogenous, entrainable oscillation of about 24 hours

italopram An SSRI, sold under the brand name Celexa

lock genes Genes that oscillate on a 24-hour basis, providing circadian rhythms for the cell's functions

cognition/cognitive ability Thinking, knowing, perceiving

co-morbid The presence of one or more additional conditions co-occurring with a primary condition

cortex A thin sheet-like structure

corticosterone The primary glucocorticoid in rodents

cortisol The primary glucocorticoid in primates (including humans)

CRH Corticotrophin releasing hormone

cross-sectional study A study of a population taking its data at one point in time

CRP C-reactive protein; indicates a recent or on-going state of inflammation

Cymbalta Brand name for duloxetine, an SNRI

cytokines Chemical messengers that signal to elements of the immune system

cytoplasm The part of a cell that is outside the nucleus

DBT Dialectical behavior therapy, a form of psychotherapy

default mode network The large-scale intrinsic neural network involved in self-referential thinking; usually most active when the subject is at rest

dendrite The part of a nerve cell that receives and combines inputs from other nerve cells

depolarization The sudden increase in positive charge in a neuron that generates an action potential

DHA Docosahexaenoic acid, an omega-3 polyunsaturated fatty acid

disorder An irregularity, disturbance, or interruption of normal functions serious enough to interrupt normal functioning for a significant amount of time

DNA Deoxyribonucleic acid, the information molecule containing the genetic code of an individual

DNA methylation A biological process by which methyl groups are added to the DNA molecule, repressing its transcription

dopamine	A monoamine neurotransmitter synthesized from the amino acid tyrosine, associated with reward and motivation
dopaminergic	Applies to neurons and other system components involved in the release of dopamine
dorsal	As a brain coordinate, refers to the direction of the backbone
dorsolateral prefrontal cortex	An area of the brain associated with executive control, working memory, and inhibitory control over behavior
DSM	The Diagnostics and Statistical Manual of Mental Disorders
dysphoria	A state of unease or generalized dissatisfaction
dysthymia	A chronic form of depression, a low mood occurring for at least two years, along with at least two other symptoms of depression
ECT	Electroconvulsive therapy
EEG	Electroencephalogram
EEG-NF	Neurofeedback using an electroencephalogram (EEG) for cueing information
effectiveness	The performance of an intervention or therapy under real-world conditions
Effexor	The brand name for venlafaxine, an SNRI
efficacy	The performance of an intervention or therapy under ideal and controlled circumstances
electroconvulsive shock	Electric current applied to a lab animal to induce a convulsion
endocrine	Refers to the hormone system
endothelial cell	A cell layer that lines all blood vessels and regulates exchanges between the bloodstream and the surrounding tissues
enzyme	A substance created in a cell to bring about a specific biochemical reaction
EPA	Eicosapentaenoic acid, an omega-3 polyunsaturated fatty acid
epidemiology/ epidemiological	A branch of medicine dealing with the incidence, distribution, and control of diseases across populations
epigenetic	A modification to inherited characteristics that takes place on top of the genome
epinephrine	A neurotransmitter and hormone produced by the adrenal glands; also known as adrenaline
esketamine	The S-enantiomer of ketamine, sold as an antidepressant under the brand name Spravato
essential amino acid	A building block of a protein that cannot be manufactured by the body but must come from the diet

excitatory	Tends to build up an electrical charge to fire an action potential
excitotoxicity	A pathological process by which neurons are damaged and killed by the over-activations of receptors for the excitatory neurotransmitter glutamate
expression	Refers to the transcription and translation of a gene to produce its protein
extracellular	In the fluid outside a cell
FDA	U.S. Food and Drug Administration
"fight or flight" response	An automatic physiological reaction to an event that is perceived as stressful or frightening; an activation of the sympathetic nervous system from perception of a threat, triggering an acute stress response that prepares the body to fight or flee
FKBP5	The gene that codes for the glucocorticoid receptor
fluoxetine	An antidepressant sold under the name Prozac, an SSRI
fMRI	Functional magnetic resonance imaging
folate	Vitamin B9
free radicals	Molecules, atoms, or ions with one or more unpaired electrons; highly reactive and damaging to proteins, lipids, and DNA
functional connectivity	Time-correlated neurophysiological index measured in different brain areas, indicates brain areas associated in common in a defined task
functional imaging	Technology that shows indications of different activity levels in the brain through changes in blood flow, oxygenation, or glucose metabolism
GABA	Gamma-aminobutyric acid, the primary inhibitory neurotransmitter
GABAergic	Related to neurons that release and/or respond to GABA
GDNF	Glial-derived growth factor
gene expression	The transcription and translation of a gene to produce its protein
glia/glial cell	A type of brain cell that performs support functions for neurons and neuronal signaling
glucocorticoid	A class of steroid hormone involved in the stress response
glucocorticoid resistance	The faulty response of glucocorticoid receptors in the presence of high concentrations of cortisol
glutamate	The primary excitatory neurotransmitter in the brain

glutamatergic	Refers to neurons and components that release and/or respond to glutamate
glutamine	An amino acid precursor for glutamate and GABA
GR	Glucocorticoid receptor
gray matter	Neuronal cell bodies, dendrites, unmyelinated axons, and glial cells
gut microbiome/ microbiota	The totality of microorganisms, bacteria, viruses, protozoa, and fungi, and their collective genetic material present in the gastrointestinal tract
gut-brain axis	Bidirectional communication between the central and the enteric nervous systems, linking emotional and cognitive centers of the brain with intestinal functions
GWAS	Genome-wide association study
hallucinogens/ hallucinogenic	A drug that causes hallucinations
hippocampus	A brain structure in the temporal lobe involved in the formation of new memories and associated with learning and emotions
histamine	A compound released by cells in response to injury and in allergic and inflammatory reactions to cause the body the expel the noxious factor
histone acetylation	An epigenetic mechanism that initiates expression of a gene
histone methylation	An epigenetic mechanism that blocks expression of a gene
hormone	A chemical messenger created by an endocrine gland and released into the bloodstream
HPA axis	Hypothalamus-pituitary-adrenal axis, the central stress response system
hyperconnectivity	Involves additional areas of time-correlated activity, and/or the correlations are within a tighter timeframe, and/or fewer areas of negative correlation than normal
hypercortisolemia	Over-secretion of cortisol
hyperpolarization	A state in which the electrical potential across a neuronal membrane has a greater negative charge than in the resting potential
hypersomnia	A tendency to sleep too much, more than normal
hypoconnectivity	Involves fewer areas of time-correlated activity, the correlations are within a looser timeframe, and/or more areas of negative correlation than normal
hypocortisolemia	Under-secretion of cortisol

hypomania	A mental state of elevated, expansive, or irritable mood and persistently increased level of activity or energy; of a lesser degree than mania
hypothalamus	A part of the brain specialized for releasing hormones, controlling body temperature, sexual behavior, and appetite, and modulating the stress response
ICD	International Classification of Diseases, international diagnostic classification standard for all clinical and research purposes
IDO	Indoleamine 2,3-dioxygenase, an enzyme that initiates the breakdown of the brain's tryptophan supply along a non-serotonin pathway
IFN-a	Interferon-alpha, a signaling protein (cytokine) specializing in alerting the immune system to the presence of a virus
IL-10	Interleukin-10, an anti-inflammatory signaling protein (cytokine)
IL-6	Interleukin-6, a pro-inflammatory signaling protein (cytokine)
imipramine	The first tricyclic antidepressant, sold under the brand name Tofranil
immunotherapy	A treatment that uses a person's own immune system to fight cancer or hepatitis
inferior	As a brain coordinate, refers to lower or below
inhibitory	Tends to increase negative charge in a neuron so as to keep the neuron from firing an action potential
insomnia	The inability to sleep; troubled, disrupted sleep
insular cortex/insula	A portion of the cerebral cortex folded deep within each hemisphere of the brain, specialized to link sensory experience and emotional valence
ion channel	A pore on a neuronal membrane used to shuttle ions into or out of the cell in order to change the electric charge potential across the membrane
ionotropic receptor	A receptor that sits directly on an ion channel so that activating the receptor opens the channel for ion flow
iproniazid	The first monoamine oxidase inhibitor, an antidepressant medication
IPT	Interpersonal therapy, a form of psychotherapy
ISNPR	International Society for Nutritional Psychiatry Research
isoniazid	An anti-tuberculosis drug, related to iproniazid
IV	Intravenous

ketamine	A drug commonly used as an anesthetic
kynurenine	A metabolite of the amino acid tryptophan, often synthesized in response to immune activation
large-scale intrinsic networks	A set of brain areas/structures that work together to accomplish some function
lateral	As a brain coordinate, refers to the outer side
light therapy	A form of psychiatric treatment in which light is shown near the eyes to affect circadian clocks
limbic system	The ancient part of the brain, involved with emotion and instinct
lithium	A mineral commonly used in the treatment of bipolar disorder and sometimes used as an adjunctive treatment in major depressive disorder
locus coeruleus	A structure in the brainstem where the brain's norepinephrine supply is synthesized
longitudinal study	An observational study in which a number of individuals are monitored over a long period of time to see their disease process
LSD	Lysergic acid diethylamide, a hallucinogenic drug
LTP	Long-term potentiation, a process by which neuronal connections are strengthened through repeated use, key to learning and memory
major depressive disorder	A psychiatric condition involving discrete episodes of at least 2 weeks' duration involving clear-cut changes in affect, cognition, and neurovegetative functions and inter-episode remissions. A major depressive episode is characterized by sad, empty, or irritable mood, accompanied by somatic and cognitive changes that significantly affect the individual's capacity to function
major depressive episode	A period of at least 2 weeks' duration involving sad, empty, or irritable mood, accompanied by somatic and cognitive changes
mania	A mental state of elevated, expansive, or irritable mood and persistently increased level of activity or energy
MAOI	Monoamine oxidase inhibitor, an early class of antidepressant drug
MAP	Mycobacterium avium paratuberculosis
MBCT	Mindfulness-based cognitive therapy, a form of psychotherapy
MBSR	Mindfulness-based stress reduction, a form of psychotherapy
medial	As a brain coordinate, refers to the direction of the centerline

melancholic features/ depression	A subtype of depression in which the sufferer loses appetite and sleep, and experiences a loss of pleasure in all, or almost all, activities or a lack of reaction to usually pleasurable stimuli, and even something good happening cannot improve his mood. These features are typically at their worst in the morning and can improve as the day goes on
melancholy/ melancholia	A term for depression used until the early 20th century
melatonin	A hormone produced in the brain in response to darkness that affects timing of circadian rhythms and sleep
meta-analysis	A statistical procedure for combining data from multiple studies
metabolite	A product of the conversion of a substance through the actions of an enzyme
metabolized	Converted to another substance through actions of an enzyme
metabotropic receptor	A receptor in which the binding of its neurotransmitter releases a set of messengers inside the neuron that trigger a cascade of biochemical signaling pathways
methylation	An epigenetic process in which a methyl group is added to DNA, RNA, a histone, or a protein to block access for transcription
microbiome	The totality of microorganisms, bacteria, viruses, protozoa, and fungi, and their collective genetic material
microbiota	A collective term for the micro-organisms that live in or on the body
microbiota-gut-brain axis	Bidirectional communication between the central and enteric nervous systems and the gut's resident microorganisms, linking emotional and cognitive centers of the brain with intestinal functions as they are affected by the products of the microorganisms
microglia	A type of glial cell specialized to carry out immune system functions in the brain
mineralocorticoid	A class of steroid hormone involved in the absorption of minerals
mirtazapine	A tetracyclic antidepressant medication sold under the brand name Remeron, considered an atypical antidepressant
mitochondria	The energy-producing organelle in a cell
monoamine	A compound having a single amine group in its molecule, especially one that is a neurotransmitter (e.g., serotonin, norepinephrine)

monoamine deficiency hypothesis	A theory proposing that the underlying biological basis for depression is a lack of neurotransmission of serotonin and/or norepinephrine, and that antidepressant treatments restore normal function by targeting this deficiency
monoamine oxidase	An enzyme that breaks down a monoamine into a waste product
monoaminergic	Applies to neurons and other system components involved in the release of any monoamine
mood disorders	Psychiatric disorders involving mood, usually major depressive disorder, dysthymia, and bipolar disorders
MR	Mineralocorticoid receptor
MRI	Magnetic resonance imaging
mRNA	Messenger ribonucleic acid, a temporary template for a protein created by the transcription of a gene
MRS	Magnetic resonance spectroscopy
myelin sheath	An insulating layer formed around the axon of a neuron, composed of protein and fatty substances
NAC	N-acetyl-cysteine
NAMI	National Alliance on Mental Illness
NARI	Norepinephrine/noradrenaline reuptake inhibitor
neocortex	The outermost area of the cerebral cortex, involved in higher-order brain functions such as sensory perception, cognition, generation of motor commands, spatial reasoning and language
neural networks	A set of brain areas/structures that work together to accomplish some function
neurodegeneration	The progressive atrophy and loss of function of neurons
neurofeedback	A therapeutic intervention that provides immediate feedback from a computer-based program on a patient's brainwave activity. The program then uses sound or visual signals to train the patient to reorganize or retrain these brain signals
neurogenesis	The birth of new brain cells
neuroinflammation	A state of inflammation in the brain
neuronal membrane	The cell membrane around a neuron
neuropathic	Involves damage to neurons or glia
neuroplasticity	The set of adaptive changes in the brain, including the birth of new brain cells, changes in cell function, reduction or creation of synaptic connections, and changes in structure of cells

neurotrophin	A growth factor, helps brain cells grow, differentiate, migrate, and form connections to other brain cells
NIH	U.S. National Institutes of Health
NIMH	U.S. National Institute of Mental Health
NMDA	N-methyl-D-aspartate, a type of glutamate receptor
noradrenaline	U.K. nomenclature for norepinephrine, a monoamine neurotransmitter that controls the sympathetic nervous system
noradrenergic	Applies to the system of neurons involved in norepinephrine (noradrenaline in the U.K.) signaling
norepinephrine	A monoamine neurotransmitter that controls the sympathetic nervous system
nucleus	The cellular structure that holds a cell's genetic material
nucleus accumbens	A small structure in the brain, considered the center of reward and pleasure
nutraceuticals	A dietary supplement administered as treatment for a health condition
oligodendrocyte	A type of glial cell specialized for wrapping an appendage around axons of neurons to protect them and speed the signal to its destination
oxytocin	A hormone secreted by the posterior lobe of the pituitary gland, released when people bond socially
paroxetine	An SSRI, sold under the brand name Paxil
pathogen	A bacterium, virus, or other microorganism that can cause disease
pathology	The study of the causes and effects of disease or injury
PET	Positron emission tomography
phagocyte	A cell that engulfs and destroys another cell or cell part
phenelzine	An MAOI, sold under the brand name Nardil
pituitary	A small pea-sized gland that plays a major role in regulating vital body functions and general wellbeing through release of hormones
posterior	As a brain coordinate, refers to the direction of the tail
posterior cingulate cortex	The rear portion of the cingulate cortex; has a central role in supporting internally-directed thoughts and shows increased activity when people retrieve autobiographical memories or plan for the future
postpartum depression	Depression with onset during pregnancy or within 4 weeks of the birth

postsynaptic A neuron that receives the neurotransmitter after it has crossed the synapse

prebiotic Compound in food that induce the growth or activity of beneficial microorganisms such as bacteria and fungi

precuneus Part of the parietal lobe, a brain region involved in a variety of complex functions including recollection and memory, integration of information relating to perception of the environment, mental imagery strategies, episodic memory retrieval, and affective responses to pain

prefrontal cortex The anterior part of the frontal cortex, implicated in planning complex cognitive behavior, expression of personality, decision-making, executive control, and moderating social behavior

pregenual ACC The forward-most portion of the anterior cingulate cortex, involved in happy events and self-relevant tasks

presynaptic The neuron that releases a neurotransmitter into the synapse

prevalence The proportion of a population who have a specific characteristic in a given time period

Pristiq Brand name for desvenlafaxine, an SNRI

probiotic A combination of live beneficial bacteria and/or yeasts that naturally live in the body

pro-inflammatory Tends to initiate a state of inflammation

prospective study An observational study in which an at-risk population is monitored over a long period of time

Prozac Brand name for fluoxetine, an SSRI

psilocybin A naturally-occurring psychedelic drug compound produced by many species of fungus

psychedelics A class of psychoactive substances that produce changes in perception, mood and cognitive processes; also known as hallucinogens

psychopathology A mental or behavioral disorder

psychosocial stress A perceived threat to one's social status, social esteem, respect, and/or acceptance within a group; threat to self-worth; or a threat that feels uncontrollable

psychotic Marked by delusions or hallucinations

psychotic depression/features Major depressive disorder involving delusions and/or hallucinations

psychotropic Any drug that affects behavior, mood, thoughts, or perception

PTSD Post-traumatic stress disorder

PUFA	Polyunsaturated fatty acid
quinolinic acid	A downstream product of kynurenine metabolism, thought to be toxic to cells
raphe nuclei	A structure in the brainstem in which the brain's serotonin supply is synthesized
remission	In depression, a patient is in remission (or remits) when their depression scale scores reach a low, nominal level
repolarization	The change in membrane potential that returns it to a negative value just after the depolarization phase of an action potential, which has changed the membrane potential to a positive value
respiratory chain	Structures within a mitochondrial membrane that produce the cell's energy substrate, ATP
response	In depression, response means that a patient's depression scale scores are reduced to half of their starting or highest value
resting potential	The difference in electrical charge between the inside and outside of a neuron when it is at rest, usually about -70 millivolts
retinoid	A derivative of vitamin A
retrograde amnesia	Inability to remember something that happened in the past
reuptake	Transport of a neurotransmitter from the synapse back into the neuron that released it
RNA	Ribonucleic acid, a form of genetic information created when a gene is transcribed
rostral	As a brain coordinate, refers to the direction toward the nose
rTMS	Repetitive transcranial magnetic stimulation
salience network	A large-scale brain network involved in detecting and filtering salient stimuli in order to recruit the relevant functional network
SCFA	Short-chain fatty acid
schizo-affective disorder	A psychiatric disorder in which mood episode and the active-phase symptoms of schizophrenia occur together and were preceded or are followed by at least 2 weeks of delusions or hallucinations without prominent mood symptoms
schizophrenia	A psychiatric disorder defined by abnormalities in one or more of the following domains: delusions, hallucinations, disorganized thinking (speech), grossly disorganized or abnormal motor behavior (including catatonia), and negative symptoms (diminished emotional expression and avolition)

SCN	Suprachiasmatic nucleus, the brain's master clock
seasonal affective disorder	A form of major depression in which there has been a regular temporal relationship between the onset of major depressive episodes and a particular time of the year (e.g., in the fall or winter), and remission also occurs along a seasonal pattern
serotonergic	Applies to the system of neurons involved in serotonin signaling
serotonin	A monoamine neurotransmitter, broadly involved in mood, sensory perception, behavior, and memory
sertraline	An SSRI, sold under the brand name Zoloft
sham-controlled/sham treatment	A control condition, like a placebo, but used when the treatment is a procedure rather than a medication
sleep deprivation therapy	Also known as wake therapy, attempts to correct a circadian rhythms dysfunction by altering the sleep schedule
SNP	Single nucleotide polymorphism
SNRI	Serotonin and norepinephrine reuptake inhibitor, a type of antidepressant medication
somatic	Refers to the body
Spravato	Brand name for esketamine, a rapid-acting antidepressant delivered as a nasal spray
SSRI	Selective serotonin reuptake inhibitor, a type of antidepressant medication
STAR*D	The Sequenced Treatment Alternatives to Relieve Depression study
statistical significance	In this book, results are noted as "statistically significant" only if the confidence interval bounding the results does not cross or include 1.0, and the p value is less than 0.05
structural imaging	Approaches specialized for visualization and analysis of anatomical properties of the brain
subgenual ACC	The most ventral part of the anterior cingulate cortex, activated in the memory of negative events and sadness tasks
suicidal behaviors	Any degree of thought or action towards taking one's own life, from thoughts (suicidal ideation) to completed suicide
suicidal ideation	Thoughts about taking action to end one's life, including identifying a method, having a plan, and/or having intent to act
suicidality	Any degree of thought or action towards taking one's own life, from thoughts (suicidal ideation) to completed suicide
suicide attempt	A potentially self-injurious behavior associated with at least some intent to die

superior	As a brain coordinate, refers to above or top
sympathetic nervous systems	An extensive network of neurons that controls aspects of the body related to the flight-or-fight response, such as mobilizing fat reserves, increasing the heart rate, and releasing epinephrine
synapse	A junction between two neurons, consisting of a minute gap across which impulses pass by diffusion of a neurotransmitter
synaptic plasticity	The activity-dependent modification of the strength or efficacy of synaptic transmission at pre-existing synapses
TCA	Tricyclic antidepressant
tDCS	Transcranial direct current stimulation
terminals	Structures at the end of an axon from which neurotransmitters are released
TES	Transcranial electrical stimulation
tetracyclic antidepressants	A class of antidepressants containing four rings in their structure, closely related to tricyclic antidepressants
thalamus	A brain structure that processes most of the sensory information from the rest of the central nervous system, directing it to areas in the cerebral cortex
threshold potential	The critical level to which a membrane potential must be depolarized to initiate an action potential
TMS	Transcranial magnetic stimulation
TNF	Tumor necrosis factor, a type of pro-inflammatory cytokine
transcription	The process of reading a DNA sequence and form the corresponding RNA nucleotide sequence
transporter	A protein situated on a presynaptic neuron terminal, specialized to catch some of the released neurotransmitter and store it back in the presynaptic neuron
trazodone	An atypical antidepressant, sold under the brand name Desyrel
treatment-resistant depression	Failed to respond to at least two different antidepressant medications administered for enough time at adequate dosage
tryptophan	An essential amino acid, precursor to serotonin
tryptophan hydroxylase	An enzyme used in the synthesis of serotonin from tryptophan
tyrosine	An essential amino acid, precursor to dopamine and norepinephrine
UCC	University College Cork

unipolar depression	Refers to major depression or dysthymia, excludes bipolar disorders
vagus nerve	The longest and most complex of the 12 pairs of cranial nerves that emanate from the brain; transmits information to or from the brain to tissues and organs elsewhere in the body, including the gut
vasopressin	A hormone secreted by the hypothalamus, involved in the stress response and male-typical social behaviors, including aggression, pair-bond formation, scent marking, and courtship
venlafaxine	An SNRI, sold under the brand name Effexor
ventral striatum	Part of the basal ganglia, holds the nucleus accumbens (the pleasure and reward center of the brain)
ventromedial prefrontal cortex	A part of the cerebral cortex with extensive connections to the limbic system, part of the "emotion circuit"
visceral	Refers to the gut
wake therapy	Also known as sleep deprivation therapy, attempts to correct a circadian rhythms dysfunction by altering the sleep schedule
Wellbutrin	Brand name for bupropion, an atypical antidepressant
white matter	Tracts of axons in the brain, named for the pale-colored myelin sheaths surrounding and protecting them
WHO	World Health Organization
Xanax	Brand name for an anti-anxiety medication
Zoloft	Brand name for sertraline, an SSRI

NOTES

Introduction AWAKENING

1 Hasin, D. S., Goodwin, R. D., Stinson, F. S., & Grant, B. F. (2005). Epidemiology of major depressive disorder: results from the National Epidemiologic Survey on Alcoholism and Related Conditions. *Archives of general psychiatry*, 62(10), 1097-1106.
Kessler, R. C., Berglund, P., Demler, O., Jin, R., Koretz, D., Merikangas, K. R., ... & Wang, P. S. (2003). The epidemiology of major depressive disorder: results from the National Comorbidity Survey Replication (NCS-R). *Journal of the American Medical Association*, 289(23), 3095-3105.
2 World Health Organization. (2017, February 23). Depression and Other Common Mental Disorders. Retrieved from https://www.who.int/mental_health/management/depression/prevalence_global_health_estimates/en/
3 American Psychiatric Association. (1952). *Diagnostic and statistical manual of mental disorders*. Washington DC: American Psychiatric Association, pg. v-viii.
4 American Psychiatric Association. (2013). *Diagnostic and statistical manual of mental disorders, 5th ed.* (DSM-5). Arlington: American Psychiatric Association, pg. 155, 160-171.
5 DSM-5, pg. 132-136.
6 Begley, S. (2013, July 17). DSM-5 Finally Unveiled. Retrieved March 17, 2019, from https://www.huffpost.com/entry/dsm-5-unveiled-changes-disorders-_n_3290212
7 Nasrallah, H. A. (2009). Diagnosis 2.0 are mental illnesses diseases, disorders, or syndromes? A major challenge for the DSM-V committees as they revise the diagnostic" bible" of psychiatric disorders is to determine whether mental illnesses are diseases, disorders, or syndromes. *Current Psychiatry*, 8(1), 14-16.
8 Arsenault-Lapierre, G., Kim, C., & Turecki, G. (2004). Psychiatric diagnoses in 3275 suicides: a meta-analysis. *BMC psychiatry*, 4(1), 37.
9 Pratt, L. A., Druss, B. G., Manderscheid, R. W., & Walker, E. R. (2016). Excess mortality due to depression and anxiety in the United States: results from a nationally representative survey. *General hospital psychiatry*, 39, 39-45.

Chapter One THE MONOAMINES

1 Baptista, R. J. (2009, October 13). Geigy-A Historic Dyestuffs Company. Retrieved June 16, 2019, from http://www.colorantshistory.org/Geigy.html
2 Steinberg, H., & Himmerich, H. (2012). Roland Kuhn–100th Birthday of an Innovator of Clinical Psychopharmacology. *Psychopharmacology bulletin*, 45(1), 48.
3 Cahn, C. (2006). Roland Kuhn, 1912–2005. *Neuropsychopharmacology*, 31(5), 1096.
4 Kuhn, R. (1958). The treatment of depressive states with G 22355 (imipramine hydrochloride). *American Journal of Psychiatry*, 115(5), 459-464.
5 Lehmann, H. E., Cahn, C. H., & De Verteuil, R. L. (1958). The treatment of depressive conditions with imipramine (G 22355). *Canadian Psychiatric Association Journal*, 3(4), 155-164.
Mann, A. M., & MacPherson, A. S. (1959). Clinical experience with imipramine (G 22355) in the treatment of depression. *Canadian Psychiatric Association Journal*, 4(1), 38-47.
6 Edelstein, S. (2012). The Great American Mow Down. Retrieved from https://envisioningtheamericandream.com/2012/05/10/the-great-american-mow-down/
7 Robitzek, E. H., & Selikoff, I. J. (1952).

Hydrazine derivatives of isonicotinic acid (Rimifon, Marsilid) in the treatment of active progressive caseous-pneumonic tuberculosis: a preliminary report. *American review of tuberculosis*, 65(4), 402-428.
8 Dally, P. J. (1958). Indications for use of iproniazid in psychiatric practice. *British medical journal*, 1(5083), 1338.
9 Pare, C. M. B., & Sandler, M. (1959). A clinical and biochemical study of a trial of iproniazid in the treatment of depression. *Journal of neurology, neurosurgery, and psychiatry*, 22(3), 247.
10 Mayo Foundation for Medical Education and Research. (2019, October 8). Tricyclic antidepressants (TCAs). Retrieved from https://www.mayoclinic.org/diseases-conditions/depression/in-depth/antidepressants/art-20046983
11 Mayo Foundation for Medical Education and Research. (2019, September 12). An option if other antidepressants haven't helped. Retrieved from https://www.mayoclinic.org/diseases-conditions/depression/in-depth/maois/art-20043992
12 Kandel, E. R. (2012). *Principles of neural science, 5th ed*. NY, NY: McGraw-Hill Medical, pg. 22-24, 27.
13 McCoy, A. N., & Tan, Y. S. (2014). Otto Loewi (1873–1961): Dreamer and Nobel laureate. *Singapore medical journal*, 55(1), 3.
14 Kandel 2012, pg. 289-297
Lodish, H. (2016). *Molecular cell biology, 8th ed*. New York: W. H. Freeman and Co, pg. 673-676
15 Kandel 2012, pg. 29-31
16 Purves, D., Augustine, G. J., Fitzpatrick, D., Hall, W. C., LaMantia, A.-S., & White, L. E. (2001). *Neuroscience, 2nd Edition*. Sunderland, Massachusetts.: Sinauer Associates, Inc, Available from: https://www.ncbi.nlm.nih.gov/books/NBK10799/
17 Purves et al 2001
18 Kandel 2012, pg. 186, 236-237, 255-257, 290
19 Kandel 2012, pg. 295-301
20 Mulinari, S. (2012). Monoamine theories of depression: historical impact on biomedical research. *Journal of the History of the Neurosciences*, 21(4), 366-392.
21 Mulinari 2012
Schildkraut, J. J. (1965). The catecholamine hypothesis of affective disorders: a review of supporting evidence. *American journal of Psychiatry*, 122(5), 509-522.
22 Mulinari 2012
Joseph J. Schildkraut. (2008, October 23). Faculty of Medicine - Memorial Minute. Retrieved from https://news.harvard.edu/gazette/story/2008/10/joseph-j-schildkraut/
23 Kandel 2012, pg. 291-293
24 Mulinari 2012
25 Coppen, A. (1967). The biochemistry of affective disorders. *The British Journal of Psychiatry*, 113(504), 1237-1264.
26 Mulinari 2012
27 Hirschfeld, R. M. (2000). History and evolution of the monoamine hypothesis of depression. *The Journal of clinical psychiatry*, 61, 4.
28 Kandel 2012, pg. 291, 1398, 1105
29 Mishra, A., Singh, S., & Shukla, S. (2018). Physiological and functional basis of dopamine receptors and their role in neurogenesis: possible implication for Parkinson's disease. *Journal of experimental neuroscience*, 12, 1179069518779829.
30 Gold, P. W. (2015). The organization of the stress system and its dysregulation in depressive illness. *Molecular psychiatry*, 20(1), 32.
31 Ressler, K. J., & Nemeroff, C. B. (2000). Role of serotonergic and noradrenergic systems in the pathophysiology of depression and anxiety disorders. *Depression and anxiety*, 12(S1), 2-19.
Kandel 2012, pg. 1410-1411
Maletic, V., Eramo, A., Gwin, K., Offord, S. J., & Duffy, R. A. (2017). The role of norepinephrine and its α-adrenergic receptors in the pathophysiology and treatment of major depressive disorder and schizophrenia: a systematic review. *Frontiers in psychiatry*, 8, 42.
32 Nichols, D. E., & Nichols, C. D. (2008). Serotonin receptors. *Chemical reviews*, 108(5), 1614-1641.
33 Kandel 2012, pg. 1410-1411; Nichols et al, 2008
34 Nichols et al, pg. 2008

35 Nichols et al, pg. 2008
36 Nichols et al, pg. 2008
37 Coppen 1967
38 Walsh, B. T., Seidman, S. N., Sysko, R., & Gould, M. (2002). Placebo response in studies of major depression: variable, substantial, and growing. *JAMA, 287*(14), 1840-1847.
39 Charney, D. S., Menkes, D. B., & Heninger, G. R. (1981). Receptor sensitivity and the mechanism of action of antidepressant treatment: Implications for the etiology and therapy of depression. *Archives of General Psychiatry, 38*(10), 1160-1180.
40 Ruhé, H. G., Mason, N. S., & Schene, A. H. (2007). Mood is indirectly related to serotonin, norepinephrine and dopamine levels in humans: a meta-analysis of monoamine depletion studies. *Molecular psychiatry, 12*(4), 331-359.
41 Warden, D., Rush, A. J., Trivedi, M. H., Fava, M., & Wisniewski, S. R. (2007). The STAR* D Project results: a comprehensive review of findings. *Current psychiatry reports, 9*(6), 449-459.
42 Kaufman, J., DeLorenzo, C., Choudhury, S., & Parsey, R. V. (2016). The 5-HT1A receptor in major depressive disorder. *European Neuropsychopharmacology, 26*(3), 397-410.

Chapter Two
STRESS

1 Selye, H. (1936). A syndrome produced by diverse nocuous agents. *Nature, 138*(3479), 32-32.
2 Szabo, S., Tache, Y., & Somogyi, A. (2012). The legacy of Hans Selye and the origins of stress research: a retrospective 75 years after his landmark brief "letter" to the editors of *Nature*. *Stress, 15*(5), 472-478.
3 Brown, T. M., & Fee, E. (2002). Walter Bradford Cannon: Pioneer physiologist of human emotions. *American Journal of Public Health, 92*(10), 1594-1595.
4 Tan, S. Y., & Yip, A. (2018). Hans Selye (1907–1982): Founder of the stress theory. *Singapore medical journal, 59*(4), 170.
5 Selye 1936
6 Tan & Yip 2018
7 Tan & Yip 2018
8 Selye 1936
9 Szabo et al 2012
10 Gold 2015
11 Kandel 1074, 1409; Gold 2015
12 Gold 2015
13 Gold 2015
14 Sachar, E. J., Hellman, L., Roffwarg, H. P., Halpern, F. S., Fukushima, D. K., & Gallagher, T. F. (1973). Disrupted 24-hour patterns of cortisol secretion in psychotic depression. *Archives of General Psychiatry, 28*(1), 19-24.
15 Sachar et al 1973
16 Sachar et al 1973
17 Lovallo, W. R. (2006). Cortisol secretion patterns in addiction and addiction risk. *International Journal of Psychophysiology, 59*(3), 195-202.
18 Cowen, P. J. (2002). Cortisol, serotonin and depression: all stressed out?. *The British Journal of Psychiatry, 180*(2), 99-100.
19 Anacker, C., Zunszain, P. A., Carvalho, L. A., & Pariante, C. M. (2011). The glucocorticoid receptor: pivot of depression and of antidepressant treatment?. *Psychoneuroendocrinology, 36*(3), 415-425.
Herbert, J., Goodyer, I.M., Grossman, A.B., Hastings, M.H., de Kloet, E.R., Lightman, S.L., Lupien, S.J., Roozendaal, B. & Seckl J.R. (2006). Do Corticosteroids Damage the Brain? *Journal of Neuroendocrinology 18*, 393–411.
20 DSM-5, pg. 185-186
21 Gold, P. W., & Chrousos, G. P. (2002). Organization of the stress system and its dysregulation in melancholic and atypical depression: high vs low CRH/NE states. *Molecular psychiatry, 7*(3), 254-275.
Interview with Dr. Gold, Jun 4, 2019
22 Lamers, F., Vogelzangs, N., Merikangas, K. R., De Jonge, P., Beekman, A. T. F., & Penninx, B. W. J. H. (2013). Evidence for a differential role of HPA-axis function, inflammation and metabolic syndrome in melancholic versus atypical depression. *Molecular psychiatry, 18*(6), 692-699.
23 Dinan, T. G. (1994). Glucocorticoids and the genesis of depressive illness a psychobiological

model. *The British Journal of Psychiatry, 164*(3), 365-371.4
24 Pariante, C. M., & Miller, A. H. (2001). Glucocorticoid receptors in major depression: relevance to pathophysiology and treatment. *Biological psychiatry, 49*(5), 391-404. Anacker et al 2011
25 Pariante 2001, Pariante, C. M., & Lightman, S. L. (2008). The HPA axis in major depression: classical theories and new developments. *Trends in neurosciences, 31*(9), 464-468.

Chapter Three
INFLAMMATION

1 Dantzer, R., Wollman, E., Vitkovic, L., & Yirmiya, R. (1999). Cytokines and depression: fortuitous or causative association?. *Molecular Psychiatry, 4*(4), 328-332.
2 Raison, C. L., Capuron, L., & Miller, A. H. (2006). Cytokines sing the blues: inflammation and the pathogenesis of depression. *Trends in immunology, 27*(1), 24-31.
3 Raison, C. L., & Miller, A. H. (2011). Is depression an inflammatory disorder?. *Current psychiatry reports, 13*(6), 467-475.
4 Dowlati, Y., Herrmann, N., Swardfager, W., Liu, H., Sham, L., Reim, E. K., & Lanctôt, K. L. (2010). A meta-analysis of cytokines in major depression. *Biological psychiatry, 67*(5), 446-457.
5 Haapakoski, R., Ebmeier, K. P., Alenius, H., & Kivimäki, M. (2016). Innate and adaptive immunity in the development of depression: an update on current knowledge and technological advances. *Progress in Neuro-Psychopharmacology and Biological Psychiatry, 66*, 63-72.
6 Lodish 2016, pg. 1084-1087
7 Lodish 2016, pg. 1084-1087
8 Lodish 2016, pg. 1084-1087
9 Kandel 2012, pg. 1565-1568
10 Yirmiya, R., Rimmerman, N., & Reshef, R. (2015). Depression as a microglial disease. *Trends in neurosciences, 38*(10), 637-658.
11 Rajkowska, G., & Miguel-Hidalgo, J. J. (2007). *Gliogenesis and glial pathology in depression. CNS & Neurological Disorders-Drug Targets (Formerly Current Drug Targets-CNS & Neurological Disorders)*, 6(3), 219-233.
12 Maier, S. F., & Watkins, L. R. (1998). Cytokines for psychologists: implications of bidirectional immune-to-brain communication for understanding behavior, mood, and cognition. *Psychological review, 105*(1), 83.
Steiner, J., Bogerts, B., Sarnyai, Z., Walter, M., Gos, T., Bernstein, H. G., & Myint, A. M. (2012). Bridging the gap between the immune and glutamate hypotheses of schizophrenia and major depression: potential role of glial NMDA receptor modulators and impaired blood–brain barrier integrity. *The World Journal of Biological Psychiatry, 13*(7), 482-492.
Louveau, A., Smirnov, I., Keyes, T. J., Eccles, J. D., Rouhani, S. J., Peske, J. D., ... & Harris, T. H. (2015). Structural and functional features of central nervous system lymphatics. *Nature*, 523(7560), 337-341.
13 Maes, M., Yirmyia, R., Noraberg, J., Brene, S., Hibbeln, J., Perini, G., ... & Maj, M. (2009). The inflammatory & neurodegenerative (I&ND) hypothesis of depression: leads for future research and new drug developments in depression. *Metabolic brain disease, 24*(1), 27-53.
14 Raison et al 2006
15 Gimeno, D., Kivimäki, M., Brunner, E. J., Elovainio, M., De Vogli, R., Steptoe, A., ... & Ferrie, J. E. (2009). Associations of C-reactive protein and interleukin-6 with cognitive symptoms of depression: 12-year follow-up of the Whitehall II study. *Psychological medicine*, 39(3), 413-423.
16 Duivis, H. E., de Jonge, P., Penninx, B. W., Na, B. Y., Cohen, B. E., & Whooley, M. A. (2011). Depressive symptoms, health behaviors, and subsequent inflammation in patients with coronary heart disease: prospective findings from the heart and soul study. *American Journal of Psychiatry, 168*(9), 913-920.
17 Yirmiya et al 2015
18 Yirmiya et al 2015
19 Yirmiya et al 2015
20 Raison & Miller 2011
21 Raison & Miller 2011
22 Dantzer, R. (2009). Cytokine, sickness behavior, and depression. *Immunology and*

Allergy Clinics, 29(2), 247-264.
23 Raison, C. L., & Miller, A. H. (2017). Pathogen–host defense in the evolution of depression: insights into epidemiology, genetics, bioregional differences and female preponderance. *Neuropsychopharmacology*, 42(1), 5-27.

Chapter Four PANDORA'S BOX

1 Kandel 2012, pg. 8-10
2 Kandel 2012, pg. 339
3 Kandel 2012, pg. 8-10, 342-343
4 Kandel 2012, pg. 8-10, 342-343
5 Harmon-Jones, E., Gable, P. A., & Peterson, C. K. (2010). The role of asymmetric frontal cortical activity in emotion-related phenomena: A review and update. *Biological psychology*, *84*(3), 451-462.
6 Sackeim, H. A., Greenberg, M. S., Weiman, A. L., Gur, R. C., Hungerbuhler, J. P., & Geschwind, N. (1982). Hemispheric asymmetry in the expression of positive and negative emotions: Neurologic evidence. *Archives of neurology*, 39(4), 210-218.
7 Sackeim et al 1982
8 U.S. Department of Health and Human Services. (n.d.). Magnetic Resonance Imaging (MRI). Retrieved from https://www.nibib.nih.gov/science-education/science-topics/magnetic-resonance-imaging-mri
9 Shin, D. (n.d.). What is fMRI? Retrieved March 28, 2019, from cfmriweb.ucsd.edu/Research/whatisfmri.html
Berger, A. (2003). How does it work?: Positron emission tomography. BMJ: British Medical Journal, 326(7404), 1449.
National Institutes of Health. (n.d.). Nuclear Medicine Fact Sheet. Retrieved from https://www.nibib.nih.gov/sites/default/files/Nuclear Medicine Fact Sheet.pdf
10 Savitz, J., & Drevets, W. C. (2009). Bipolar and major depressive disorder: neuroimaging the developmental-degenerative divide. *Neuroscience & Biobehavioral Reviews*, 33(5), 699-771.
11 Savitz & Drevets 2009
Grimm, S., Beck, J., Schuepbach, D., Hell, D., Boesiger, P., Bermpohl, F., ... & Northoff, G. (2008). Imbalance between left and right dorsolateral prefrontal cortex in major depression is linked to negative emotional judgment: an fMRI study in severe major depressive disorder. *Biological psychiatry*, 63(4), 369-376.
Koenigs, M., & Grafman, J. (2009). The functional neuroanatomy of depression: distinct roles for ventromedial and dorsolateral prefrontal cortex. *Behavioural brain research*, *201*(2), 239-243.
12 Drevets, W. C., Price, J. L., Simpson, J. R., Todd, R. D., Reich, T., Vannier, M., & Raichle, M. E. (1997). Subgenual prefrontal cortex abnormalities in mood disorders. *Nature*, *386*(6627), 824-827.
Savitz & Drevets 2009
13 Yu, C., Zhou, Y., Liu, Y., Jiang, T., Dong, H., Zhang, Y., & Walter, M. (2011). Functional segregation of the human cingulate cortex is confirmed by functional connectivity based neuroanatomical parcellation. *Neuroimage*, 54(4), 2571-2581.
14 Savitz & Drevets 2009
15 Savitz & Drevets 2009; Gold 2015
16 Videbech, P., & Ravnkilde, B. (2004). Hippocampal volume and depression: a meta-analysis of MRI studies. *American Journal of Psychiatry*, *161*(11), 1957-1966.

Chapter Five NEUROPLASTICITY

1 Liu, B., Liu, J., Wang, M., Zhang, Y., & Li, L. (2017). From serotonin to neuroplasticity: evolvement of theories for major depressive disorder. *Frontiers in cellular neuroscience*, *11*, 305.
2 Pittenger, C., & Duman, R. S. (2008). Stress, depression, and neuroplasticity: a convergence of mechanisms. *Neuropsychopharmacology*, 33(1), 88-109.
3 Gross, C. G. (2000). Neurogenesis in the adult brain: death of a dogma. *Nature Reviews Neuroscience*, *1*(1), 67-73.
4 Ohira, K., Takeuchi, R., Shoji, H.,

& Miyakawa, T. (2013). Fluoxetine-induced cortical adult neurogenesis. *Neuropsychopharmacology, 38*(6), 909-920.
5 Duman, R. S. (2004). Depression: a case of neuronal life and death?. *Biological psychiatry, 56*(3), 140-145.
Spalding, K. L., Bergmann, O., Alkass, K., Bernard, S., Salehpour, M., Huttner, H. B., ... & Possnert, G. (2013). Dynamics of hippocampal neurogenesis in adult humans. *Cell, 153*(6), 1219-1227.
6 Kandel 2012, pg. 21
7 Kempermann, G., Gage, F. H., Aigner, L., Song, H., Curtis, M. A., Thuret, S., ... & Gould, E. (2018). Human adult neurogenesis: evidence and remaining questions. *Cell stem cell, 23*(1), 25-30.
8 Jacobs, B. L., Van Praag, H., & Gage, F. H. (2000). Adult brain neurogenesis and psychiatry: a novel theory of depression. *Molecular psychiatry, 5*(3), 262-269.
9 Cameron, H. A., & Gould, E. (1994). Adult neurogenesis is regulated by adrenal steroids in the dentate gyrus. *Neuroscience, 61*(2), 203-209.
10 Duman 2004
Snyder, J. S., Soumier, A., Brewer, M., Pickel, J., & Cameron, H. A. (2011). Adult hippocampal neurogenesis buffers stress responses and depressive behaviour. *Nature, 476*(7361), 458-461.
11 Kempermann, G., & Kronenberg, G. (2003). Depressed new neurons?–Adult hippocampal neurogenesis and a cellular plasticity hypothesis of major depression. *Biological psychiatry, 54*(5), 499-503.
12 Kandel 2012, pg. 1205
Anderson, G., & Maes, M. (2014). Oxidative/nitrosative stress and immuno-inflammatory pathways in depression: treatment implications. *Current pharmaceutical design, 20*(23), 3812-3847.
13 Sheline, Y. I. (2011). Depression and the hippocampus: cause or effect?. *Biological psychiatry, 70*(4), 308.
14 Sapolsky, R. M., Krey, L. C., & McEwen, B. S. (1986). The neuroendocrinology of stress and aging: the glucocorticoid cascade hypothesis. *Endocrine reviews, 7*(3), 284-301.
15 Anderson & Maes 2014
16 Rajkowska & Miguel-Hidalgo 2007
17 Liu et al 2017
18 Park, H., & Poo, M. M. (2013). Neurotrophin regulation of neural circuit development and function. *Nature Reviews Neuroscience, 14*(1), 7-23.
19 Kandel 2012, pg. 283-284
20 Spitzer, N. C. (2015). Neurotransmitter switching? No surprise. *Neuron, 86*(5), 1131-1144.
21 Duman, R. S., & Monteggia, L. M. (2006). A neurotrophic model for stress-related mood disorders. *Biological psychiatry, 59*(12), 1116-1127.
22 Duman, R. S., & Li, N. (2012). A neurotrophic hypothesis of depression: role of synaptogenesis in the actions of NMDA receptor antagonists. *Philosophical Transactions of the Royal Society B: Biological Sciences, 367*(1601), 2475-2484.
23 Park & Poo 2013; Duman & Monteggia 2006
24 Groves, J. O. (2007). Is it time to reassess the BDNF hypothesis of depression?. *Molecular psychiatry, 12*(12), 1079-1088.

Chapter Six
GLUTAMATE AND GABA

1 Berman, R. M., Cappiello, A., Anand, A., Oren, D. A., Heninger, G. R., Charney, D. S., & Krystal, J. H. (2000). Antidepressant effects of ketamine in depressed patients. *Biological psychiatry, 47*(4), 351-354.
2 Mount Sinai was involved in the research that led to the development of this new treatment method for treatment-resistant depression and receives financial remuneration from the manufacturer of SPRAVATO (esketmaine). Mount Sinai's Dean is a co-inventor of patents related to this new treatment method and as such receives remuneration through Mount Sinai from the manufacturer. For more information about these financial interests and Mount Sinai's leadership role in SPRAVATO, please visit bit.ly/esketamine-development.
In the past 5 years, Dr. Murrough has provided

consultation services and/or served on advisory boards for Allergan, Boehreinger Ingelheim, Clexio Biosciences, Fortress Biotech, FSV7, Global Medical Education (GME), Impel Neuropharma, Janssen Research and Development, Medavante-Prophase, Novartis, Otsuka, and Sage Therapeutics. Dr. Murrough is named on a patent pending for neuropeptide Y as a treatment for mood and anxiety disorders and on a patent pending for the use of ezogabine and other KCNQ channel openers to treat depression and related conditions. The Icahn School of Medicine (employer of Dr. Murrough) is named on a patent and has entered into a licensing agreement and will receive payments related to the use of ketamine or esketamine for the treatment of depression. The Icahn School of Medicine is also named on a patent related to the use of ketamine for the treatment of PTSD. Dr. Murrough is not named on these patents and will not receive any payments.

3 Murrough, J. W., Perez, A. M., Pillemer, S., Stern, J., Parides, M. K., aan het Rot, M., ... & Iosifescu, D. V. (2013). Rapid and longer-term antidepressant effects of repeated ketamine infusions in treatment-resistant major depression. *Biological psychiatry*, 74(4), 250-256.

4 Katalinic, N., Lai, R., Somogyi, A., Mitchell, P. B., Glue, P., & Loo, C. K. (2013). Ketamine as a new treatment for depression: a review of its efficacy and adverse effects. *Australian & New Zealand Journal of Psychiatry*, 47(8), 710-727.

5 Sanacora, G., Treccani, G., & Popoli, M. (2012). Towards a glutamate hypothesis of depression: an emerging frontier of neuropsychopharmacology for mood disorders. *Neuropharmacology*, 62(1), 63-77.

6 Sanacora et al 2012; Kandel 2012, pg. 213

7 Sanacora et al 2012; Kandel 2012, pg. 213-225

8 Kandel 2012, pg. 30-31, 213-225

9 Hertz, L. (2013). The glutamate–glutamine (GABA) cycle: importance of late postnatal development and potential reciprocal interactions between biosynthesis and degradation. *Frontiers in endocrinology*, 4, 59. Hamidi, M., Drevets, W. C., & Price, J. L. (2004). Glial reduction in amygdala in major depressive disorder is due to oligodendrocytes. *Biological psychiatry*, 55(6), 563-569.

10 Murrough, J. W., Abdallah, C. G., & Mathew, S. J. (2017). Targeting glutamate signalling in depression: progress and prospects. *Nature Reviews Drug Discovery*, 16(7), 472.
Park, H., Popescu, A., & Poo, M. M. (2014). Essential role of presynaptic NMDA receptors in activity-dependent BDNF secretion and corticostriatal LTP. *Neuron*, 84(5), 1009-1022.
Kandel 2012, pg. 213-215

11 Kandel 2012, pg. 222-225

12 Luscher, B., Shen, Q., & Sahir, N. (2011). The GABAergic deficit hypothesis of major depressive disorder. *Molecular psychiatry*, 16(4), 383-406.

13 Sanacora et al 2012

14 Tognarelli, J. M., Dawood, M., Shariff, M. I., Grover, V. P., Crossey, M. M., Cox, I. J., ... & McPhail, M. J. (2015). Magnetic resonance spectroscopy: principles and techniques: lessons for clinicians. Journal of clinical and experimental hepatology, 5(4), 320-328.

15 Sanacora et al 2012

16 Luscher et al 2011

17 Paul, I. A., Nowak, G., Layer, R. T., Popik, P., & Skolnick, P. (1994). Adaptation of the N-methyl-D-aspartate receptor complex following chronic antidepressant treatments. *Journal of Pharmacology and Experimental Therapeutics*, 269(1), 95-102.

18 Sanacora et al 2012; Luscher et al 2011

19 Abdallah, C. G., Sanacora, G., Duman, R. S., & Krystal, J. H. (2018). The neurobiology of depression, ketamine and rapid-acting antidepressants: Is it glutamate inhibition or activation?. *Pharmacology & therapeutics*, 190, 148-158.

20 Gerhard, D. M., Wohleb, E. S., & Duman, R. S. (2016). Emerging treatment mechanisms for depression: focus on glutamate and synaptic plasticity. *Drug discovery today*, 21(3), 454-464.

21 Monteggia, L. M., & Zarate Jr, C. (2015). Antidepressant actions of ketamine: from molecular mechanisms to clinical practice. *Current opinion in neurobiology*, 30, 139-143.
Interview with Dr. Monteggia, 11-14-19

Chapter Seven GLIAL PATHOLOGY

1 Rajkowska, G., Miguel-Hidalgo, J. J., Wei, J., Dilley, G., Pittman, S. D., Meltzer, H. Y., ... & Stockmeier, C. A. (1999). Morphometric evidence for neuronal and glial prefrontal cell pathology in major depression. *Biological psychiatry*, 45(9), 1085-1098.
2 Öngür, D., Drevets, W. C., & Price, J. L. (1998). Glial reduction in the subgenual prefrontal cortex in mood disorders. *Proceedings of the National Academy of Sciences*, 95(22), 13290-13295.
3 Rajkowska & Miguel-Hidalgo 2007; Hamidi et al 2004
4 Yirmiya et al 2015
5 Yirmiya et al 2015
6 Rajkowska & Miguel-Hidalgo 2007
7 Ongur et al 1998; Rajkowska & Miguel-Hidalgo 2007
8 Choudary, P. V., Molnar, M., Evans, S. J., Tomita, H., Li, J. Z., Vawter, M. P., ... & Jones, E. G. (2005). Altered cortical glutamatergic and GABAergic signal transmission with glial involvement in depression. *Proceedings of the National Academy of Sciences*, 102(43), 15653-15658.
9 Sanacora, G., & Banasr, M. (2013). From pathophysiology to novel antidepressant drugs: glial contributions to the pathology and treatment of mood disorders. *Biological psychiatry*, 73(12), 1172-1179.
10 Rajkowska & Miguel-Hidalgo 2007
11 Savitz & Drevets 2009
12 Hamidi et al 2004
13 Edgar, N., & Sibille, E. (2012). A putative functional role for oligodendrocytes in mood regulation. *Translational psychiatry*, 2(5), e109-e109.
14 Rajkowska & Miguel-Hidalgo 2007

Chapter Eight BIOENERGETICS

1 Lodish 2016, pg. 523 - 527
2 DiMauro, S. (2004). Mitochondrial diseases. *Biochimica et Biophysica Acta (BBA)-Bioenergetics*, 1658(1-2), 80-88.
3 Boles, R. G., Burnett, B. B., Gleditsch, K., Wong, S., Guedalia, A., Kaariainen, A., ... & Brumm, V. (2005). A high predisposition to depression and anxiety in mothers and other matrilineal relatives of children with presumed maternally inherited mitochondrial disorders. *American Journal of Medical Genetics Part B: Neuropsychiatric Genetics*, 137(1), 20-24.
4 Burnett, B. B., Gardner, A., & Boles, R. G. (2005). Mitochondrial inheritance in depression, dysmotility and migraine?. *Journal of affective disorders*, 88(1), 109-116.
5 Bergemann, E. R., & Boles, R. G. (2010). Maternal inheritance in recurrent early-onset depression. *Psychiatric genetics*, 20(1), 31-34.
6 Lodish 2016, pg. 520-522
7 Klinedinst, N. J., & Regenold, W. T. (2015). A mitochondrial bioenergetic basis of depression. *Journal of bioenergetics and biomembranes*, 47(1-2), 155-171.
8 Gardner, A., & Boles, R. G. (2011). Beyond the serotonin hypothesis: mitochondria, inflammation and neurodegeneration in major depression and affective spectrum disorders. *Progress in Neuro-Psychopharmacology and Biological Psychiatry*, 35(3), 730-743.
9 Hyder, F., Rothman, D. L., & Bennett, M. R. (2013). Cortical energy demands of signaling and nonsignaling components in brain are conserved across mammalian species and activity levels. *Proceedings of the National Academy of Sciences*, 110(9), 3549-3554.
10 Ferrari, F., & Villa, R. F. (2017). The neurobiology of depression: an integrated overview from biological theories to clinical evidence. *Molecular neurobiology*, 54(7), 4847-4865.
11 Lodish 2016, pg. 513-515, 553-555
12 Kandel 2012, pg. 428 - 432
13 DiMauro 2004
Moylan, S., Berk, M., Dean, O. M., Samuni, Y., Williams, L. J., O'Neil, A., ... & Maes, M. (2014). Oxidative & nitrosative stress in depression: why so much stress?. *Neuroscience & Biobehavioral Reviews*, 45, 46-62.
14 Shao, L., Martin, M. V., Watson, S. J., Schatzberg, A., Akil, H., Myers, R. M., ...

& Vawter, M. P. (2008). Mitochondrial involvement in psychiatric disorders. *Annals of medicine*, 40(4), 281-295.
Ben-Shachar, D., & Karry, R. (2008). Neuroanatomical pattern of mitochondrial complex I pathology varies between schizophrenia, bipolar disorder and major depression. *PloS one*, 3(11).
15 Abdallah, C. G., Jiang, L., De Feyter, H. M., Fasula, M., Krystal, J. H., Rothman, D. L., ... & Sanacora, G. (2014). Glutamate metabolism in major depressive disorder. *American Journal of Psychiatry*, 171(12), 1320-1327.
16 Harper, D. G., Jensen, J. E., Ravichandran, C., Perlis, R. H., Fava, M., Renshaw, P. F., & Iosifescu, D. V. (2017). Tissue type-specific bioenergetic abnormalities in adults with major depression. *Neuropsychopharmacology*, 42(4), 876-885.
17 Lodish 2016, pg. 518-520, 550-557, 1017-1019
18 Anderson & Maes 2014
19 Anderson & Maes 2014
20 Anderson & Maes 2014
21 Moylan et al 2014
22 Palta, P., Samuel, L. J., Miller III, E. R., & Szanton, S. L. (2014). Depression and oxidative stress: results from a meta-analysis of observational studies. *Psychosomatic medicine*, 76(1), 12.
Liu, T., Zhong, S., Liao, X., Chen, J., He, T., Lai, S., & Jia, Y. (2015). A meta-analysis of oxidative stress markers in depression. *PloS one*, 10(10).

Chapter Nine CONNECTIVITY

1 Shin, D. (n.d.). What is fMRI? Retrieved March 28, 2019, from cfmriweb.ucsd.edu/Research/whatisfmri.html
2 Raichle, M. E. (2015). The restless brain: how intrinsic activity organizes brain function. *Philosophical Transactions of the Royal Society B: Biological Sciences*, 370(1668), 20140172.
3 Shulman, G. L., Fiez, J. A., Corbetta, M., Buckner, R. L., Miezin, F. M., Raichle, M. E., & Petersen, S. E. (1997). Common blood flow changes across visual tasks: II. Decreases in cerebral cortex. *Journal of cognitive neuroscience*, 9(5), 648-663.
4 Raichle, M. E., MacLeod, A. M., Snyder, A. Z., Powers, W. J., Gusnard, D. A., & Shulman, G. L. (2001). A default mode of brain function. *Proceedings of the National Academy of Sciences*, 98(2), 676-682.
5 Raichle et al 2001
6 Biswal, B., Zerrin Yetkin, F., Haughton, V. M., & Hyde, J. S. (1995). Functional connectivity in the motor cortex of resting human brain using echo-planar MRI. *Magnetic resonance in medicine*, 34(4), 537-541.
7 Biswal et al 1995
8 Biswal et al 1995
9 Menon, V. (2011). Large-scale brain networks and psychopathology: a unifying triple network model. *Trends in cognitive sciences*, 15(10), 483-506.
10 Bullmore, E., & Sporns, O. (2009). Complex brain networks: graph theoretical analysis of structural and functional systems. *Nature reviews neuroscience*, 10(3), 186-198.
11 Leech, R., & Sharp, D. J. (2014). The role of the posterior cingulate cortex in cognition and disease. *Brain*, 137(1), 12-32.
12 Menon 2011
13 Sridharan, D., Levitin, D. J., & Menon, V. (2008). A critical role for the right fronto-insular cortex in switching between central-executive and default-mode networks. *Proceedings of the National Academy of Sciences*, 105(34), 12569-12574.
14 Chen, A. C., Oathes, D. J., Chang, C., Bradley, T., Zhou, Z. W., Williams, L. M., ... & Etkin, A. (2013). Causal interactions between fronto-parietal central executive and default-mode networks in humans. *Proceedings of the National Academy of Sciences*, 110(49), 19944-19949.
15 Sridharan et al 2008
16 Menon 2011
17 Kaiser, R. H., Andrews-Hanna, J. R., Wager, T. D., & Pizzagalli, D. A. (2015). Large-scale network dysfunction in major depressive disorder: a meta-analysis of resting-state functional connectivity. *JAMA psychiatry*, 72(6), 603-611.
18 Bar, M. (2009). A cognitive neuroscience

hypothesis of mood and depression. *Trends in cognitive sciences*, 13(11), 456-463.
19 Nolen-Hoeksema, S., Wisco, B. E., & Lyubomirsky, S. (2008). Rethinking rumination. *Perspectives on psychological science*, 3(5), 400-424.
20 Greicius, M. D., Flores, B. H., Menon, V., Glover, G. H., Solvason, H. B., Kenna, H., ... & Schatzberg, A. F. (2007). Resting-state functional connectivity in major depression: abnormally increased contributions from subgenual cingulate cortex and thalamus. *Biological psychiatry*, 62(5), 429-437.
21 Berman, M. G., Peltier, S., Nee, D. E., Kross, E., Deldin, P. J., & Jonides, J. (2011). Depression, rumination and the default network. *Social cognitive and affective neuroscience*, 6(5), 548-555.
22 Sheline, Y. I., Price, J. L., Yan, Z., & Mintun, M. A. (2010). Resting-state functional MRI in depression unmasks increased connectivity between networks via the dorsal nexus. *Proceedings of the National Academy of Sciences*, 107(24), 11020-11025.
23 Lythe, K. E., Moll, J., Gethin, J. A., Workman, C. I., Green, S., Ralph, M. A. L., ... & Zahn, R. (2015). Self-blame–selective hyperconnectivity between anterior temporal and subgenual cortices and prediction of recurrent depressive episodes. *JAMA psychiatry*, 72(11), 1119-1126.
24 Scharinger, C., Rabl, U., Kasess, C. H., Meyer, B. M., Hofmaier, T., Diers, K., ... & Hartinger, B. (2014). Platelet serotonin transporter function predicts default-mode network activity. *PLoS One*, 9(3).
Hahn, A., Wadsak, W., Windischberger, C., Baldinger, P., Höflich, A. S., Losak, J., ... & Mitterhauser, M. (2012). Differential modulation of the default mode network via serotonin-1A receptors. *Proceedings of the National Academy of Sciences*, 109(7), 2619-2624.
Tomasi, D., Volkow, N. D., Wang, R., Telang, F., Wang, G. J., Chang, L., ... & Fowler, J. S. (2009). Dopamine transporters in striatum correlate with deactivation in the default mode network during visuospatial attention. *PloS one*, 4(6).
25 Kapogiannis, D., Reiter, D. A., Willette, A. A., & Mattson, M. P. (2013). Posteromedial cortex glutamate and GABA predict intrinsic functional connectivity of the default mode network. *Neuroimage*, 64, 112-119.

Chapter Ten CIRCADIAN RHYTHMS

1 Nutt, D., Wilson, S., & Paterson, L. (2008). Sleep disorders as core symptoms of depression. *Dialogues in clinical neuroscience*, 10(3), 329.
2 Vyazovskiy, V. V., Walton, M. E., Peirson, S. N., & Bannerman, D. M. (2017). Sleep homeostasis, habits and habituation. *Current opinion in neurobiology*, 44, 202-211.
3 Borbély, A. A., & Wirz-Justice, A. (1982). Sleep, sleep deprivation and depression. *Hum Neurobiol*, 1(205), 10.
Borbély, A. A., Daan, S., Wirz-Justice, A., & Deboer, T. (2016). The two-process model of sleep regulation: a reappraisal. *Journal of sleep research*, 25(2), 131-143.
4 Kandel 2012, pg. 1141-1144
5 Kandel 2012, pg. 1141
6 Nutt et al 2008
7 Ohayon, M. M., & Roth, T. (2003). Place of chronic insomnia in the course of depressive and anxiety disorders. *Journal of psychiatric research*, 37(1), 9-15.
8 Baglioni, C., Battagliese, G., Feige, B., Spiegelhalder, K., Nissen, C., Voderholzer, U., ... & Riemann, D. (2011). Insomnia as a predictor of depression: a meta-analytic evaluation of longitudinal epidemiological studies. *Journal of affective disorders*, 135(1-3), 10-19.
9 Srinivasan, V., Pandi-Perumal, S. R., Trakht, I., Spence, D. W., Hardeland, R., Poeggeler, B., & Cardinali, D. P. (2009). Pathophysiology of depression: role of sleep and the melatonergic system. *Psychiatry research*, 165(3), 201-214.
10 Wirz-Justice, A., & Benedetti, F. (2019). Perspectives in affective disorders: Clocks and sleep. *European Journal of Neuroscience*, 00, 1-20.
11 McClung, C. A. (2013). How might circadian rhythms control mood? Let me count the ways... *Biological psychiatry*, 74(4), 242-249.
12 Bunney, W. E., & Bunney, B. G. (2000). Molecular clock genes in man and

lower animals: possible implications for circadian abnormalities in depression. *Neuropsychopharmacology, 22*(4), 335-345.
McClung, C. R. (2006). Plant circadian rhythms. *The Plant Cell, 18*(4), 792-803.
13 Lall, G. S., Atkinson, L. A., Corlett, S. A., Broadbridge, P. J., & Bonsall, D. R. (2012). Circadian entrainment and its role in depression: a mechanistic review. *Journal of neural transmission, 119*(10), 1085-1096.
14 Srinivasan et al 2009
15 McClung 2013
16 Archer, S. N., Laing, E. E., Möller-Levet, C. S., van der Veen, D. R., Bucca, G., Lazar, A. S., ... & Smith, C. P. (2014). Mistimed sleep disrupts circadian regulation of the human transcriptome. *Proceedings of the National Academy of Sciences, 111*(6), E682-E691.
17 Bunney, B. G., Li, J. Z., Walsh, D. M., Stein, R., Vawter, M. P., Cartagena, P., ... & Akil, H. (2015). Circadian dysregulation of clock genes: clues to rapid treatments in major depressive disorder. *Molecular psychiatry, 20*(1), 48-55.
18 Spitzer, N. C. (2012). Activity-dependent neurotransmitter respecification. *Nature Reviews Neuroscience, 13*(2), 94-106.
Dulcis, D., Jamshidi, P., Leutgeb, S., & Spitzer, N. C. (2013). Neurotransmitter switching in the adult brain regulates behavior. *Science, 340*(6131), 449-453.
19 McClung 2013
20 Lambert, G. W., Reid, C., Kaye, D. M., Jennings, G. L., & Esler, M. D. (2002). Effect of sunlight and season on serotonin turnover in the brain. *The Lancet, 360*(9348), 1840-1842.
21 Praschak-Rieder, N., Willeit, M., Wilson, A. A., Houle, S., & Meyer, J. H. (2008). Seasonal variation in human brain serotonin transporter binding. *Archives of general psychiatry, 65*(9), 1072-1078.
22 Archer et al 2014
23 Li, J. Z., Bunney, B. G., Meng, F., Hagenauer, M. H., Walsh, D. M., Vawter, M. P., ... & Schatzberg, A. F. (2013). Circadian patterns of gene expression in the human brain and disruption in major depressive disorder. *Proceedings of the National Academy of Sciences, 110*(24), 9950-9955.
Bunney et al 2015
24 Orozco-Solis, R., Montellier, E., Aguilar-Arnal, L., Sato, S., Vawter, M. P., Bunney, B. G., ... & Sassone-Corsi, P. (2017). A circadian genomic signature common to ketamine and sleep deprivation in the anterior cingulate cortex. *Biological psychiatry, 82*(5), 351-360.
25 Duncan Jr, W. C., Slonena, E., Hejazi, N. S., Brutsche, N., Kevin, C. Y., Park, L., ... & Zarate Jr, C. A. (2017). Motor-activity markers of circadian timekeeping are related to ketamine's rapid antidepressant properties. *Biological psychiatry, 82*(5), 361-369.
26 Benedetti, F., Dallaspezia, S., Fulgosi, M. C., Barbini, B., Colombo, C., & Smeraldi, E. (2007). Phase advance is an actimetric correlate of antidepressant response to sleep deprivation and light therapy in bipolar depression. *Chronobiology international, 24*(5), 921-937.
27 Wirz-Justice & Benedetti 2019

Chapter Eleven
THE ROAD TAKEN

1 Kendler, K. S., Gardner, C. O., & Prescott, C. A. (2002). Toward a comprehensive developmental model for major depression in women. *American Journal of Psychiatry, 159*(7), 1133-1145.
2 Kendler et al 2002
3 Kendler et al 2002
4 Hasin, D. S., Goodwin, R. D., Stinson, F. S., & Grant, B. F. (2005). Epidemiology of major depressive disorder: results from the National Epidemiologic Survey on Alcoholism and Related Conditions. *Archives of general psychiatry, 62*(10), 1097-1106.
5 Kendler, K. S., Gardner, C. O., & Prescott, C. A. (2006). Toward a comprehensive developmental model for major depression in men. *American Journal of Psychiatry, 163*(1), 115-124.
6 Eaton, W. W., Shao, H., Nestadt, G., Lee, B. H., Bienvenu, O. J., & Zandi, P. (2008). Population-based study of first onset and chronicity in major depressive disorder. *Archives of general psychiatry, 65*(5), 513-520.

7 Mattisson, C., Bogren, M., Horstmann, V., Munk-Jörgensen, P., & Nettelbladt, P. (2007). The long-term course of depressive disorders in the Lundby Study. *Psychological Medicine, 37*(6), 883-891.

Chapter Twelve
GENETICS

1 Sullivan, P. F., Neale, M. C., & Kendler, K. S. (2000). Genetic epidemiology of major depression: review and meta-analysis. *American Journal of Psychiatry, 157*(10), 1552-1562.
2 Sullivan et al 2000
3 Sullivan et al 2000
4 Sullivan et al 2000
5 Baselmans, B. M., Yengo, L., van Rheenen, W., & Wray, N. R. (2020). Risk in relatives, heritability, SNP-based heritability and genetic correlations in psychiatric disorders: a review. *Biological Psychiatry*.
6 Kandel 2012, pg. 42-43
7 Lodish 2016, pg. 167-170, 176-183, 301-302
8 Lodish 2016, pg. 179-183
9 Lesch, K. P., Bengel, D., Heils, A., Sabol, S. Z., Greenberg, B. D., Petri, S., ... & Murphy, D. L. (1996). Association of anxiety-related traits with a polymorphism in the serotonin transporter gene regulatory region. *Science, 274*(5292), 1527-1531.
10 Lesch et al 1996
11 Smeraldi, E., Zanardi, R., Benedetti, F., Di Bella, D., Perez, J., & Catalano, M. (1998). Polymorphism within the promoter of the serotonin transporter gene and antidepressant efficacy of fluvoxamine. *Molecular psychiatry, 3*(6), 508-511.
12 Mann, J. J., Huang, Y. Y., Underwood, M. D., Kassir, S. A., Oppenheim, S., Kelly, T. M., ... & Arango, V. (2000). A serotonin transporter gene promoter polymorphism (5-HTTLPR) and prefrontal cortical binding in major depression and suicide. *Archives of general psychiatry, 57*(8), 729-738.
13 Anguelova, M., Benkelfat, C., & Turecki, G. (2003). A systematic review of association studies investigating genes coding for serotonin receptors and the serotonin transporter: I. Affective disorders. *Molecular psychiatry, 8*(6), 574-591.
14 Caspi, A., Sugden, K., Moffitt, T. E., Taylor, A., Craig, I. W., Harrington, H., ... & Poulton, R. (2003). Influence of life stress on depression: moderation by a polymorphism in the 5-HTT gene. *Science, 301*(5631), 386-389.
15 Caspi et al 2003
16 Caspi et al 2003
17 Karg, K., Burmeister, M., Shedden, K., & Sen, S. (2011). The serotonin transporter promoter variant (5-HTTLPR), stress, and depression meta-analysis revisited: evidence of genetic moderation. *Archives of general psychiatry, 68*(5), 444-454.
Fergusson, D. M., Horwood, L. J., Miller, A. L., & Kennedy, M. A. (2011). Life stress, 5-HTTLPR and mental disorder: findings from a 30-year longitudinal study. *The British Journal of Psychiatry, 198*(2), 129-135.
18 Zimmermann, P., Brückl, T., Nocon, A., Pfister, H., Binder, E. B., Uhr, M., ... & Ising, M. (2011). Interaction of FKBP5 gene variants and adverse life events in predicting depression onset: results from a 10-year prospective community study. *American Journal of Psychiatry, 168*(10), 1107-1116.
19 Zimmermann et al 2011
Wang, Q., Shelton, R. C., & Dwivedi, Y. (2018). Interaction between early-life stress and FKBP5 gene variants in major depressive disorder and post-traumatic stress disorder: A systematic review and meta-analysis. *Journal of affective disorders, 225*, 422-428.
20 Gatt, J. M., Nemeroff, C. B., Dobson-Stone, C., Paul, R. H., Bryant, R. A., Schofield, P. R., ... & Williams, L. M. (2009). Interactions between BDNF Val66Met polymorphism and early life stress predict brain and arousal pathways to syndromal depression and anxiety. *Molecular psychiatry, 14*(7), 681-695.
21 Dunn, E. C., Brown, R. C., Dai, Y., Rosand, J., Nugent, N. R., Amstadter, A. B., & Smoller, J. W. (2015). Genetic determinants of depression: recent findings and future directions. *Harvard review of psychiatry, 23*(1), 1.
22 Ripke, S., Wray, N. R., Lewis, C. M.,

Hamilton, S. P., Weissman, M. M., Breen, G., ... & Heath, A. C. (2013). A mega-analysis of genome-wide association studies for major depressive disorder. *Molecular psychiatry, 18*(4), 497.
23 Dunn et al 2015
24 Wray, N. R., Ripke, S., Mattheisen, M., Trzaskowski, M., Byrne, E. M., Abdellaoui, A., ... & Bacanu, S. A. (2018). Genome-wide association analyses identify 44 risk variants and refine the genetic architecture of major depression. *Nature genetics, 50*(5), 668.
25 Wray et al 2018

Chapter Thirteen
EARLY LIFE ADVERSITY

1 Briere, J., & Runtz, M. (1988). Symptomatology associated with childhood sexual victimization in a nonclinical adult sample. *Child abuse & neglect, 12*(1), 51-59.
2 Children's Bureau. (2017, January 19). Child Maltreatment 2015. Retrieved from https://www.acf.hhs.gov/cb/resource/child-maltreatment-2015
3 Felitti, V. J., Anda, R. F., Nordenberg, D., Williamson, D. F., Spitz, A. M., Edwards, V., ... & Marks, J. S. (1998). Relationship of childhood abuse and household dysfunction to many of the leading causes of death in adults: The Adverse Childhood Experiences (ACE) Study. *American journal of preventive medicine, 56*(6), 774-786.
4 Chapman, D. P., Whitfield, C. L., Felitti, V. J., Dube, S. R., Edwards, V. J., & Anda, R. F. (2004). Adverse childhood experiences and the risk of depressive disorders in adulthood. *Journal of affective disorders, 82*(2), 217-225.
5 Chapman et al 2004
6 Chapman et al 2004
7 Norman, R. E., Byambaa, M., De, R., Butchart, A., Scott, J., & Vos, T. (2012). The long-term health consequences of child physical abuse, emotional abuse, and neglect: a systematic review and meta-analysis. *PLoS medicine, 9*(11).
Lindert, J., von Ehrenstein, O. S., Grashow, R., Gal, G., Braehler, E., & Weisskopf, M. G. (2014). Sexual and physical abuse in childhood is associated with depression and anxiety over the life course: systematic review and meta-analysis. *International journal of public health, 59*(2), 359-372.
8 Agid, O. B. J. M. B. H. T. M. U. B., Shapira, B., Zislin, J., Ritsner, M., Hanin, B., Murad, H., ... & Lerer, B. (1999). Environment and vulnerability to major psychiatric illness: a case control study of early parental loss in major depression, bipolar disorder and schizophrenia. *Molecular psychiatry, 4*(2), 163-172.
9 Vythilingam, M., Heim, C., Newport, J., Miller, A. H., Anderson, E., Bronen, R., ... & Nemeroff, C. B. (2002). Childhood trauma associated with smaller hippocampal volume in women with major depression. *American Journal of Psychiatry, 159*(12), 2072-2080.
10 Toga, A. W., Thompson, P. M., & Sowell, E. R. (2006). Mapping brain maturation. *Focus, 29*(3), 148-390.
11 Toga et al 2006
Andersen, S. L., & Teicher, M. H. (2008). Stress, sensitive periods and maturational events in adolescent depression. *Trends in neurosciences, 31*(4), 183-191.
12 Andersen, S. L., Tomada, A., Vincow, E. S., Valente, E., Polcari, A., & Teicher, M. H. (2008). Preliminary evidence for sensitive periods in the effect of childhood sexual abuse on regional brain development. *The Journal of neuropsychiatry and clinical neurosciences, 20*(3), 292-301.
13 Heim, C., & Binder, E. B. (2012). Current research trends in early life stress and depression: Review of human studies on sensitive periods, gene–environment interactions, and epigenetics. *Experimental neurology, 233*(1), 102-111.
14 Yu, M., Linn, K. A., Shinohara, R. T., Oathes, D. J., Cook, P. A., Duprat, R., ... & Fava, M. (2019). Childhood trauma history is linked to abnormal brain connectivity in major depression. *Proceedings of the National Academy of Sciences, 116*(17), 8582-8590.
15 Yu et al 2019

Chapter Fourteen **EPIGENETICS**

1 Waddington, C. H. (1939). Development as an epigenetic process. *An introduction to modern genetics.*
2 Zimmer, C. (2018, January 31). The Famine Ended 70 Years Ago, but Dutch Genes Still Bear Scars. Retrieved from https://www.nytimes.com/2018/01/31/science/dutch-famine-genes.html
Heijmans, B. T., Tobi, E. W., Stein, A. D., Putter, H., Blauw, G. J., Susser, E. S., ... & Lumey, L. H. (2008). Persistent epigenetic differences associated with prenatal exposure to famine in humans. *Proceedings of the National Academy of Sciences, 105*(44), 17046-17049.
Lumey, L. H., Stein, A. D., Kahn, H. S., Van der Pal-de Bruin, K. M., Blauw, G. J., Zybert, P. A., & Susser, E. S. (2007). Cohort profile: the Dutch Hunger Winter families study. *International journal of epidemiology, 36*(6), 1196-1204.
3 Lodish 2016, pg. 302
4 Lodish 2016, pg. 327-332, 404-405
5 Heijmans, B. T., Kremer, D., Tobi, E. W., Boomsma, D. I., & Slagboom, P. E. (2007). Heritable rather than age-related environmental and stochastic factors dominate variation in DNA methylation of the human IGF2/H19 locus. *Human molecular genetics, 16*(5), 547-554.
6 Heijmans et al 2008
7 Francis, D., Diorio, J., Liu, D., & Meaney, M. J. (1999). Nongenomic transmission across generations of maternal behavior and stress responses in the rat. *Science, 286*(5442), 1155-1158.
8 Francis et al 1999
9 Weaver, I. C., Cervoni, N., Champagne, F. A., D'Alessio, A. C., Sharma, S., Seckl, J. R., ... & Meaney, M. J. (2004). Epigenetic programming by maternal behavior. *Nature neuroscience, 7*(8), 847-854.
10 McGowan, P. O., Sasaki, A., D'alessio, A. C., Dymov, S., Labonté, B., Szyf, M., ... & Meaney, M. J. (2009). Epigenetic regulation of the glucocorticoid receptor in human brain associates with childhood abuse. *Nature neuroscience, 12*(3), 342.
11 Turecki, G., & Meaney, M. J. (2016). Effects of the social environment and stress on glucocorticoid receptor gene methylation: a systematic review. *Biological psychiatry, 79*(2), 87-96.
12 Klengel, T., Mehta, D., Anacker, C., Rex-Haffner, M., Pruessner, J. C., Pariante, C. M., ... & Nemeroff, C. B. (2013). Allele-specific FKBP5 DNA demethylation mediates gene–childhood trauma interactions. *Nature neuroscience, 16*(1), 33.
13 Stankiewicz, A. M., Swiergiel, A. H., & Lisowski, P. (2013). Epigenetics of stress adaptations in the brain. *Brain Research Bulletin, 98*, 76-92.
Wilkinson, M. B., Xiao, G., Kumar, A., LaPlant, Q., Renthal, W., Sikder, D., ... & Nestler, E. J. (2009). Imipramine treatment and resiliency exhibit similar chromatin regulation in the mouse nucleus accumbens in depression models. *Journal of Neuroscience, 29*(24), 7820-7832.

Chapter Fifteen **DIET**

1 Sarris, J., Logan, A. C., Akbaraly, T. N., Amminger, G. P., Balanzá-Martínez, V., Freeman, M. P., ... & Nanri, A. (2015). Nutritional medicine as mainstream in psychiatry. *The Lancet Psychiatry*, 2(3), 271-274.
2 Lai, J. S., Hiles, S., Bisquera, A., Hure, A. J., McEvoy, M., & Attia, J. (2014). A systematic review and meta-analysis of dietary patterns and depression in community-dwelling adults. *The American journal of clinical nutrition*, 99(1), 181-197.
3 Sánchez-Villegas, A., Henríquez-Sánchez, P., Ruiz-Canela, M., Lahortiga, F., Molero, P., Toledo, E., & Martínez-González, M. A. (2015). A longitudinal analysis of diet quality scores and the risk of incident depression in the SUN Project. *BMC medicine*, 13(1), 197.
4 Sánchez-Villegas et al 2015
5 Lai et al 2014
6 Gangwisch, J. E., Hale, L., Garcia, L., Malaspina, D., Opler, M. G., Payne, M. E., ...

& Lane, D. (2015). High glycemic index diet as a risk factor for depression: analyses from the Women's Health Initiative. *The American journal of clinical nutrition, 102*(2), 454-463.
7 Hibbeln, J. R. (1998). Fish consumption and major depression. *The Lancet, 351*(9110), 1213.
8 Oregon State University. (n.d.). Essential Fatty Acids. Retrieved from https://lpi.oregonstate.edu/mic/other-nutrients/essential-fatty-acids
9 Bazinet, R. P., & Layé, S. (2014). Polyunsaturated fatty acids and their metabolites in brain function and disease. *Nature Reviews Neuroscience, 15*(12), 771-785.
10 Stanley, W. C., Khairallah, R. J., & Dabkowski, E. R. (2012). Update on lipids and mitochondrial function: impact of dietary n-3 polyunsaturated fatty acids. *Current opinion in clinical nutrition and metabolic care, 15*(2), 122.
11 Calder, P. C. (2009). Polyunsaturated fatty acids and inflammatory processes: new twists in an old tale. *Biochimie, 91*(6), 791-795.
12 Sánchez-Villegas, A., Álvarez-Pérez, J., Toledo, E., Salas-Salvadó, J., Ortega-Azorín, C., Zomeño, M. D., ... & López-Miranda, J. (2018). Seafood consumption, omega-3 fatty acids intake, and life-time prevalence of depression in the PREDIMED-plus trial. *Nutrients, 10*(12), 2000.
13 Mihrshahi, S., Dobson, A. J., & Mishra, G. D. (2015). Fruit and vegetable consumption and prevalence and incidence of depressive symptoms in mid-age women: results from the Australian longitudinal study on women's health. *European journal of clinical nutrition, 69*(5), 585-591. Table 4
14 Sánchez-Villegas, A., Delgado-Rodríguez, M., Alonso, A., Schlatter, J., Lahortiga, F., Majem, L. S., & Martínez-González, M. A. (2009). Association of the Mediterranean dietary pattern with the incidence of depression: the Seguimiento Universidad de Navarra/University of Navarra follow-up (SUN) cohort. *Archives of general psychiatry, 66*(10), 1090-1098.
15 Gangwisch et al 2015
16 Gea, A., Martinez-Gonzalez, M. A., Toledo, E., Sanchez-Villegas, A., Bes-Rastrollo, M., Nuñez-Cordoba, J. M., ... & Beunza, J. J. (2012). A longitudinal assessment of alcohol intake and incident depression: the SUN project. *BMC public health, 12*(1), 954.
17 Bremner, J. D., & McCaffery, P. (2008). The neurobiology of retinoic acid in affective disorders. *Progress in Neuro-Psychopharmacology and Biological Psychiatry, 32*(2), 315-331.
Misner, D. L., Jacobs, S., Shimizu, Y., De Urquiza, A. M., Solomin, L., Perlmann, T., ... & Evans, R. M. (2001). Vitamin A deprivation results in reversible loss of hippocampal long-term synaptic plasticity. *Proceedings of the National Academy of Sciences, 98*(20), 11714-11719.
Etchamendy, N., Enderlin, V., Marighetto, A., Vouimba, R. M., Pallet, V., Jaffard, R., & Higueret, P. (2001). Alleviation of a selective age-related relational memory deficit in mice by pharmacologically induced normalization of brain retinoid signaling. *Journal of Neuroscience, 21*(16), 6423-6429.
18 Wray et al 2018
19 Gilbody, S., Lightfoot, T., & Sheldon, T. (2007). Is low folate a risk factor for depression? A meta-analysis and exploration of heterogeneity. *Journal of Epidemiology & Community Health, 61*(7), 631-637.
Kim, J. M., Stewart, R., Kim, S. W., Yang, S. J., Shin, I. S., & Yoon, J. S. (2008). Predictive value of folate, vitamin B 12 and homocysteine levels in late-life depression. *The British Journal of Psychiatry, 192*(4), 268-274.
20 Anglin, R. E., Samaan, Z., Walter, S. D., & McDonald, S. D. (2013). Vitamin D deficiency and depression in adults: systematic review and meta-analysis. *The British journal of psychiatry, 202*(2), 100-107.
21 Gowda, U., Mutowo, M. P., Smith, B. J., Wluka, A. E., & Renzaho, A. M. (2015). Vitamin D supplementation to reduce depression in adults: meta-analysis of randomized controlled trials. *Nutrition, 31*(3), 421-429.
22 Chasapis, C. T., Loutsidou, A. C., Spiliopoulou, C. A., & Stefanidou, M. E. (2012). Zinc and human health: an update. *Archives of toxicology, 86*(4), 521-534.
23 Swardfager, W., Herrmann, N., Mazereeuw,

G., Goldberger, K., Harimoto, T., & Lanctôt, K. L. (2013). Zinc in depression: a meta-analysis. *Biological psychiatry, 74*(12), 872-878.
24 Lai, J., Moxey, A., Nowak, G., Vashum, K., Bailey, K., & McEvoy, M. (2012). The efficacy of zinc supplementation in depression: systematic review of randomised controlled trials. *Journal of affective disorders, 136*(1-2), e31-e39.
25 Jacka, F. N., O'Neil, A., Opie, R., Itsiopoulos, C., Cotton, S., Mohebbi, M., ... & Brazionis, L. (2017). A randomised controlled trial of dietary improvement for adults with major depression (the 'SMILES'trial). *BMC medicine*, 15(1), 1-13.

Chapter Sixteen
THE GUT MICROBIOME

1 Goldsmith, W. B. (1885). Syphilis and Insanity. *The Boston Medical and Surgical Journal, 113*(19), 433-435.
2 PLEASURE, HYMAN. "Psychiatric and neurological side-effects of isoniazid and iproniazid." *AMA Archives of Neurology & Psychiatry* 72.3 (1954): 313-320.
3 O'Brien, M. E. R., et al. "A randomized phase II study of SRL172 (Mycobacterium vaccae) combined with chemotherapy in patients with advanced inoperable non-small-cell lung cancer and mesothelioma." *British Journal of Cancer* 83.7 (2000): 853-857.
4 Reber, S. O., Siebler, P. H., Donner, N. C., Morton, J. T., Smith, D. G., Kopelman, J. M., ... & Greenwood, B. N. (2016). Immunization with a heat-killed preparation of the environmental bacterium Mycobacterium vaccae promotes stress resilience in mice. *Proceedings of the National Academy of Sciences, 113*(22), E3130-E3139.
5 Dinan, T. G., & Cryan, J. F. (2017). Gut instincts: microbiota as a key regulator of brain development, ageing and neurodegeneration. *The Journal of physiology, 595*(2), 489-503.
Qin, J., Li, R., Raes, J., Arumugam, M., Burgdorf, K. S., Manichanh, C., ... & Mende, D. R. (2010). A human gut microbial gene catalogue established by metagenomic sequencing. *Nature, 464*(7285), 59-65.
6 Dinan & Cryan 2017
Dominguez-Bello, M. G., Costello, E. K., Contreras, M., Magris, M., Hidalgo, G., Fierer, N., & Knight, R. (2010). Delivery mode shapes the acquisition and structure of the initial microbiota across multiple body habitats in newborns. *Proceedings of the National Academy of Sciences*, 107(26), 11971-11975.
7 Moloney, R. D., Desbonnet, L., Clarke, G., Dinan, T. G., & Cryan, J. F. (2014). The microbiome: stress, health and disease. *Mammalian Genome*, 25(1-2), 49-74.
Huh, S. Y., Rifas-Shiman, S. L., Zera, C. A., Edwards, J. W. R., Oken, E., Weiss, S. T., & Gillman, M. W. (2012). Delivery by caesarean section and risk of obesity in preschool age children: a prospective cohort study. *Archives of disease in childhood*, 97(7), 610-616.
8 Claesson, M. J., Jeffery, I. B., Conde, S., Power, S. E., O'connor, E. M., Cusack, S., ... & Fitzgerald, G. F. (2012). Gut microbiota composition correlates with diet and health in the elderly. *Nature, 488*(7410), 178-184.
9 Dinan & Cryan 2017
10 Böbel, T. S., Hackl, S. B., Langgartner, D., Jarczok, M. N., Rohleder, N., Rook, G. A., ... & Reber, S. O. (2018). Less immune activation following social stress in rural vs. urban participants raised with regular or no animal contact, respectively. *Proceedings of the National Academy of Sciences*, 115(20), 5259-5264.
11 Rook, G. A. (2012). Hygiene hypothesis and autoimmune diseases. *Clinical reviews in allergy & immunology*, 42(1), 5-15.
12 Senator H. Über ein Fall von Hydrothionamie und über Selbstinfektion durch abnorme Verdauungsvorgänge. *Berliner klinische Wochenschrift*. 1868;5:254–256.
13 Townsend, A. A. (1905). Mental Depression and Melancholia considered in regard to Auto-intoxication, with special Reference to the presence of Indoxyl in the Urine and its Clinical Significance; Essay for which was awarded the Bronze Medal of the Medico-Psychological Association, 1904. *Journal of Mental Science*, 51(212), 51-62.
14 Bested, A. C., Logan, A. C., & Selhub, E. M. (2013). Intestinal microbiota, probiotics and

mental health: from Metchnikoff to modern advances: Part I–autointoxication revisited. *Gut pathogens*, 5(1), 5.
15 O'Mahony, S. M., Marchesi, J. R., Scully, P., Codling, C., Ceolho, A. M., Quigley, E. M., ... & Dinan, T. G. (2009). Early life stress alters behavior, immunity, and microbiota in rats: implications for irritable bowel syndrome and psychiatric illnesses. *Biological psychiatry*, 65(3), 263-267.
16 Sudo, N., Chida, Y., Aiba, Y., Sonoda, J., Oyama, N., Yu, X. N., ... & Koga, Y. (2004). Postnatal microbial colonization programs the hypothalamic–pituitary–adrenal system for stress response in mice. *The Journal of physiology*, 558(1), 263-275.
17 Sudo et al 2004
18 Heijtz, R. D., Wang, S., Anuar, F., Qian, Y., Björkholm, B., Samuelsson, A., ... & Pettersson, S. (2011). Normal gut microbiota modulates brain development and behavior. *Proceedings of the National Academy of Sciences*, 108(7), 3047-3052.
19 Clarke, G., Grenham, S., Scully, P., Fitzgerald, P., Moloney, R. D., Shanahan, F., ... & Cryan, J. F. (2013). The microbiome-gut-brain axis during early life regulates the hippocampal serotonergic system in a sex-dependent manner. *Molecular psychiatry*, 18(6), 666-673.
Schéle, E., Grahnemo, L., Anesten, F., Hallén, A., Bäckhed, F., & Jansson, J. O. (2013). The gut microbiota reduces leptin sensitivity and the expression of the obesity-suppressing neuropeptides proglucagon (Gcg) and brain-derived neurotrophic factor (Bdnf) in the central nervous system. *Endocrinology*, 154(10), 3643-3651.
Ogbonnaya, E. S., Clarke, G., Shanahan, F., Dinan, T. G., Cryan, J. F., & O'Leary, O. F. (2015). Adult hippocampal neurogenesis is regulated by the microbiome. *Biological psychiatry*, 78(4), e7-e9.
20 Fung, T. C., Olson, C. A., & Hsiao, E. Y. (2017). Interactions between the microbiota, immune and nervous systems in health and disease. *Nature neuroscience*, 20(2), 145.
21 Erny, D., de Angelis, A. L. H., Jaitin, D., Wieghofer, P., Staszewski, O., David, E., ... & Schwierzeck, V. (2015). Host microbiota constantly control maturation and function of microglia in the CNS. *Nature neuroscience*, 18(7), 965.
22 Desbonnet, L., Clarke, G., Traplin, A., O'Sullivan, O., Crispie, F., Moloney, R. D., ... & Cryan, J. F. (2015). Gut microbiota depletion from early adolescence in mice: implications for brain and behaviour. *Brain, behavior, and immunity*, 48, 165-173.
23 Cryan, J. F., O'Riordan, K. J., Cowan, C. S., Sandhu, K. V., Bastiaanssen, T. F., Boehme, M., ... & Guzzetta, K. E. (2019). The microbiota-gut-brain axis. *Physiological reviews*, 99(4), 1877-2013.
24 Cryan et al 2019
25 Bravo, J. A., Forsythe, P., Chew, M. V., Escaravage, E., Savignac, H. M., Dinan, T. G., ... & Cryan, J. F. (2011). Ingestion of Lactobacillus strain regulates emotional behavior and central GABA receptor expression in a mouse via the vagus nerve. *Proceedings of the National Academy of Sciences*, 108(38), 16050-16055.
26 Cryan et al 2019
27 O'Mahony, S. M., Clarke, G., Borre, Y. E., Dinan, T. G., & Cryan, J. F. (2015). Serotonin, tryptophan metabolism and the brain-gut-microbiome axis. *Behavioural brain research*, 277, 32-48.
28 Clarke et al 2013
29 Zheng, P., Zeng, B., Zhou, C., Liu, M., Fang, Z., Xu, X., ... & Zhang, X. (2016). Gut microbiome remodeling induces depressive-like behaviors through a pathway mediated by the host's metabolism. *Molecular psychiatry*, 21(6), 786-796.
30 Fung et al 2017; Qin et al 2010
Luna, R. A., & Foster, J. A. (2015). Gut brain axis: diet microbiota interactions and implications for modulation of anxiety and depression. *Current opinion in biotechnology*, 32, 35-41.
31 Oriach, C. S., Robertson, R. C., Stanton, C., Cryan, J. F., & Dinan, T. G. (2016). Food for thought: The role of nutrition in the microbiota-gut–brain axis. *Clinical Nutrition Experimental*, 6, 25-38.
Dinan & Cryan 2017

32 Messaoudi, M., Lalonde, R., Violle, N., Javelot, H., Desor, D., Nejdi, A., ... & Cazaubiel, J. M. (2011). Assessment of psychotropic-like properties of a probiotic formulation (Lactobacillus helveticus R0052 and Bifidobacterium longum R0175) in rats and human subjects. *British Journal of Nutrition, 105*(5), 755-764.
33 Tillisch, K., Labus, J., Kilpatrick, L., Jiang, Z., Stains, J., Ebrat, B., ... & Mayer, E. A. (2013). Consumption of fermented milk product with probiotic modulates brain activity. *Gastroenterology, 144*(7), 1394-1401.
34 Oriach et al 2015
35 Schmidt, K., Cowen, P. J., Harmer, C. J., Tzortzis, G., Errington, S., & Burnet, P. W. (2015). Prebiotic intake reduces the waking cortisol response and alters emotional bias in healthy volunteers. *Psychopharmacology, 232*(10), 1793-1801.

Chapter Seventeen UNSEEN

1 Brody, D. J. (2018, February 13). Prevalence of Depression Among Adults Aged 20 and Over: United States, 2013–2016. Retrieved from https://www.cdc.gov/nchs/products/databriefs/db303.htm
2 Weissman, M. M., Bland, R. C., Canino, G. J., Faravelli, C., Greenwald, S., Hwu, H. G., ... & Lépine, J. P. (1996). Cross-national epidemiology of major depression and bipolar disorder. *JAMA, 276*(4), 293-299.
3 Hasin et al 2005
4 Hasin et al 2005
5 Cahill, L. (2006). Why sex matters for neuroscience. *Nature reviews neuroscience, 7*(6), 477-484.
6 Cahill 2006
7 Heinrichs, M., von Dawans, B., & Domes, G. (2009). Oxytocin, vasopressin, and human social behavior. *Frontiers in neuroendocrinology, 30*(4), 548-557.
8 McQuaid, R. J., McInnis, O. A., Abizaid, A., & Anisman, H. (2014). Making room for oxytocin in understanding depression. *Neuroscience & Biobehavioral Reviews, 45*, 305-322.
Heinrichs et al 2009
9 Kessler, R. C., & Bromet, E. J. (2013). The epidemiology of depression across cultures. *Annual review of public health, 34*, 119-138. Demyttenaere, K., Bruffaerts, R., Posada-Villa, J., Gasquet, I., Kovess, V., Lepine, J., ... & Kikkawa, T. (2004). Prevalence, severity, and unmet need for treatment of mental disorders in the World Health Organization World Mental Health Surveys. *JAMA, 291*(21), 2581-2590.
10 Kessler & Bromet 2013
11 Liu, Q., He, H., Yang, J., Feng, X., Zhao, F., & Lyu, J. (2020). Changes in the global burden of depression from 1990 to 2017: Findings from the Global Burden of Disease study. *Journal of psychiatric research, 126*, 134-140.
12 Hasin et al 2005
13 Wu, Z., & Schimmele, C. M. (2005). The healthy migrant effect on depression: variation over time?. *Canadian Studies in Population [ARCHIVES]*, 271-295.
14 Breslau, J., Borges, G., Hagar, Y., Tancredi, D., & Gilman, S. (2009). Immigration to the USA and risk for mood and anxiety disorders: variation by origin and age at immigration. *Psychological medicine, 39*(7), 1117-1127.
15 Abe-Kim, J., Takeuchi, D. T., Hong, S., Zane, N., Sue, S., Spencer, M. S., ... & Alegría, M. (2007). Use of mental health–related services among immigrant and US-born Asian Americans: results from the National Latino and Asian American study. *American journal of public health, 97*(1), 91-98.
16 Kessler, R. C., Chiu, W. T., Demler, O., & Walters, E. E. (2005). Prevalence, severity, and comorbidity of 12-month DSM-IV disorders in the National Comorbidity Survey Replication. *Archives of general psychiatry, 62*(6), 617-627.
17 Abe-Kim et al 2007.
18 Kessler & Bromet 2013
19 World Health Organization. (2017, February 23). Depression and Other Common Mental Disorders. Retrieved from https://www.who.int/mental_health/management/depression/prevalence_global_health_estimates/en/
20 Pratt et al 2016

Chapter Eighteen SUICIDE

1 Arsenault-Lapierre, G., Kim, C., & Turecki, G. (2004). Psychiatric diagnoses in 3275 suicides: a meta-analysis. *BMC psychiatry*, 4(1), 37.

2 Turecki, G., & Brent, D. A. (2016). Suicide and suicidal behaviour. *The Lancet*, 387(10024), 1227-1239.

3 Borges, G., Nock, M. K., Abad, J. M. H., Hwang, I., Sampson, N. A., Alonso, J., ... & Bruffaerts, R. (2010). Twelve month prevalence of and risk factors for suicide attempts in the WHO World Mental Health Surveys. *The Journal of clinical psychiatry*, 71(12), 1617.
World Health Organization. (2018, April 5). World Health Statistics data visualizations dashboard. Retrieved from https://www.who.int/mental_health/prevention/suicide/estimates/en/

4 Bostwick, J. M., & Pankratz, V. S. (2000). Affective disorders and suicide risk: a reexamination. *American Journal of Psychiatry*, 157(12), 1925-1932.

5 Bostwick & Pankratz 2000

6 Bostwick, J. M., Pabbati, C., Geske, J. R., & McKean, A. J. (2016). Suicide attempt as a risk factor for completed suicide: even more lethal than we knew. *American journal of psychiatry*, 173(11), 1094-1100.

7 Nock, M. K., Borges, G., Bromet, E. J., Alonso, J., Angermeyer, M., Beautrais, A., ... & De Graaf, R. (2008). Cross-national prevalence and risk factors for suicidal ideation, plans and attempts. *The British Journal of Psychiatry*, 192(2), 98-105.
World Health Organization. (2017, February 23). Depression and Other Common Mental Disorders. Retrieved from https://www.who.int/mental_health/management/depression/prevalence_global_health_estimates/en/

8 Centers for Disease Control and Prevention, National Center for Health Statistics. Underlying Cause of Death 1999-2018 on CDC WONDER Online Database, released in 2020. Data are from the Multiple Cause of Death Files, 1999-2018, as compiled from data provided by the 57 vital statistics jurisdictions through the Vital Statistics Cooperative Program. Accessed at http://wonder.cdc.gov/ucd-icd10.html on May 21, 2020.
World Health Organization. (2018, April 5). World Health Statistics data visualizations dashboard. Retrieved from https://www.who.int/mental_health/prevention/suicide/estimates/en/

9 Nock et al 2008

10 Centers for Disease Control and Prevention. (n.d.). Underlying Cause of Death, 1999-2018 Request. Retrieved from https://wonder.cdc.gov/ucd-icd10.html on May 21, 2020.

11 Centers for Disease Control and Prevention. (n.d.). Underlying Cause of Death, 1999-2018 Request. Retrieved from https://wonder.cdc.gov/ucd-icd10.html on May 21, 2020.

12 Centers for Disease Control and Prevention. (n.d.). Underlying Cause of Death, 1999-2018 Request. Retrieved from https://wonder.cdc.gov/ucd-icd10.html on May 21, 2020.

13 Cash, S. J., & Bridge, J. A. (2009). Epidemiology of youth suicide and suicidal behavior. *Current opinion in pediatrics*, 21(5), 613.

14 Brent, D. A., & Mann, J. J. (2005). Family genetic studies, suicide, and suicidal behavior. In *American Journal of Medical Genetics Part C: Seminars in Medical Genetics*. 133(1), 13-24.

15 Brent & Mann 2005

16 Ernst, C., Mechawar, N., & Turecki, G. (2009). Suicide neurobiology. *Progress in neurobiology*, 89(4), 315-333.

17 Turecki & Brent 2016
McGirr, A., Alda, M., Séguin, M., Cabot, S., Lesage, A., & Turecki, G. (2009). Familial aggregation of suicide explained by cluster B traits: a three-group family study of suicide controlling for major depressive disorder. *American Journal of Psychiatry*, 166(10), 1124-1134.

18 Coon, H., Darlington, T. M., DiBlasi, E., Callor, W. B., Ferris, E., Fraser, A., ... & Chen, D. (2018). Genome-wide significant regions in 43 Utah high-risk families implicate multiple genes involved in risk for completed suicide. *Molecular psychiatry*, 1-14.

19 Docherty, A., Shabalin, A. A., DiBlasi, E., Monsen, E., Mullins, N., Adkins, D. E., ... &

Gray, D. (2019). Genome-wide association study of suicide death and polygenic prediction of clinical antecedents. *bioRxiv*, 234674.
20 Mann, J. J., Currier, D., Stanley, B., Oquendo, M. A., Amsel, L. V., & Ellis, S. P. (2006). Can biological tests assist prediction of suicide in mood disorders?. *International Journal of Neuropsychopharmacology*, 9(4), 465-474.
21 Mann, J. J. (2013). The serotonergic system in mood disorders and suicidal behaviour. *Philosophical Transactions of the Royal Society B: Biological Sciences*, 368(1615), 20120537.
22 Black, C., & Miller, B. J. (2015). Meta-analysis of cytokines and chemokines in suicidality: distinguishing suicidal versus nonsuicidal patients. *Biological psychiatry*, 78(1), 28-37.
23 Haws, C. A., Gray, D. D., Yurgelun-Todd, D. A., Moskos, M., Meyer, L. J., & Renshaw, P. F. (2009). The possible effect of altitude on regional variation in suicide rates. *Medical hypotheses*, 73(4), 587-590.
24 World Health Organization. (2018, April 5). World Health Statistics data visualizations dashboard. Retrieved from https://www.who.int/mental_health/prevention/suicide/estimates/en/
25 Reno, E., Brown, T. L., Betz, M. E., Allen, M. H., Hoffecker, L., Reitinger, J., ... & Honigman, B. (2018). Suicide and high altitude: An integrative review. *High altitude medicine & biology*, 19(2), 99-108.
26 Katz, I. R. (1982). Is there a hypoxic affective syndrome?. *Psychosomatics: Journal of Consultation and Liaison Psychiatry.*
27 Young, S. N. (2013). Elevated incidence of suicide in people living at altitude, smokers and patients with chronic obstructive pulmonary disease and asthma: possible role of hypoxia causing decreased serotonin synthesis. *Journal of psychiatry & neuroscience: JPN*, 38(6), 423.
28 Harmatz, M. G., Well, A. D., Overtree, C. E., Kawamura, K. Y., Rosal, M., & Ockene, I. S. (2000). Seasonal variation of depression and other moods: a longitudinal approach. *Journal of biological rhythms*, 15(4), 344-350.
29 Centers for Disease Control and Prevention. (n.d.). Underlying Cause of Death, 1999-2018 Request. Retrieved from https://wonder.cdc.gov/ucd-icd10.html on May 21, 2020
30 Lambert, G., Reid, C., Kaye, D., Jennings, G., & Esler, M. (2003). Increased suicide rate in the middle-aged and its association with hours of sunlight. *American Journal of Psychiatry*, 160(4), 793-795.
31 Ernst et al 2009
32 Ernst et al 2009
33 Fong, C. W. (2014). Statins in therapy: understanding their hydrophilicity, lipophilicity, binding to 3-hydroxy-3-methylglutaryl-CoA reductase, ability to cross the blood brain barrier and metabolic stability based on electrostatic molecular orbital studies. *European journal of medicinal chemistry*, 85, 661-674.
34 Björkhem, I., & Meaney, S. (2004). Brain cholesterol: long secret life behind a barrier. *Arteriosclerosis, thrombosis, and vascular biology*, 24(5), 806-815.
35 Wu, S., Ding, Y., Wu, F., Xie, G., Hou, J., & Mao, P. (2016). Serum lipid levels and suicidality: a meta-analysis of 65 epidemiological studies. *Journal of psychiatry & neuroscience: JPN*, 41(1), 56.
36 Ernst et al 2009
37 Turecki & Brent 2016
38 Zalsman, G., Hawton, K., Wasserman, D., van Heeringen, K., Arensman, E., Sarchiapone, M., ... & Purebl, G. (2016). Suicide prevention strategies revisited: 10-year systematic review. *The Lancet Psychiatry*, 3(7), 646-659.
39 Zalsman et al 2016

Chapter Nineteen
THE WARNING

1 Hammad, T. A. (2004, August 16). Relationship between psychotropic drugs and pediatric suicidality. Retrieved from https://www.fda.gov/ohrms/dockets/ac/04/briefing/2004-4065b1-10-TAB08-Hammads-Review.pdf
2 Libby, A. M., Orton, H. D., & Valuck, R. J. (2009). Persisting decline in depression

treatment after FDA warnings. *Archives of General Psychiatry, 66*(6), 633-639.
3 Laughren, T. P. (2006, November 16). Memorandum to Members of PDAC. Retrieved from https://wayback.archive-it.org/7993/20170405070114/https://www.fda.gov/ohrms/dockets/ac/06/briefing/2006-4272b1-01-FDA.pdf
4 Laughren 2006 pg. 112 of 140. Medicines and Healthcare products Regulatory Agency (MHRA). (2008, January). Antidepressants and suicidal thoughts and behaviour. Retrieved from https://www.hma.eu/fileadmin/dateien/Human_Medicines/CMD_h_/Product_Information/PhVWP_Recommendations/Antidepressants/PAR_suicidal_thoughts.pdf, 7
5 MHRA 2008
6 Libby et al 2009
7 Sparks, J. A., & Duncan, B. L. (2013). Outside the black box: Re-assessing pediatric antidepressant prescription. *Journal of the Canadian Academy of Child and Adolescent Psychiatry, 22*(3), 240.
8 Walsh et al 2002
9 Benedetti, F., Mayberg, H. S., Wager, T. D., Stohler, C. S., & Zubieta, J. K. (2005). Neurobiological mechanisms of the placebo effect. *Journal of Neuroscience, 25*(45), 10390-10402.
10 Brown, W. A., & Brown, W. A. (2013). *The placebo effect in clinical practice*. Oxford University Press.
11 Walsh et al 2002
12 Kann, L. (2016, June 10). Youth Risk Behavior Surveillance - United States, 2015. Retrieved from https://www.cdc.gov/mmwr/volumes/65/ss/ss6506a1.htm
13 Hammad Report 2004
14 Brent, D., Melhem, N., & Turecki, G. (2010). Pharmacogenomics of suicidal events. *Pharmacogenomics, 11*(6), 793-807.
15 Gibbons, R. D., Coca Perraillon, M., Hur, K., Conti, R. M., Valuck, R. J., & Brent, D. A. (2015). Antidepressant treatment and suicide attempts and self-inflicted injury in children and adolescents. *Pharmacoepidemiology and drug safety, 24*(2), 208-214.
16 Hammad, T. A., Laughren, T., & Racoosin, J. (2006). Suicidality in pediatric patients treated with antidepressant drugs. *Archives of general psychiatry, 63*(3), 332-339.
17 Simon, G. E., Savarino, J., Operskalski, B., & Wang, P. S. (2006). Suicide risk during antidepressant treatment. *American Journal of Psychiatry, 163*(1), 41-47.
18 Simon, G. E., & Savarino, J. (2007). Suicide attempts among patients starting depression treatment with medications or psychotherapy. *American Journal of Psychiatry, 164*(7), 1029-1034.
19 Price, R. B., Nock, M. K., Charney, D. S., & Mathew, S. J. (2009). Effects of intravenous ketamine on explicit and implicit measures of suicidality in treatment-resistant depression. *Biological psychiatry, 66*(5), 522-526. Murrough, J. W., Soleimani, L., DeWilde, K. E., Collins, K. A., Lapidus, K. A., Iacoviello, B. M., ... & Price, R. B. (2015). Ketamine for rapid reduction of suicidal ideation: a randomized controlled trial. *Psychological medicine, 45*(16), 3571-3580.
Duman, R. S. (2018). Ketamine and rapid-acting antidepressants: a new era in the battle against depression and suicide. *F1000Research, 7*.
20 Gibbons, R. D., Brown, C. H., Hur, K., Davis, J. M., & Mann, J. J. (2012). Suicidal thoughts and behavior with antidepressant treatment: reanalysis of the randomized placebo-controlled studies of fluoxetine and venlafaxine. *Archives of general psychiatry, 69*(6), 580-587.
21 Hammad Report 2004
22 Centers for Disease Control and Prevention. (n.d.). Underlying Cause of Death, 1999-2017 Request. Retrieved from https://wonder.cdc.gov/ucd-icd10.html on Aug 9, 2019.
23 Centers for Disease Control and Prevention. (n.d.). Underlying Cause of Death, 1999-2017 Request. Retrieved from https://wonder.cdc.gov/ucd-icd10.html on Aug 16, 2019.
24 Ayers, J. W., Althouse, B. M., Leas, E. C., Dredze, M., & Allem, J. P. (2017). Internet searches for suicide following the release of 13 Reasons Why. *JAMA internal medicine, 177*(10), 1527-1529.

25 Centers for Disease Control and Prevention. (n.d.). Underlying Cause of Death, 1999-2018 Request. Retrieved from https://wonder.cdc.gov/ucd-icd10.html on Aug 16, 2019.

Chapter Twenty
THE REAL WORLD

1 Warden, D., Rush, A. J., Trivedi, M. H., Fava, M., & Wisniewski, S. R. (2007). The STAR* D Project results: a comprehensive review of findings. *Current psychiatry reports*, 9(6), 449-459.
2 Warden et al 2007

Chapter Twenty-One
PSYCHOTHERAPY

1 Etkin, A., Pittenger, C., Polan, H. J., & Kandel, E. R. (2005). Toward a neurobiology of psychotherapy: basic science and clinical applications. *The Journal of Neuropsychiatry and Clinical Neurosciences*, 17(2), 145-158.
2 Beck, A. T. (2005). The current state of cognitive therapy: a 40-year retrospective. *Archives of general psychiatry*, 62(9), 953-959.
3 Beck 2005
4 Longmore, R. J., & Worrell, M. (2007). Do we need to challenge thoughts in cognitive behavior therapy?. *Clinical psychology review*, 27(2), 173-187.
Jacobson, N. S., Dobson, K. S., Truax, P. A., Addis, M. E., Koerner, K., Gollan, J. K., ... & Prince, S. E. (1996). A component analysis of cognitive-behavioral treatment for depression. *Journal of consulting and clinical psychology*, 64(2), 295.
5 Cuijpers, P., Berking, M., Andersson, G., Quigley, L., Kleiboer, A., & Dobson, K. S. (2013). A meta-analysis of cognitive-behavioural therapy for adult depression, alone and in comparison with other treatments. *The Canadian Journal of Psychiatry*, 58(7), 376-385.
Hollon, S. D., Stewart, M. O., & Strunk, D. (2006). Enduring effects for cognitive behavior therapy in the treatment of depression and anxiety. *Annu. Rev. Psychol.*, 57, 285-315.
6 Beck 2005
7 Markowitz, J. C., & Weissman, M. M. (2012). Interpersonal psychotherapy: past, present and future. *Clinical psychology & psychotherapy*, 19(2), 99-105.
8 Markowitz & Weissman 2012
9 Markowitz & Weissman 2012
10 Markowitz, J. C., & Weissman, M. M. (2004). Interpersonal psychotherapy: principles and applications. *World Psychiatry*, 3(3), 136.
11 de Mello, M. F., de Jesus Mari, J., Bacaltchuk, J., Verdeli, H., & Neugebauer, R. (2005). A systematic review of research findings on the efficacy of interpersonal therapy for depressive disorders. *European archives of psychiatry and clinical neuroscience*, 255(2), 75-82.
12 Keng, S. L., Smoski, M. J., & Robins, C. J. (2011). Effects of mindfulness on psychological health: A review of empirical studies. *Clinical psychology review*, 31(6), 1041-1056.
13 Khoury, B., Lecomte, T., Fortin, G., Masse, M., Therien, P., Bouchard, V., ... & Hofmann, S. G. (2013). Mindfulness-based therapy: a comprehensive meta-analysis. Clinical psychology review, 33(6), 763-771.
14 Dimeff, L., & Linehan, M. M. (2001). Dialectical behavior therapy in a nutshell. *The California Psychologist*, 34(3), 10-13.
15 Sussex Publishers. (n.d.). Dialectical Behavior Therapy. Retrieved from https://www.psychologytoday.com/us/therapy-types/dialectical-behavior-therapy
16 Keng et al 2011
17 Cuijpers et al 2013
18 Aldao, A., Nolen-Hoeksema, S., & Schweizer, S. (2010). Emotion-regulation strategies across psychopathology: A meta-analytic review. *Clinical psychology review*, 30(2), 217-237.
19 Aldao et al 2010
20 Aldao et al 2010
21 Messina, I., Sambin, M., Palmieri, A., & Viviani, R. (2013). Neural correlates of psychotherapy in anxiety and depression: a meta-analysis. *PloS one*.
22 Goldapple, K., Segal, Z., Garson, C., Lau, M., Bieling, P., Kennedy, S., & Mayberg, H. (2004). Modulation of cortical-limbic pathways in

major depression: treatment-specific effects of cognitive behavior therapy. *Archives of general psychiatry*, *61*(1), 34-41.

Chapter Twenty-Two
EXTERNAL REBALANCING

1 Rossini, P. M., Burke, D., Chen, R., Cohen, L. G., Daskalakis, Z., Di Iorio, R., ... & Hallett, M. (2015). Non-invasive electrical and magnetic stimulation of the brain, spinal cord, roots and peripheral nerves: basic principles and procedures for routine clinical and research application. An updated report from an IFCN Committee. *Clinical Neurophysiology*, *126*(6), 1071-1107.
Burt, T., Lisanby, S. H., & Sackeim, H. A. (2002). Neuropsychiatric applications of transcranial magnetic stimulation: a meta analysis. *International Journal of Neuropsychopharmacology*, *5*(1), 73-103.
2 Höflich, G., Kasper, S., Hufnagel, A., Ruhrmann, S., & Möller, H. J. (1993). Application of transcranial magnetic stimulation in treatment of drug-resistant major depression–a report of two cases. *Human Psychopharmacology: Clinical and Experimental*, *8*(5), 361-365.
3 Burt et al 2002
4 Rossini et al 2015
5 Rossini et al 2015
6 Lefaucheur, J. P., André-Obadia, N., Antal, A., Ayache, S. S., Baeken, C., Benninger, D. H., ... & Devanne, H. (2014). Evidence-based guidelines on the therapeutic use of repetitive transcranial magnetic stimulation (rTMS). *Clinical Neurophysiology*, *125*(11), 2150-2206.
7 Gershon, A. A., Dannon, P. N., & Grunhaus, L. (2003). Transcranial magnetic stimulation in the treatment of depression. *American Journal of Psychiatry*, *160*(5), 835-845.
8 Lefaucheur et al 2014
9 Rossini et al 2015
10 McClintock, S. M., Reti, I. M., Carpenter, L. L., McDonald, W. M., Dubin, M., Taylor, S. F., ... & Krystal, A. D. (2018). Consensus recommendations for the clinical application of repetitive transcranial magnetic stimulation (rTMS) in the treatment of depression. *The Journal of clinical psychiatry*, *79*(1).
11 Lefaucheur et al 2014
12 Rossini et al 2015; Lefaucheur et al 2014
13 McClintock et al 2018
14 Lefaucheur, J. P., Aleman, A., Baeken, C., Benninger, D. H., Brunelin, J., Di Lazzaro, V., ... & Jääskeläinen, S. K. (2020). Evidence-based guidelines on the therapeutic use of repetitive transcranial magnetic stimulation (rTMS): An update (2014–2018). *Clinical Neurophysiology*.
15 Utz, K. S., Dimova, V., Oppenländer, K., & Kerkhoff, G. (2010). Electrified minds: transcranial direct current stimulation (tDCS) and galvanic vestibular stimulation (GVS) as methods of non-invasive brain stimulation in neuropsychology–a review of current data and future implications. *Neuropsychologia*, *48*(10), 2789-2810.
Parent, A. (2004). Giovanni Aldini: from animal electricity to human brain stimulation. *Canadian journal of neurological sciences*, *31*(4), 576-584.
16 Brunoni, A. R., Ferrucci, R., Fregni, F., Boggio, P. S., & Priori, A. (2012). Transcranial direct current stimulation for the treatment of major depressive disorder: a summary of preclinical, clinical and translational findings. *Progress in Neuro-Psychopharmacology and Biological Psychiatry*, *39*(1), 9-16.
17 Brunoni et al 2012
18 Lefaucheur, J. P., Antal, A., Ayache, S. S., Benninger, D. H., Brunelin, J., Cogiamanian, F., ... & Marangolo, P. (2017). Evidence-based guidelines on the therapeutic use of transcranial direct current stimulation (tDCS). *Clinical Neurophysiology*, *128*(1), 56-92.
19 Brunoni, A. R., Moffa, A. H., Fregni, F., Palm, U., Padberg, F., Blumberger, D. M., ... & Loo, C. K. (2016). Transcranial direct current stimulation for acute major depressive episodes: meta-analysis of individual patient data. *The British Journal of Psychiatry*, *208*(6), 522-531.
20 Lefaucheur et al 2017
21 Linden, D. E. (2014). Neurofeedback and networks of depression. *Dialogues in clinical neuroscience*, *16*(1), 103.

22 Linden 2014
23 Linden 2014
Yuan, H., Young, K. D., Phillips, R., Zotev, V., Misaki, M., & Bodurka, J. (2014). Resting-state functional connectivity modulation and sustained changes after real-time functional magnetic resonance imaging neurofeedback training in depression. *Brain connectivity*, 4(9), 690-701.
24 Emmert, K., Kopel, R., Sulzer, J., Brühl, A. B., Berman, B. D., Linden, D. E., ... & Johnston, S. (2016). Meta-analysis of real-time fMRI neurofeedback studies using individual participant data: How is brain regulation mediated?. *Neuroimage*, 124, 806-812.
25 Kong, J., Fang, J., Park, J., Li, S., & Rong, P. (2018). Treating depression with transcutaneous auricular vagus nerve stimulation: state of the art and future perspectives. *Frontiers in psychiatry*, 9, 20.
26 Bonaz, B., Sinniger, V., & Pellissier, S. (2017). The vagus nerve in the neuro-immune axis: implications in the pathology of the gastrointestinal tract. *Frontiers in immunology*, 8, 1452.
27 Bonaz et al 2017
Ben-Menachem, E., Revesz, D., Simon, B. J., & Silberstein, S. (2015). Surgically implanted and non-invasive vagus nerve stimulation: a review of efficacy, safety and tolerability. *European journal of neurology*, 22(9), 1260-1268.
Aaronson, S. T., Sears, P., Ruvuna, F., Bunker, M., Conway, C. R., Dougherty, D. D., ... & Zajecka, J. M. (2017). A 5-year observational study of patients with treatment-resistant depression treated with vagus nerve stimulation or treatment as usual: comparison of response, remission, and suicidality. *American Journal of Psychiatry*, 174(7), 640-648.
28 Kong et al 2018

Chapter Twenty-Three MEDICINES

1 Duman, R. S., Aghajanian, G. K., Sanacora, G., & Krystal, J. H. (2016). Synaptic plasticity and depression: new insights from stress and rapid-acting antidepressants. *Nature medicine*, 22(3), 238-249.
Castrén, E., & Hen, R. (2013). Neuronal plasticity and antidepressant actions. *Trends in neurosciences*, 36(5), 259-267.
2 Van Schaik, D. J., Klijn, A. F., Van Hout, H. P., Van Marwijk, H. W., Beekman, A. T., De Haan, M., & Van Dyck, R. (2004). Patients' preferences in the treatment of depressive disorder in primary care. General hospital psychiatry, 26(3), 184-189.
3 Kraus, C., Castrén, E., Kasper, S., & Lanzenberger, R. (2017). Serotonin and neuroplasticity–links between molecular, functional and structural pathophysiology in depression. *Neuroscience & Biobehavioral Reviews*, 77, 317-326.
4 Ingelman-Sundberg, M., Sim, S. C., Gomez, A., & Rodriguez-Antona, C. (2007). Influence of cytochrome P450 polymorphisms on drug therapies: pharmacogenetic, pharmacoepigenetic and clinical aspects. *Pharmacology & therapeutics*, 116(3), 496-526.
5 Millan, M. J. (2006). Multi-target strategies for the improved treatment of depressive states: conceptual foundations and neuronal substrates, drug discovery and therapeutic application. *Pharmacology & therapeutics*, 110(2), 135-370.
WW Hasselmann, H. (2014). Ketamine as antidepressant? Current state and future perspectives. *Current neuropharmacology*, 12(1), 57-70.
6 Millan 2006
Baghai, T. C., Blier, P., Baldwin, D. S., Bauer, M., Goodwin, G. M., Fountoulakis, K. N., ... & Versiani, M. (2011). General and comparative efficacy and effectiveness of antidepressants in the acute treatment of depressive disorders: a report by the WPA section of pharmacopsychiatry. *European archives of psychiatry and clinical neuroscience*, 261(3), 207-245.
7 Millan 2006; Hasselmann 2014
8 Millan 2006
9 Millan 2006
Kent, J. M. (2000). SNaRIs, NaSSAs, and NaRIs: new agents for the treatment of depression.

The Lancet, 355(9207), 911-918.
10 Millan 2006
11 Millan 2006
12 Millan 2006
Office of the Commissioner. (2018, November 20). FDA warns marketers of products labeled as dietary supplements that contain tianeptine for making unproven claims to treat serious conditions, including opioid use disorder. Retrieved from https://www.fda.gov/news-events/press-announcements/fda-warns-marketers-products-labeled-dietary-supplements-contain-tianeptine-making-unproven-claims
Centers for Disease Control and Prevention. (2018, August 3). Characteristics of Tianeptine Exposures Reported to the National Poison Data System - United States, 2000–2017. Retrieved from https://www.cdc.gov/mmwr/volumes/67/wr/mm6730a2.htm
13 McEwen, B. S., Chattarji, S., Diamond, D. M., Jay, T. M., Reagan, L. P., Svenningsson, P., & Fuchs, E. (2010). The neurobiological properties of tianeptine (Stablon): from monoamine hypothesis to glutamatergic modulation. *Molecular psychiatry*, 15(3), 237-249.
14 Baghai et al 2011
De Bodinat, C., Guardiola-Lemaitre, B., Mocaër, E., Renard, P., Muñoz, C., & Millan, M. J. (2010). Agomelatine, the first melatonergic antidepressant: discovery, characterization and development. *Nature reviews Drug discovery*, 9(8), 628-642.
15 Millan 2006; Duman et al 2016
16 Gerhard et al 2016
17 Sarris, J., Murphy, J., Mischoulon, D., Papakostas, G. I., Fava, M., Berk, M., & Ng, C. H. (2016). Adjunctive nutraceuticals for depression: a systematic review and meta-analyses. *American Journal of Psychiatry*, 173(6), 575-587.
18 Appleton, K. M., Rogers, P. J., & Ness, A. R. (2010). Updated systematic review and meta-analysis of the effects of n− 3 long-chain polyunsaturated fatty acids on depressed mood. *The American journal of clinical nutrition*, 91(3), 757-770.
19 Mocking, R. J. T., Harmsen, I., Assies, J., Koeter, M. W. J., Ruhé, H., & Schene, A. H. (2016). Meta-analysis and meta-regression of omega-3 polyunsaturated fatty acid supplementation for major depressive disorder. *Translational psychiatry*, 6(3), e756-e756.
20 Sarris et al 2016
21 Morgan, A. J., & Jorm, A. F. (2008). Self-help interventions for depressive disorders and depressive symptoms: a systematic review. *Annals of general psychiatry*, 7(1), 13.
Papakostas, G. I., Cassiello, C. F., & Iovieno, N. (2012). Folates and S-adenosylmethionine for major depressive disorder. The Canadian Journal of Psychiatry, 57(7), 406-413.
22 Fernandes, B. S., Dean, O. M., Dodd, S., Malhi, G. S., & Berk, M. (2016). N-acetylcysteine in depressive symptoms and functionality: a systematic review and meta-analysis. *The Journal of clinical psychiatry*, 77(4), 457-466.
23 Wang, S. M., Han, C., Lee, S. J., Patkar, A. A., Masand, P. S., & Pae, C. U. (2014). A review of current evidence for acetyl-l-carnitine in the treatment of depression. *Journal of psychiatric research*, 53, 30-37.
24 Sarris, J. (2018). Herbal medicines in the treatment of psychiatric disorders: 10-year updated review. *Phytotherapy Research*, 32(7), 1147-1162.
25 Gaster, B., & Holroyd, J. (2000). St John's wort for depression: a systematic review. *Archives of internal medicine*, 160(2), 152-156.
Apaydin, E. A., Maher, A. R., Shanman, R., Booth, M. S., Miles, J. N., Sorbero, M. E., & Hempel, S. (2016). A systematic review of St. John's wort for major depressive disorder. *Systematic reviews*, 5(1), 148.
26 Morgan & Jorm 2008
27 Sarris 2018
28 Rucker, J. J., Jelen, L. A., Flynn, S., Frowde, K. D., & Young, A. H. (2016). Psychedelics in the treatment of unipolar mood disorders: a systematic review. *Journal of Psychopharmacology*, 30(12), 1220-1229.
29 dos Santos, R. G., Osório, F. L., Crippa, J. A. S., Riba, J., Zuardi, A. W., & Hallak, J. E. (2016). Antidepressive, anxiolytic, and antiaddictive

effects of ayahuasca, psilocybin and lysergic acid diethylamide (LSD): a systematic review of clinical trials published in the last 25 years. *Therapeutic advances in psychopharmacology, 6*(3), 193-213.
30 Carhart-Harris, R. L., Bolstridge, M., Day, C. M. J., Rucker, J., Watts, R., Erritzoe, D. E., ... & Rickard, J. A. (2018). Psilocybin with psychological support for treatment-resistant depression: six-month follow-up. *Psychopharmacology, 235*(2), 399-408.
31 dos Santos et al 2016

Chapter Twenty-Four
SEIZURE THERAPY

1 Gazdag, G., Bitter, I., Ungvari, G. S., Baran, B., & Fink, M. (2009). Laszlo Meduna's pilot studies with camphor inductions of seizures: the first 11 patients. *The journal of ECT, 25*(1), 3-11.
2 Fink, M. (2000). Electroshock revisited: Electroconvulsive therapy, once vilified, is slowly receiving greater interest and use in the treatment of mental illness. *American Scientist, 88*(2), 162-167.
3 Gazdag et al 2009
Payne, N. A., & Prudic, J. (2009a). Electroconvulsive therapy Part I: a perspective on the evolution and current practice of ECT. *Journal of psychiatric practice, 15*(5), 346.
4 Payne & Pudic 2009a
5 Payne, N. A., & Prudic, J. (2009b). Electroconvulsive therapy part II: A biopsychosocial perspective. *Journal of psychiatric practice, 15*(5), 369.
6 Payne & Prudic 2009b
7 Husain, M. M., Rush, A. J., Fink, M., Knapp, R., Petrides, G., Rummans, T., ... & Zhao, W. (2004). Speed of response and remission in major depressive disorder with acute electroconvulsive therapy (ECT): a Consortium for Research in ECT (CORE) report. *The Journal of clinical psychiatry*, 65(4), 485.
8 Payne & Prudic 2009b
9 Payne & Pudic 2009a
10 Payne & Pudic 2009a
11 Lee, W. H., Lisanby, S. H., Laine, A. F., & Peterchev, A. V. (2016). Comparison of electric field strength and spatial distribution of electroconvulsive therapy and magnetic seizure therapy in a realistic human head model. *European Psychiatry, 36*, 55-64.
12 Payne & Pudic 2009a
Bolwig, T. G. (2011). How does electroconvulsive therapy work? Theories on its mechanism. *The Canadian Journal of Psychiatry, 56*(1), 13-18.
13 Sackeim, H. A., Prudic, J., Nobler, M. S., Fitzsimons, L., Lisanby, S. H., Payne, N., ... & Devanand, D. P. (2008). Effects of pulse width and electrode placement on the efficacy and cognitive effects of electroconvulsive therapy. *Brain stimulation, 1*(2), 71-83.
14 Sackeim, H. A., Prudic, J., Fuller, R., Keilp, J., Lavori, P. W., & Olfson, M. (2007). The cognitive effects of electroconvulsive therapy in community settings. *Neuropsychopharmacology*, 32(1), 244-254.
Kellner, C. H., Knapp, R., Husain, M. M., Rasmussen, K., Sampson, S., Cullum, M., ... & Bailine, S. H. (2010). Bifrontal, bitemporal and right unilateral electrode placement in ECT: randomised trial. *The British Journal of Psychiatry*, 196(3), 226-234.
Sackeim et al 2008
15 Spellman, T., Peterchev, A. V., & Lisanby, S. H. (2009). Focal electrically administered seizure therapy: a novel form of ECT illustrates the roles of current directionality, polarity, and electrode configuration in seizure induction. *Neuropsychopharmacology*, 34(8), 2002-2010.
16 Lee et al 2017
Lisanby, S. H., Schlaepfer, T. E., Fisch, H. U., & Sackeim, H. A. (2001). Magnetic seizure therapy of major depression. *Archives of General Psychiatry*, 58(3), 303-305.
Lisanby, S. H., Luber, B., Schlaepfer, T. E., & Sackeim, H. A. (2003). Safety and feasibility of magnetic seizure therapy (MST) in major depression: randomized within-subject comparison with electroconvulsive therapy. *Neuropsychopharmacology*, 28(10), 1852-1865.

Chapter Twenty-Five
EXERCISE

1 Kvam, S., Kleppe, C. L., Nordhus, I. H., & Hovland, A. (2016). Exercise as a treatment for depression: a meta-analysis. *Journal of affective disorders, 202*, 67-86.
2 Kvam et al 2016
3 Kvam et al 2016
Silveira, H., Moraes, H., Oliveira, N., Coutinho, E. S. F., Laks, J., & Deslandes, A. (2013). Physical exercise and clinically depressed patients: a systematic review and meta-analysis. *Neuropsychobiology, 67*(2), 61-68.
4 Christie, B. R., Eadie, B. D., Kannangara, T. S., Robillard, J. M., Shin, J., & Titterness, A. K. (2008). Exercising our brains: how physical activity impacts synaptic plasticity in the dentate gyrus. *Neuromolecular medicine, 10*(2), 47.
5 Lopresti, A. L., Hood, S. D., & Drummond, P. D. (2013). A review of lifestyle factors that contribute to important pathways associated with major depression: diet, sleep and exercise. *Journal of affective disorders, 148*(1), 12-27.
6 Saraulli, D., Costanzi, M., Mastrorilli, V., & Farioli-Vecchioli, S. (2017). The long run: neuroprotective effects of physical exercise on adult neurogenesis from youth to old age. *Current neuropharmacology, 15*(4), 519-533.
7 Szuhany, K. L., Bugatti, M., & Otto, M. W. (2015). A meta-analytic review of the effects of exercise on brain-derived neurotrophic factor. *Journal of psychiatric research, 60*, 56-64.
8 Christie et al 2008
9 Lopresti et al 2012
10 Lopresti et al 2012
11 Eyre, H., & Baune, B. T. (2012). Neuroimmunological effects of physical exercise in depression. *Brain, behavior, and immunity, 26*(2), 251-266.
12 Vukovic, J., Colditz, M. J., Blackmore, D. G., Ruitenberg, M. J., & Bartlett, P. F. (2012). Microglia modulate hippocampal neural precursor activity in response to exercise and aging. *Journal of Neuroscience, 32*(19), 6435-6443.

Chapter Twenty-Six
CHRONOTHERAPIES

1 Hemmeter, U. M., Hemmeter-Spernal, J., & Krieg, J. C. (2010). Sleep deprivation in depression. *Expert review of neurotherapeutics, 10*(7), 1101-1115.
2 Hemmeter et al 2010
Dopierała, E., & Rybakowski, J. (2015). Sleep deprivation as a method of chronotherapy in the treatment of depression. *Psychiatr Pol, 49*(3), 423-433.
3 Wirz-Justice & Benedetti 2019
Wu, J. C., Kelsoe, J. R., Schachat, C., Bunney, B. G., DeModena, A., Shahrokh, G., Gillin, J. C., Potkin, S. G., Bunney, W. E. (2009). Rapid and Sustained Antidepressant Response with Sleep Deprivation and Chronotherapy in Bipolar Disorder. *Biological Psychiatry, 66* (3), 298-301.
4 Wirz-Justice & Benedetti 2019
5 Lam, R. W., Levitt, A. J., Levitan, R. D., Michalak, E. E., Cheung, A. H., Morehouse, R., ... & Tam, E. M. (2016). Efficacy of bright light treatment, fluoxetine, and the combination in patients with nonseasonal major depressive disorder: a randomized clinical trial. *JAMA psychiatry, 73*(1), 56-63.
6 Bunney, B. G., & Bunney, W. E. (2012). Rapid-acting antidepressant strategies: mechanisms of action. *International Journal of Neuropsychopharmacology, 15*(5), 695-713.
7 Pail, G., Huf, W., Pjrek, E., Winkler, D., Willeit, M., Praschak-Rieder, N., & Kasper, S. (2011). Bright-light therapy in the treatment of mood disorders. *Neuropsychobiology, 64*(3), 152-162.
8 Wirz-Justice & Benedetti 2019
9 Engelmann, W. (1973). A slowing down of circadian rhythms by lithium ions. *Zeitschrift für Naturforschung C, 28*(11-12), 733-736.
Kafka, M. S., Wirz-Justice, A., Naber, D., Marangos, P. J., O'Donohue, T. L., & Wehr, T. A. (1982). Effect of lithium on circadian neurotransmitter receptor rhythms. *Neuropsychobiology, 8*(1), 41-50.
10 Wirz-Justice & Benedetti 2019
11 Wirz-Justice & Benedetti 2019

CPSIA information can be obtained
at www.ICGtesting.com
Printed in the USA
JSHW021005030621
15471JS00003B/12

9 781735 845432